Taxing Choices for Managing Natural Resources, the Environment, and Global Climate Change

Anwar Shah
Editor

Taxing Choices for Managing Natural Resources, the Environment, and Global Climate Change

Fiscal Systems Reform Perspectives

palgrave
macmillan

Editor
Anwar Shah
Brookings Institution
Washington, DC, USA

ISBN 978-3-031-22605-2   ISBN 978-3-031-22606-9 (eBook)
https://doi.org/10.1007/978-3-031-22606-9

© The Editor(s) (if applicable) and The Author(s), under exclusive license to Springer Nature Switzerland AG 2023
This work is subject to copyright. All rights are solely and exclusively licensed by the Publisher, whether the whole or part of the material is concerned, specifically the rights of translation, reprinting, reuse of illustrations, recitation, broadcasting, reproduction on microfilms or in any other physical way, and transmission or information storage and retrieval, electronic adaptation, computer software, or by similar or dissimilar methodology now known or hereafter developed.
The use of general descriptive names, registered names, trademarks, service marks, etc. in this publication does not imply, even in the absence of a specific statement, that such names are exempt from the relevant protective laws and regulations and therefore free for general use.
The publisher, the authors, and the editors are safe to assume that the advice and information in this book are believed to be true and accurate at the date of publication. Neither the publisher nor the authors or the editors give a warranty, expressed or implied, with respect to the material contained herein or for any errors or omissions that may have been made. The publisher remains neutral with regard to jurisdictional claims in published maps and institutional affiliations.

This Palgrave Macmillan imprint is published by the registered company Springer Nature Switzerland AG
The registered company address is: Gewerbestrasse 11, 6330 Cham, Switzerland

The editor is pleased to dedicate this book to the following scholars who inspired his research interests in fiscal system reform options managing natural resources, local environmental protection and global climate change.

Professor Robin Boadway, Queen's University, Canada, inspired his work on fiscal federalism dimensions of natural resource management, environmental protection, and global climate change.

Professor Melville McMillan of the University of Alberta, Canada, introduced him to public finance of the environment and natural resource management.

Professor Lawrence Summers, Harvard University, motivated his work on carbon taxes and removing energy subsidies as tools for fiscal system reforms. He convincingly argued in 1980s that such policies make eminent economic sense even in the unlikely event that "global warming turned out to be the swine flu epidemic of the 1960s".

Professor John Whalley, University of Western Ontario, Canada introduced him to the economics of global climate change in the 1980s and was the principal advisor in the editor's advocacy to the World Bank to undertake a major research project on economic instruments to combat global climate change.

About This Book

This book presents fiscal system reform perspectives to protect the local and global environment and preserve and sustain natural resources. Strategies and practices to manage and sharing of revenues from natural resources are highlighted. Alternative economic instruments such as carbon taxes, elimination of energy subsidies, and tradable permits to combat global climate change are examined. In addition, the roles of various orders of government in managing, taxing, and sharing natural resources in selected federal countries are documented to highlight the impact of such division of responsibilities in preserving natural resources and the environment. Finally, reform options to achieve integrity in oil and gas operations are highlighted.

Contents

Part I Introduction and Overview

1 Overview 3
   Anwar Shah

Part II The Taxation of Natural Resources and the Environment, Revenue Sharing and Revenue Fund Management

2 The Taxation of Natural Resources: Principles and Policy Issues 17
   Robin Boadway and Frank Flatters

3 Green Taxes and Policies for Environmental Protection 83
   Neil Bruce and Gregory Ellis

4 Revenue Sharing from Natural Resources: Principles and Practices 121
   Baoyun Qiao and Anwar Shah

5 Non-renewable Resource Revenue Funds: Critical Issues in Design and Management 153
   Anwar Shah

Part III   Environmental Federalism

6 Green Federalism: Principles and Practice in Mature Federations and the European Union   173
   Anwar Shah

7 Environmental Federalism in Brazil   211
   Jorge Jatobá

Part IV   Combating Global Climate Change

8 Carbon Tax as a Tool for Tax Reform and Protecting Local and Global Environments   255
   Bjorn Larsen and Anwar Shah

9 Worldwide Energy Subsidies and the Impact of Their Removal on Economic Welfare and Global Climate Change   325
   Bjorn Larsen, Angie Nga Le, and Anwar Shah

Part V   Combating Corruption

10 Combating Corruption in the Oil and Gas Sector   387
    Anwar Shah

Author Index   407

Subject Index   413

Notes on Contributors

Robin Boadway is Emeritus Professor of Economics at Queen's University, Canada. He is past Editor of the *Journal of Public Economics* and *Canadian Journal of Economics* and past President of the International Institute of Public Finance and of the Canadian Economics Association. He was Distinguished Fellow of the Centre for Economic Studies at the University of Munich. He is an Officer of the Order of Canada and a Fellow of the Royal Society of Canada.

Neil Bruce (now deceased) was Professor Emeritus, University of Washington Victoria. He previously held faculty positions at Queen's University, the University of Colorado. He published widely on taxation, public goods, social security, and welfare in a broad range of professional journals and books. Much of his work was with his distinguished colleagues such as Robin Boadway, Michael Waldman, Richard Harris, Robert Halvorsen, and Stephen Turnovsky, appearing in the most prestigious journals such as *The American Economic Review*, *Quarterly Journal of Economics*, *Journal of Public Economics*, and *Journal of Political Economy*. His book and monograph publications included his widely adopted book, coauthored with Robin Boadway, *Welfare Economics* (Blackwell, New York, 1984) and more recently *Public Finance and the American Economy* (Boston: Addison-Wesley, 1998, 2002), which considers the government's role in a market economy. Among his professional associations and service, Neil served as a member of the Washington State Governor's Council of Economic Advisors, the Prosperity Partnership Steering Committee on

Tax Reform, the Washington State Tax Structure Study Committee, and the American Economic Association and the National Tax Association.

Gregory Ellis (now deceased) was Principal Teaching Professor at the University of Washington. He received his Ph.D. in 1992 from the University of California, Berkeley, with a dissertation entitled, "Environmental Regulation with Incomplete Information and Imperfect Monitoring." He received the Buechel Award for Distinguished Undergraduate Teaching in Economics in 2004 and again in 2013. In addition, he received the Buechel Award for Outstanding Contribution to the Economics Undergraduate Program in 2008. He has widely published on natural resources and environmental economics issues.

Frank Flatters is Professor Emeritus of Economics at Queen's University, Canada. His academic work centered on aspects of applied welfare economics including international trade, fiscal decentralization, cost-benefit analysis, and corruption. Following early retirement from Queen's, he worked as a policy economist—as a researcher, advisor, commentator, and mentor. He has provided policy advice to a number of countries in Asia and Africa on a variety of issues, including the design, evaluation and reform of trade, regulatory, fiscal, and industrial policies. He has published widely on fiscal federalism, tax reform, tax incentives, natural resources management, trade, and regulatory policy issues. His contributions to these fields have been widely quoted and reprinted.

Jorge Jatobá is currently a partner in the Consulting and Economic Planning (CEPLAN), Brazil. He previously retired as Professor in Economics at the UFPE, Brazil. He received his Ph.D. in Economics from Vanderbilt University (USA) and Post-Doctorate in Labor Economics and Industrial Relations at the University of Wisconsin-Madison. He also was a Visiting Scholar in the Department of Economics at UFPE Economic Growth Center, Yale University, and Visiting Professor at Brown University. He coordinated Getúlio Vargas Foundation Regional office in the Northeast, Brazil. He also previously served as the Secretary for Employment and Wages of the Ministry of Labor; Head of the Special Advisory Office of the Ministry of Labor and Secretary of Finance of Pernambuco; Consultant for the International Labor Organization (ILO), World Bank (IBRD), Inter-American Development Bank (IDB), Economic Commission for Latin America and the Caribbean (ECLAC), Federal and State

Governments, and for the private sector organizations in Brazil. He has published several works in Brazil and abroad.

Bjorn Larsen is an environmental economist and consultant to the World Bank, governments, and research institutions. His areas of research and interests have over the last decades been economic assessment and cost-benefit analysis of environmental health issues including household air pollution, ambient air pollution, water-sanitation-hygiene, and lead and arsenic exposure. He has received worldwide recognition for his work on local and global environmental issues. In early 1990s, he, in collaboration with his colleague Anwar Shah at the World Bank, did pioneering work on global climate change issues including quantifying the impact of carbon taxes and energy pricing reform options on tax reform, economic welfare, health costs, and local pollution and global carbon emissions for developing countries.

Angie Nga Le is a visiting postdoctoral associate in the Department of Public Policy and Administration, Steven J. Green School of International and Public Affairs, Florida International University (FIU). Her research interest focuses on public administration, public policy, and public finance. She previously served as a researcher for the Vietnamese Government's Ministry of Planning and Investment. She assisted the International Labor Organization's research team in developing Vietnam's National Employment Strategy and was actively involved in the Vietnam National Assembly's project on public budgetary and financial supervision. She received her Ph.D. degree in Public Affairs from FIU and masters degrees from the KDI School of Public Policy and Management (in development policy) and Université Libre de Bruxelles (in public management and economics).

Baoyun Qiao is Professor of Economics, China Academy of Public Finance and Policy, Central University of Finance and Economics. His research has mainly focused on public finance, macroeconomics, development economics, and china economy. He is an author of more than a dozen books. He published numerous scholarly papers in leading economic journals. He has advised many countries such as China and Thailand on public finance, fiscal decentralization, taxation, budget management, subnational debt management, local governance, and economic development. He has directed research projects on China Public Finance Reform, China Green Taxes, China Subnational Debt

Management, etc., for the World Bank and the Asian Development Bank. He holds Ph.D. in economics from Georgia State University, USA.

Anwar Shah is a Non-resident Senior Fellow, Brookings Institution, Washington, DC, USA, and a Fellow, Institute of Public Economics, Edmonton, Canada. He has previously served as the Director, Centre for Public Economics, SWUFE, Chengdu/Wenjiang, China, and also the World Bank, Asian Development Bank, Canadian Ministry of Finance, the Government of Alberta Urban Advisory Group, and the UN Intergovernmental Panel on Climate Change. He has published more than two dozen books and numerous articles in leading journals. His books include *Fiscal Federalism* (with Robin Boadway), by the Cambridge University Press, and *Local Public Finance and Economics* (with Harry Kitchen and Melville McMillan); *Local Public, Fiscal and Financial Governance* (with Brian Drollery, Harry Kitchen and Melville McMillan) by Palgrave McMillan/Springer Nature Press and *Fiscal Incentives for Investment and Innovation* by Oxford University Press.

List of Figures

Fig. 3.1	Taxes versus standards—standard preferred	104
Fig. 3.2	Taxes versus standards—tax preferred	105
Fig. 8.1	Coal releases the most CO_2 emissions, natural gas the least	257
Fig. 9.1	Ratios of domestic prices to world prices	338
Fig. 9.2	Impact of subsidy removal (in subsidizing country)	341
Fig. 9.3	Impact of subsidy removal (World Market)	341
Fig. 9.4	Welfare gain from subsidy removal	368
Fig. 9.5	Welfare effect of subsidy removal	374
Fig. 10.1	Inconsistency of weights across countries: Transparency International's Corruption Perception Index—2005 weights	397
Fig. 10.2	Inconsistency of weights across countries used by the Worldwide Governance Indicators (WGIs)	398
Fig. 10.3	Inconsistency of weights over time—Albania as an example	399

List of Tables

Table 4.1	Resource Tax Assignment: Brazil vs Canada—A comparative perspective	128
Table 4.2	Brazil—tax sharing of oil and gas royalties and special participation in federating units—2009	129
Table 4.3	A bird's eye view of resource revenue sharing systems	131
Table 4.4	Natural resource tax sharing in Indonesia—2012 (% share)	137
Table 4.5	Statutory assignment of selected natural resource revenues in the Russian Federation (%)	141
Table 4.6	Resource revenue sharing in Nigeria	143
Table 5.1	Non-renewable Resource Revenue Funds—Examples of large funds	156
Table 6.1	Australia: assignment of environmental functions	192
Table 6.2	Canada: assignment of environmental functions	194
Table 6.3	USA: assignment of environmental functions	200
Table 7.1	Division of responsibilities for environmental protection among federal, state and local government	222
Table 7.2	Environmental indications for Brazil compared to Latin America and the Caribbean region and upper middle income countries	229
Table 8.1	CO_2 content and carbon tax rates for fossil fuels @US$49/metric ton of Carbon Dioxide equivalent (metric ton CO_2)	257
Table 8.2	Net transfer by country under alternate carbon tax regimes (using UN national Accounts GDP)	264

Table 8.3	Net transfer by country under alternate global carbon tax regimes (using PENN GDP)	270
Table 8.4	Revenue potential of a US $10/ton domestic carbon tax (using UN National Accounts GDP)	276
Table 8.5	Carbon tax ($10/ton) incidence—Pakistan (Carbon taxes [TAX as percent of monthly income (Y) or expenditure (EXP)])	280
Table 8.6	Incidence of personal and corporate income taxes in Pakistan under alternate approaches	282
Table 8.7	Carbon emissions, Carbon prices and energy taxes in selected countries	287
Table 8.8	Summary of welfare effects of a $10/ton Carbon tax	288
Table 8.9	Costs and benefits of carbon taxes for selected countries	291
Table 8.10	Marginal benefits of NOx, SO_2 and PM Reductions (US$/ton)	296
Table 8.11	Impact of Carbon taxes on Pakistan industries application of Bernstein-Shah Dynamic Variable Profits Model and Shah-Baffes Dynamic Flexible Accelerator Model	301
Table 8.12	Impact of a US$10 Carbon Tax on manufacturing value added and local externalities	302
Table 9.1	Greenhouse gas emissions from fuel combustion 2019—oil	328
Table 9.2	Greenhouse gas emissions from fuel combustion 2019—gas	330
Table 9.3	Greenhouse gas emissions from fuel combustion 2019—coal	331
Table 9.4	Greenhouse gas emissions from fuel combustion 2019—total (oil, coal and natural gas)	332
Table 9.5	CO_2 emissions from fuel combustion by sector 2019—electricity and heat production	333
Table 9.6	Fossil fuel subsidies 2019, million USD	334
Table 9.7	Greenhouse gas emission reductions from removing subsidies on fossil fuels—oil (2019)	343
Table 9.8	Greenhouse gas emission reductions from removing subsidies on fossil fuels—gas	348
Table 9.9	Greenhouse gas emission reductions from removing subsidies on fossil fuels—coal (2019)	352
Table 9.10	Greenhouse gas emission reductions from removing subsidies on fossil fuels—total (oil, coal, and natural gas)	356
Table 9.11	CO_2 emission reductions from removing subsidies on fossil fuels—electricity	360

Table 9.12	Welfare impacts of subsidy removal	369
Table 9.13	Carbon tax equivalent to subsidy removal: oil, gas, coal and electricity sectors	377
Table 9.14	CO_2 emissions from fuel combustion by sector 2019—total	379
Table 9.15	IEA—fossil fuel subsidies 2019, real 2020 million USD	380
Table 10.1	Many facets of corruption in the oil and gas sector	392

PART I

Introduction and Overview

CHAPTER 1

Overview

Anwar Shah

This book presents fiscal system reform perspectives to protect the local and global environment and preserve and sustain natural resources. Strategies and practices to manage and sharing of revenues from natural resources are highlighted. Alternative economic instruments such as carbon taxes, elimination of energy subsidies, and tradable permits to combat global climate change are examined. In addition, the roles of various orders of government in managing, taxing, and sharing natural resources in selected federal countries are documented to highlight the impact of such division of responsibilities in preserving natural resources and the environment. Finally, reform options to achieve integrity in oil and gas operations are highlighted.

The book is organized into five parts. Part I provides an introductory overview of the book. Part II is concerned with the taxation of natural resources and the environment and revenue sharing and issues in the design and management of non-renewable resource revenue funds.

A. Shah (✉)
Brookings Institution, Washington, DC, USA
e-mail: shah.anwar@gmail.com

© The Author(s), under exclusive license to Springer Nature Switzerland AG 2023
A. Shah (ed.), *Taxing Choices for Managing Natural Resources, the Environment, and Global Climate Change*,
https://doi.org/10.1007/978-3-031-22606-9_1

Part III deals with environmental federalism and reviews the experiences of OECD countries and Brazil. Part IV is devoted to combating global climate change. This part discusses the potential role of carbon taxes and the elimination of energy subsidies in mitigating the greenhouse effect. Part V analyzes the susceptibility of extractive industries to grand corruption and why international efforts to combat corruption to date have been ineffective.

The following paragraphs provide an introduction to various chapters in this book.

Chapter 2 by Boadway and Flatters presents principles of taxation of natural resources and policy issues that arise in practice in applying these principles. Natural resources are typically subject both to taxation under the income tax system and to special resource taxes. Properly designed income taxes attempt to include capital income on a uniform basis. But in most countries, the income tax treats resource industries more favorably than most other industries—through favorable treatment of such capital expenses as depletion, exploration and development, and the cost of acquiring resource properties.

The case for special resource taxes is precisely to tax resource rents over and above the levies implicit in general income taxes. There are two justifications for this: (1) the efficiency-based argument that a tax on resource rents is non-distorting and complementary, and (2) the "equity" argument that the property rights to resources ought to accrue to the public at large rather than to private citizens since the rents represent the bounty nature has bestowed on the economy rather than a reward for economic effort.

If the main purpose of a resource tax is to capture rents for the public sector, the base of resource taxes should be economic rents (or their present value equivalent), contend Boadway and Flatters.

Actual resource taxes differ from rent taxes in significant ways. Unlike a general income tax—which allows the resource industries to understate capital income—resource taxes often overstate rents. This is because they typically do not offer full deductions for all costs, especially capital costs. Some systems tax revenues without allowing any deductions for costs; others allow the deduction of current costs only. As a result, they discourage investment activity in resource industries, encourage the exploitation of high-grade relative to low-grade resources, and make it difficult to impose high tax rates for fear of making the marginal tax rate higher than 100%.

Boadway and Flatters (Chapter 2) discuss three alternative "ideal" ways for the government to divert a share of rents to the public sector:

- Levy a tax on rents, ideally in the form of a cash flow tax.
- Require firms to bid for the rights to exploit resources.
- Take a share of equity in the firm.

They discuss these options in terms of their implications for the ability of firms to obtain finance, the allocation of risk, the share of rents accruing to the public sector, the extent of involvement of foreign firms, and other factors. The time has come in many countries, they say, when gains from a further refinement of imperfect existing taxes on resources are less than replacing them with simpler, more efficient forms of pure rent taxes.

Neil Bruce and Gregory Ellis, in Chapter 3, review green taxes and policies for environmental protection. They state that increasing urbanization and industrialization can exacerbate pollution problems in developing countries. Tax revenues in developing countries are too low to support adequate infrastructure for treating and disposing of waste. Still, the problem is also attributable to the classic problem of externalities in production and consumption. "Externalities" means that the costs of environmental degradation are not considered by the private decision-makers undertaking the activities that cause the problems.

Two types of policies are commonly considered to correct this market failure and improve the allocation of resources: *command and control policies* (such as emission and abatement standards) and *market-based incentives policies* (such as emissions charges, taxes on production and consumption, and marketable pollution quotas), which raise the price of such activities for the perpetrators.

Market-based incentives theoretically reduce pollution at the least cost and increase government revenues but may require costly monitoring to be effective and are usually implemented in an environment of imperfect information about the costs of abatement. Sometimes command and control policies make more economic sense in this environment.

Efficiency gains from curbing pollution in developing countries may be large. Some polluting activities are subsidized, so curtailing them brings both fiscal and environmental benefits. Taxing polluting inputs and outputs is a particularly attractive policy in developing countries, which often lack experience in administering and enforcing other types of environmental regulation. Corrective taxes make use of existing administrative structures and increase tax revenues, which can be spent on public goods to improve environmental quality (including treatment facilities for water and sewage, waste disposal, and sanitation) or can be used to reduce other taxes (which are often highly distortionary in countries with a narrow tax base).

Which goods and inputs to single out for corrective taxation depends on the main sources of pollution, which varies from country to country. Air pollution from vehicles is growing in many countries, where increased fuel taxes, perhaps coupled with improved regulations for vehicle maintenance, may be desirable. Higher taxes on high-sulfur coal would curb both industrial and household emissions of sulfur dioxide. Charges can be implemented for fixed site, easy-to-monitor industrial emissions. Subsidies to industries that cause pollution should be phased out, and those industries should be subjected to higher-than-average tax rates.

Qiao and Shah (Chapter 4) are concerned with sharing of natural resource revenues. Natural resource revenues, especially non-renewable resource revenues, present special challenges for resource management and sharing of revenues due to special features. These revenues are highly unevenly distributed and subject to uncertainty, instability, and volatility associated with large fluctuations in prices and demand. These shocks are often persistent. The uneven distribution and exhaustible nature of resources—depletion of wealth over time—have serious implications for fiscal sustainability, interjurisdictional, and intergenerational equity. Resource-rich countries also face Dutch disease with exchange rate appreciation, resulting in a lack of competitiveness of exports and an adverse impact on the industrial base and economic growth. Empirical evidence shows a negative relationship between resource revenue dependence and economic growth. Resource management and exploitation represent a complex value chain with a high probability of corruption being undetected and, when detected, even more, difficult to prosecute, and significant increases in revenues invite rent-seeking behaviors. Overall, resource-rich countries have a high incidence of corruption. Beyond these unique features, division of responsibilities for resource ownership, management, and sharing of revenues in a multi-order government represent special challenges, and what is conceptual ideal or desirable is typical, not feasible. Ideally, citizens at large and not any order of government should have resource ownership, and all net oil and gas revenues are deposited in a non-renewable resource revenue fund, i.e., a heritage fund (Norwegian style) owned by all citizens regardless of their place of residence. The assets of this fund are held in perpetuity and could not be drawn, but capital income would be available for current use. In multi-order governance, competing and conflicting goals stand in the way of ideal solutions. Political cohesion and environmental protection considerations require preferential access to resource revenues in producing regions. Economic and social union considerations require national sharing of resource wealth. Only second best solutions may be feasible in these countries. Such solutions should aim to limit adverse incentives.

Qiao and Shah (Chapter 4) show that in practice, even the second-best solutions are not politically feasible, and instead, a plethora of compromise solutions have been followed. A second-best solution is the centralization of resource rent taxes and redistribution through a federal fiscal equalization program. Alternately, decentralization of resource rent taxes accompanied by an inter-state (net) equalization program may be desirable.

Chapter 5 by Shah provides an overview of the principles and practices of non-renewable resource revenue funds. It notes the unique features of non-renewable resource revenues that motivate the establishment of these funds. These special features include the uncertainty, instability, and volatility of resource revenues, exhaustible nature of resources, vulnerability to Dutch disease, and complex value chain with a high probability of corruption being undetected. The author presents examples of large funds with the size of their assets and ratings on transparency of governance. He highlights the principal objectives of these funds. These objectives vary from economic stabilization to financing infrastructure and facilitating economic development. Norway and the State of Alaska USA share a unique goal of using these funds as a tool for common citizenship. On the effectiveness of these funds, empirical evidence is mixed in achieving economic stabilization, fiscal discipline, overcoming Dutch disease, and good governance objectives but quite positive in furthering common citizenship, wealth accumulation, intergenerational equity, and province-building goals. The author notes key considerations in significant elements of fund design-governance, transparency and oversight, rules on a total pool, and inflows and outflows.

Fund governance must be based on a legal framework with parliamentary oversight and cabinet accountability for legislative compliance.

Overall management responsibility may rest with the finance minister or treasurer with clearly specified rules on inflows and outflows, asset allocation, the threshold for risk tolerance, reporting and transparency requirements, ethics rules, and choice of internal or external management. Agency or ministry for operational management must be identified, and institutional mechanisms for operational oversight should be laid out. Various asset management options need to be carefully evaluated, and transparency and civil society oversight arrangements are elaborated. Various inflow rules are evaluated, and a choice of appropriate inflow rule is made. Similarly, outflow rules also must be critically examined to select the most appropriate rule to suit local preferences. Finally, the author draws the following lessons from the practice of non-renewable resource revenue funds (NRRFs).

- NRRFs are important tools to advance stabilization, economic diversification, wealth accumulation, common citizenship, and regional development goals.
- External contractual fund management with an independent oversight board may help maximize returns while keeping risks within tolerable limits, provided overall management responsibility and policy determination rest with the government.
- Accountable fund governance is critical to NRRFs success. Transparency, integration with the budget, and parliamentary and civil society oversight advances responsible and accountable governance.
- Citizens' right to know and access to information and external audit is critical for restraining direct or indirect raiding of funds from advancing political and bureaucratic interests.

Chapter 6 by Shah is concerned with drawing lessons from OECD experiences with environmental federalism. The chapter presents a synthesis of conceptual underpinnings in the assignment of environmental functions from the literature on federalism, public choice, political science, and neo-institutional economics. It offers a comparative perspective on environmental federalism practices in Australia, Canada, Germany, the USA, and the European Union. It concludes with the following lessons from the experiences of these mature federations.

- Cooperative or even competitive federalism is better approach to environmental management. Command and control approaches are costly and proven ineffective. Federalism principles are a valuable guide to practice. Applying "one-size-fits-all" approaches to local environmental quality leads to costly administration and jurisdictional conflict. For example, US Federal Arsenic Rule for Drinking Water, 2001, provided uniform standards that were shown to be costly and difficult to comply with by small local jurisdictions. Federal spending power could be used to have national minimum (possibly variable with for type and size of jurisdiction) standards, but unfunded mandates must be avoided.
- Decentralized governance for local environmental functions does not lead to a "race to the bottom." It may enhance economic efficiency and environmental quality through jurisdictional competition and innovation (e.g., through emission trading at the state level in the USA). Local governments typically link environmental quality with economic development, and therefore, it may lead to "a race to the top," as indicated by the experience of OECD countries. Subnational agreements may prove a less costly alternative to centralization for some cross-border externalities.
- Democratic participation ensures safeguards for environmental protection. Civil society groups help in ensuring compliance.
- Environmental federalism represents a dynamic influence of complex interactions of societal consensus and government commitment to environmental protection, constitutional-legal framework, political system, party competition, institutions of intergovernmental relations, judicial system and traditions, public interest groups, bill of rights including protection of property and international agreements and influences. Any effort to reform this system must pay attention to all these elements.

Chapter 7 by Jatobá reviews environmental federalism in Brazil—a large federal country.

The author argues that while the Brazilian Constitution has a strong emphasis on environmental protection, it does not delineate a clear division of responsibility for environmental functions among various orders of government. This lack of clarity contributes to intergovernmental conflicts and impairs the efficiency and efficacy of environmental policies. To overcome this, Brazil, in recent years, has made substantial progress in institutionalizing environmental policy and clarifying the roles of various orders/levels of government. In addition, the conception of environmental policy and the exercise of many of its instruments are democratic. In fact, there are environmental councils at all levels of government. In these councils, civil society is widely represented. As a result, environmental awareness is increasing in the country, keeping pace with the dynamism of the Brazilian environmental movement and with the maturing of environmental government institutions.

Access to environmental information and environmental statistics is easy and widespread, and the legal system is well positioned to punish environmental crimes. Further, the State and Federal Public Prosecutor's Offices have been keen observers of unlawful environmental practices and very attentive to any threats to environmental quality. Despite all this, the country faces severe environmental problems as illegal deforestation, water, soil, and air pollution, desertification, and other environmental plagues strike all regions. The contradiction posed by the institutional advances and the severity of the environmental problems is explained by the coordination failures across various orders of government in Brazil. There have been concerted efforts in recent years to overcome these failures.

The author notes that the creation of the National and State Tripartite Commissions, as well as the efforts that are being made to enact legislation that clarifies institutional assignments for the conception and implementation of environmental policies and programs, shows that governments and society at large are trying to mitigate federative conflicts that lead to waste of resources, loss of welfare, and inefficiency and inefficacy in the implementation of policies designed to better the environment and to promote sustainable development.

The author notes that the Brazilian environmental system is also vulnerable to fraud and corruption. Substantial economic interests are at stake because of the richness of the Brazilian natural resources. Reconciling economic interests with environmental protection is at the heart of sustainable development. Brazil has an immense natural capital to preserve and use as productive resources for its development. Achieving sustainable, inclusive, and equitable development in harmony with the Brazilian natural endowment is a challenge that government and the environmental movement will face for many years ahead.

Larsen and Shah, in Chapter 8, evaluate the case for carbon taxes in terms of national interest. They reach the following conclusions:

- A global carbon tax involves issues of international resource transfers and would be difficult to administer and enforce. It is thus unlikely to be implemented soon.
- National carbon taxes can raise significant revenues cost-effectively in developing countries and are not likely to be as regressive in their impact as commonly perceived. Such taxes can also enhance economic efficiency if introduced as a revenue-neutral partial replacement for corporate income taxes or in cases where subsidies are prevalent. The welfare costs of carbon taxes generally vary directly with the existing level of energy taxes, so a carbon tax should be an instrument of choice for countries such as India and Indonesia, which have few or no energy taxes.
- A carbon tax can significantly reduce local pollution and carbon dioxide emissions. Cost-benefit analysis shows countries with few, or no energy taxes substantially gain from carbon taxes in terms of an improved local environment.
- A carbon tax of $10 a ton produces minimal output losses for selected countries' industries analyzed in this paper and the output losses are fully offset by health benefits from reduced emissions of local pollutants—even ignoring the global implications of a reduced greenhouse effect
- Tradable permits are preferable to carbon taxes where the critical threshold of the stock of carbon emission beyond which temperatures would rise exponentially is known. Given our current ignorance on the costs of reducing carbon emissions and the threshold effect, a carbon tax appears to be a better and more flexible instrument for avoiding high unexpected costs.

It has been argued that economic policies to protect local and global environments should, first and foremost, remove fossil fuel subsidies. Larsen and Shah in 1992 did pioneering work on specifying a framework and quantifying world energy subsidies for oil, natural gas, and coal and their impacts on carbon emissions and economic welfare. Larsen, Le, and Shah (Chapter 9) update and extend that work to also cover the electricity subsidies. Based on a sample of 87 countries, they estimate that in 2019, world energy (oil, gas, coal, and electricity) subsidies to be more than US$1.38 trillion. They note that despite renewed heightened interest in combating global climate change, world has made little progress to get energy prices right over the last three decades and the use of energy subsidies worldwide has significantly increased over the past three decades.

Removal of these subsidies could reduce global carbon emissions by 15%, assuming no change in world fossil fuel prices, and by 9% when accounting for estimated changes in world prices. Welfare gains from subsidy removal worldwide would be more than US$172 billion assuming no change in world prices, even ignoring the benefits from curtailment of greenhouse gases emissions and abatement of local pollution. Total welfare gains from removing fossil fuel subsidies when accounting for world price changes would still be some US$148 billion in subsidizing countries. Net fossil fuel importers in Western Europe, United States, and Japan would experience a welfare gain of approximately US $30 billion in the event of subsidy removal dampening world energy prices. Equivalent reductions in carbon emissions could be achieved by an OECD carbon tax on the order of US$155–341 per ton. It should be noted that neither the subsidy removal nor an equivalent carbon tax would be sufficient to stabilize global carbon emissions at 2025 levels. To achieve that objective, stronger economic policy responses would be required.

Chapter 10 by Shah examines the grand corruption observed in oil and gas industries and discusses its root causes conceptually and empirically based upon unique features of the oil and gas industry. The oil and gas industry has a complex value chain with a high probability of corruption being undetected. Further, in dealing with corruption in resource-rich developing countries, the Western countries' strategic, security, economic, and commercial interests collide with their moral compass. This process safeguards the interests of dictatorial ruling elites in developing countries. As a result, grand corruption persists, benefits from oil and gas exploration accrue to a small group of ruling elites, and the public is deprived of opportunities for economic and social advancement. The global initiative, the so-called Extractive Industries Transparency Initiative, was undertaken to ensure transparency of payments given and received. This initiative is a welcome first small step but unlikely to significantly impact combating corruption. This is because it covers a small number of countries and a small coverage of the value chain and does not include bribe payers as members. The author argues that a more effective anti-corruption regime would provide victims' compensation, full transparency of all transactions in the oil and gas sector, and empowerment of citizens to sue their governments for non-compliance with full transparency provisions.

Shah concludes that combating corruption in the oil and gas sector is one of the most challenging tasks facing reformers worldwide. Given the enormous complexity of the underlying transactions, uncovering corruption is an enormously difficult even in the most ideal public policy environment. To assist in this effort, creating an enabling environment for people's right to know and freedom for investigative journalism to pursue all leads without fear of harassment, persecution and risks to life and liberty should be the first order of priority for policymakers. A second order of priority is to establish an authorizing environment that holds powerful individuals and entities to account for corrupt practices through timely and fair dispensation of justice. This would require difficult legislative and judicial reforms. Since corruption in the oil and gas sector, in a large part, originates from the stakeholders in industrial countries, it would also require strong judicial interventions by these countries along the lines of the US Foreign Corrupt Practices Act. International advocacy groups concerned with corruption in the oil and gas sector would be well advised to focus their activities on both the open government issues in resource-rich countries as well as creating disincentives for stakeholders in bribe-paying countries to disengage them from corrupt practices to advance their economic and political agendas.

PART II

The Taxation of Natural Resources and the Environment, Revenue Sharing and Revenue Fund Management

CHAPTER 2

The Taxation of Natural Resources: Principles and Policy Issues

Robin Boadway and Frank Flatters

2.1 Introduction

The raising of revenues from the economic activity associated with the exploitation of natural resources is virtually a universal phenomenon among the nations of the world. This can take several different forms. It may consist of taxes specific to the resources in question. It may involve special measures applicable selectively to the resource industries within more general systems of taxation (such as the corporation income tax). Or, it may consist of varying degrees of public ownership of resource property rights ranging from ownership of the resource being exploited which are sold or leased to private sector resource firms, to joint public-private ventures, to outright public ownership and operation of the resource firms themselves. Our purpose in this study is to concentrate on the use of taxation measures by the public sector to extract revenues

R. Boadway (✉) · F. Flatters
Queen's University, Kingston, ON, Canada
e-mail: boadwayr@queensu.ca

© The Author(s), under exclusive license to Springer Nature Switzerland AG 2023
A. Shah (ed.), *Taxing Choices for Managing Natural Resources, the Environment, and Global Climate Change*,
https://doi.org/10.1007/978-3-031-22606-9_2

from resource industries, especially taxes specific to the resource sector. However, we will not be able to do so in isolation from these other measures, some of which represent relatively close substitutes for taxation. In this introductory section, we set the stage for the subsequent analysis outlined by some general features of resource industries and resource taxation found across countries.

It is useful to begin with some discussion of the types of resources themselves. Natural resources consist of the various materials endowed upon a nation by nature which are useful in the production of goods and services. It is common to classify natural resources as being of two broad types, though the distinction is sometimes ambiguous. They are the following:

Renewable Resources. Renewable resources are those that can generate a continuous flow of output for an indefinite period of time. They include such things as fisheries, forests, hydro-electricity, water supplies, clean air, and agricultural land. In each case, as some of the resource is taken for economic use, the resource can replenish itself by natural or artificial means. A characteristic feature of renewable resources is that the level of flow of resource that can be sustained is an endogenous variable. It can depend upon the stock of the resource that is maintained, upon natural rates of renewal of the resources (e.g., biological rates of growth) and upon conservation and husbandry practices of those exploiting the resources (e.g., replanting of forests, regulations on the size of fish taken, fertilization practices, use of reservoirs, etc.). In some cases, it is also true that the dynamics of resource renewal are such that extinction of the stock can occur in the case of overexploitation. The tax treatment of renewable resources necessarily involves consideration of the dynamics of the resource renewal process. In some cases, the exploitation may involve a continuous flow of output (e.g., fisheries, hydro-electricity); in others, it may involve a series of cycles of extraction and replenishment, as when clearcutting is used in forests.

Non-Renewable Resources. Non-renewable resources are those such that, in principle, there is a fixed amount available for use. Two types of industries account for the most important non-renewable resources—hydro-carbon fuels (oil and gas) and mining. The latter, in turn, can be subdivided into metallic and non-metallic mining, and these can be subdivided according to type of resource. Thus, base and precious metals are often distinguished within the metal mining sector and so on. The two broad categories, oil and gas and mining, share some features in common, but they also differ in some important ways. Both are non-renewable in the sense that there is ultimately a fixed stock of the resource (ignoring the fact that hydro-carbons regenerate themselves over very long periods of time). The stock is, however, typically both of unknown size and of variable quality. Because it is of unknown size, new deposits must continually be discovered and there is an exploration industry which is devoted to that. The tax treatment of exploration activities will be of some importance for our later discussion. The variability of quality can come about because of different concentrations of the resource in a given deposit or of differing costs of extraction. Differences in quality are also important for tax policy since they result in different costs to the economy of obtaining the resource. A related characteristic of non-renewable resources is that they are typically found in impure form, that is, mixed with other elements. This implies that further processing is an important part of obtaining the resource. This, too, will have tax implications. One way in which oil and gas differ from mining is that many of its products can be used only once. Thus, natural gas and engine gasoline are burned off when used. On the other hand, the products of mining can often be re-used. This means that there can be an active recycling industry. In that sense, they approach being renewable resources.

We can summarize the above by listing the possible stages of production for non-renewable resources:

- exploration
- development
- extraction
- processing
- recycling.

Processing itself may consist of several steps including concentrating, milling, etc. At any stage beyond extraction, there may also be the holding of inventory, which involves decision-making.

Governments impose a variety of types of taxes and other levies on their resource industries. Taxes of a general broad-based sort, such as the corporate tax and general sales taxes, also apply to the resource industries. However, they often have special provisions applying to the latter. For example, corporate tax systems often allow rapid write-offs for resource activities such as exploration and development as well as special depletion allowances on non-renewable resources. There may also be investment incentives such as preferential tax rates, tax holidays, and investment tax credits. Higher sales tax rates may be levied on the consumption of oil and gas products. These broad-based taxes tend to be levied on a *residence* basis, that is, on the tax base of taxpayers resident in the country levying the tax.

Taxes specific to the resource industries are most often applied on a *source* basis, that is, on the tax base in the country where the base is generated. The simplest of these is a specific output or production tax levied on either the output or the revenues of a resource industry. In the mining industry, this is sometimes referred to as a *severance* tax. When the property rights to the resource are owned by the state, it may be referred to as a *royalty*. In the case of forestry, it is sometimes called a *stumpage fee*. The rate may be stated in *per unit* terms or in *ad valorem* terms. It may be a flat rate, or it may be graduated according to price, size or quality of deposit, etc. Production taxes may allow some costs to be deducted from them. In the simplest case, current or operating costs may be deducted. More generally, the tax can be a *profit tax* in which both capital costs and current costs are deductible. The tax treatment of capital costs is an important characteristic of resource taxes since resource industries tend to be relatively capital-intensive. Capital costs may include depreciation of installed capital, interest costs, and depletion allowances. There may also be incentives for certain types of activities such as exploration, development, and further processing. A variation on profits taxation is the so-called *rate-of-return tax* which is a tax levied on rates of return in excess of a cut-off rate.

Another very important variant of profits taxation is the so-called *cash flow tax*. The base for this tax is the real cash flows of the firm defined to be total cash (as opposed to accrued) revenues from the sale of output less total cash outlays on both current and capital inputs as they occur. The full and immediate write-off of all investment expenses implies that there is no need for costing capital on an accrual basis using depreciation and cost of capital deductions. Nor is there any need for indexing. Under this form of a cash flow tax, only real as opposed to financial transactions have tax implications. It is what the Meade Report (1978) referred to as an R-based cash flow tax. It would also be possible to treat financial purchases and sales on a cash flow basis, and there may be some merit from doing so in industries in which significant profits are generated from financial transactions, such as in financial intermediation. However, in discussing the cash flow tax in the context of the resource industries, our focus will be on the real side of the firm. There are very few instances of a pure cash flow tax, though some countries use partial variants of it. One important reason is the fact that, under a cash flow tax, firms undertaking expansionary investments will typically be in a loss position with negative tax liabilities. Symmetric treatment would require that the government make good these negative taxes, but this is rarely done. That is, *full loss offsetting* is not the rule.

This problem of the tax treatment of losses is a more general one that applies to any sort of tax allowing deductions for costs. It will be of some importance in our discussion of resource tax policy. Typically, tax systems allow *partial loss offsetting* of the following type. Firms in a loss position are allowed to carry the losses backward for a given number of years and forward for a given number of years without interest. If special investment incentives are in place, the ability to offset losses may be affected. For example, if countries offer a tax holiday in which zero taxes are payable, firms may be precluded from carrying forward losses into years in which the tax rate is positive. Naturally, the problem of loss offsetting is only relevant under tax systems in which deductions for costs are allowed from the base. Production or output tax bases could not be negative.

In many countries, resource products are traded on international markets. This gives rise to trade taxes as a form of revenue raising. In the case of an exporting country, an export tax can be used. Its effects will differ from a source-based production tax since domestic consumption of the resource is excluded from taxation. Similarly, resource-importing countries may employ tariffs on resource imports. Equivalent measures such as quotas and licenses can be used in lieu of trade taxes, although their revenues may accrue to the private rather than the public sector.

There are various non-tax measures that could be undertaken by the government to divert revenues from the resource industries to the public sector. These typically involve the direct exercise of property rights by the public sector. One common form this takes is the sale of leases from the public sector to the private sector for the exploitation of a particular resource. This is common in the oil and gas industry, in forestry, in the fishery, and in mining industries. The sale often takes the form of an auction in which competing bids are tendered. The auction itself may take various forms, including both sealed and open bidding. Depending on the resource, the lease may involve the right to explore (as in oil and gas) or the right to extract a known source of resource (as in forests and fishing grounds). The terms of leases may vary as well. An important element of a lease may be the time period over which it applies. The length of the lease or concession may affect the speed with which a non-renewable resource is extracted as well as the way in which a renewable resource is managed. Only a lease of indefinite duration would be equivalent to full private ownership of the resource property. Note that there may be an interaction between the leasing system and the subsequent taxation of the profits from the resource. The purchase of a lease is an acquisition cost which is typically treated as a cost deductible from the tax base. An alternative would be for the lease to be creditable against tax liabilities. The relationship between leases and profits taxes will be discussed further below. A related measure that can be used is licensing. Firms can be required to pay a license fee to exploit resource properties. Depending on how licenses are allotted and how their prices are set, they can have very similar effects to leases.

Direct public sector participation in resource production is another way of obtaining a share of revenues from resources. This can take the form of joint ventures in which the public sector puts up a certain share of the capital to full public ownership. This bears some analogy with cash flow taxation. As we discuss later, cash flow taxation has the effect of making the government a *silent partner* in the ownership and profits of the firm. Public share ownership makes the government an active partner. As long as the government is not in a position to exercise control of the firm, the results should be similar, with one major exception. For firms in a loss position, public share purchases will be like cash flow taxation with full loss offsetting. It will therefore differ in effect from cash flow taxation with only partial loss offsetting.

The public sector may also engage in regulatory activities which affect the behavior of resource firms without generating any revenues for the public sector. Various aspects of the resource firm's behavior may be regulated, from exploration to development through to extraction. In addition to having the disadvantage of not generating revenues for the public sector, regulation is also a discretionary form of intervention which can induce inter-firm distortions on the economy.

Before leaving this introductory section, there are three further institutional features of the resource industries which are worth highlighting. The first is that there is often a significant presence of foreign-owned firms in the resource sector, especially in developing countries. The tax treatment of such firms both by the host country and by its home government is an important determinant of the incentive to invest in the former. Typically, a foreign firm is liable for taxation both at home and in the host country. However, there may be measures in place to reduce the possibility of double taxation. Corporate tax systems typically offer partial tax credits on similar taxes paid abroad. Thus, the United States taxes the profits of foreign subsidiaries of its domestic firms when profits are repatriated and offers a tax credit up to the amount of home country tax liabilities. Similar practices are applied elsewhere. Resource taxes may not be creditable against home country taxes, in which case they may serve to discourage investment in the host country. This may be important in designing the tax system to apply to resources.

A second institutional feature of resource taxation is that, in federal economies, jurisdiction over resources may be divided between two levels of government. For example, general taxes such as corporate taxes may be levied by the central level of government, while special resource taxes may be applied at a lower level of government. This complicates the tax system considerably.

Finally, resource exploitation may give rise to environmental costs of various sorts. These costs may be external to the resource firm itself. If so, special measures may have to be taken to ensure that the external costs imposed on the environment are taken into consideration by the firm in its decision-making.

2.2 The Goals of Resource Taxation

As mentioned, governments typically tax resource industries over and above other industries, often with special taxes applying on resources alone. In this section, we consider the reasons for this practice. The most important objective of resource taxation is to obtain some share of the rents for the public sector. We begin with a discussion of the concept of resource rents and then turn to the reasons for taxing resources, one of which is to obtain a share of the rents for the public sector.

2.2.1 The Concept of Resource Rents

One of the key characteristics of natural resources is the fact that they generate *economic rents*. The rent of a stock resource is simply its ultimate economic value, or the *economic profit* from its exploitation. More specifically, the flow of rent from a given amount of resource is the difference between the real accrued revenues it generates and all real accrued costs of obtaining those revenues. It is useful to distinguish non-renewable from renewable resource rents.

2.2.1.1   Non-Renewable Resource Rents
For a non-renewable resource such as a mine, the accrued revenues result from the final sale of the mineral to a user. The accrued costs include all the current and capital costs associated with exploring for the mineral, developing the mine site, extracting the ore, and processing it to obtain the mineral in usable form.

Revenues and current costs are conceptually quite easy to account for on an accrual basis. Revenues include the sale value of the resource when the transaction occurs independent of when cash actually changes hand. Accrued revenues will differ from cash receipts by accounts receivable. The same applies for current inputs. Their accrued costs differ from cash costs by accounts payable. The valuation of accrued revenues and costs should be at their value at the time of transaction rather than actual cash receipts or disbursements. These will differ typically by implicit interest costs. This makes exact measurement difficult.

Capital costs are even more difficult to impute since all capital expenditures must be appropriately capitalized. Thus, the cost of using depreciable assets includes three components—true depreciation of the asset, the real financial costs of holding the asset whether the financing be by debt or retained earnings or new equity, and any real capital losses resulting from changes in the replacement cost of holding the asset. All of these are difficult, if not impossible, to measure since they require one to know the true rate of depreciation of the asset. For a depletable asset, similar components should be included as costs, but in this case depreciation is replaced by depletion of the asset through exploitation. Note that the acquisition cost of the depletable asset here includes the purchase price of any lease or property rights as well as exploration and development expenses. These must be capitalized appropriately as above. Any holding of inventories of goods in process or final product must also be accounted for on an accrual basis. The cost of using inventories includes the replacement cost of the inventory when used plus the real cost of holding the inventories including both financial and storage costs. Notice that if current inputs are used to produce inventories, they should not be treated as a cost until the inventory is used to produce revenues.

Finally, mining activities involve some risk and the full costs of risk-taking should be taken into account. There are various sorts of risk involved. In the exploration stage, there is the risk associated with not knowing what size of deposit will be found. There is a risk associated with future changes in the price of inputs (capital and labor) required to exploit the mine. And, there is the risk associated with uncertainty about the final price of the mineral when it is eventually sold. The measurement of the cost of risk-taking is not simple since it depends upon the extent to which risks can be pooled on capital markets. Thus, if capital markets were perfect, the only risk that needs to be a concern is the non-diversifiable risk associated with the mining activity. In principle, this component of risk may be observable as the *beta coefficient in* empirical capital asset pricing models.

2.2.1.2 Renewable Resource Rents

Similar principles apply to a renewable resource, though the emphasis will differ somewhat. Again, the economic rent from a renewable resource like a forest or a fishing ground will be the flow of accrued revenues less the flow of all accrued costs on a real basis. Accounting for revenues received and for current inputs used to produce revenues is similar to the case of non-renewable resources. Capital costs are somewhat different in nature. Any depreciable assets used in exploiting the renewable resource are treated as above. However, the asset associated with the renewable resource itself is quite different from a stock of non-renewable resource. Unlike with the former, there will typically be no exploration costs associated with discovering it. And, since it is renewable, it regenerates itself over time.

Consider a fishing ground as an example. The evolution of the stock of fish through time depends jointly upon the biological growth rate of the stock (which itself typically depends upon the stock) and the rate at which fish are taken from the fishing ground. There is usually no resource cost involved with this biological process (although fish farms may use restocking techniques). The opportunity cost of taking additional fish from the fishing ground at a point in time is the present value of the foregone flow of fish that results in the future. This is obviously a difficult thing to account for. It presumes, for example, a particular pattern of behavior into the future, ideally optimal behavior. A similar accounting difficulty arises with a forest, except here there is the additional complication that costs of reforestation must be taken into account. As with the fishing ground, there is a natural growth rate of trees, so the stock of trees depends jointly upon the frequency of cutting and the growth pattern of the species of trees. Thus, the opportunity cost of additional cutting can be treated as the cost of replanting plus the present value of the change in the value of trees harvested into the future. Again, this is a difficult thing to measure. Finally, the property used for renewable resource exploitation may have an alternative use in which case that should be part of the opportunity cost of obtaining the resource. For example, in the case of a forest, the land may have a site value independent of its use for planting trees. The capitalized value of the land ought to be part of the ongoing cost of operating the forest.

The amount of rent that a given resource will generate depends upon the behavior of the agent responsible for exploiting the resource. The agent's behavior, in turn, depends upon the institutional setting, including the way in which property rights are defined, the efficiency of capital markets, and the tax or regulatory system in place. The basic presumption is that private sector operators will maximize the present value of after-tax economic profits (rents) over the applicable time horizon. If private ownership is absolute, the time horizon will be the indefinite future. We will refer to the value of rents generated by private optimizing behavior as *private rents*. They may differ from *social rents*, which are the rents attainable from the resource from society's point of view. Private rents may differ from social rents for a variety of reasons. If taxes apply on the firm, they are part of the social return, but not of the private return. If the activities of the firm generate external costs, such as degradation of the environment, these will form part of social costs but not private costs. If the time horizon of the private sector is limited by institutional constraint, the measurement of rents from a private point of view will differ from that for society. Furthermore, potential social rents may well differ substantially from actual social rents generated by the exploitation of a resource. All of the above distortions can give rise to a pattern of exploitation which is sub-optimal from a social point of view. One of our purposes later on in this study is to consider with more precision how various taxes impinge upon the behavior of resource managers.

Naturally, the amount of rent that can be generated from a renewable or non-renewable resource depends upon the features of the resource in question. Mines with higher quality ores will generate higher rents. Resources which are found in isolated locations will be costlier to exploit and will generate lower rents. Rents will also vary with the stock of a resource. For any given resource, we can think of there being a spectrum of low rent to high rent stocks ranging from negative to positive. Only those resource stocks with non-negative rents will worth exploiting. Those resource stocks for which rents are zero will be referred to as marginal resource *stocks*. Those with positive rents will be called *inframarginal*. The location of the marginal resource stock along the spectrum will also depend upon the institutional setting. For example, if the tax system impinges upon the marginal resource, it will make the after-tax rent negative and another resource deposit will become the marginal one. Much of our later analysis will consider precisely the issue of how the tax system affects the marginal resource stock.

We have noted at several points that the measurement of rents is a difficult thing, both conceptually and practically. This is because all accounting is on an accrual basis and in real terms, and many of the costs that must be imputed are not observable and therefore hard to measure. This would seem to make the concept of rents virtually impossible to use for any policy purposes and, as we shall see, that would be very unfortunate. The concept of rents as defined above is an economically attractive one since it measures the flow of the contribution the resource makes to real economic output at any point in time as an economist would see it. However, there is an alternative measure which gives the same present value of economic rents but a different time pattern. That is the *cash flow*. It consists simply of the difference between all cash receipts from the sale of output less cash expenditures for both current and capital inputs. Because capital costs are not capitalized, costs occur much earlier in time than under an accrual accounting system. Thus, the pattern of cash flows is typically lower earlier on and higher later than for economic profits. However, in present value terms, cash flow is the same as economic profits. It also has the advantage of being much easier to measure than economic profits since all items are, in principle, observable. There is no need to measure imputed costs, nor is there any need to index. The concept of cash flow will play an important part in our analysis of tax policy options and we discuss it in more detail below.

One final important property of the concept of economic rent should be mentioned before turning to tax issues. Since rent reflects the present value of the economic profits that a resource is expected to generate into the future, the value of the resource stock in question should be precisely the present value of its future rents. That is, future rents are said to be *capitalized* into the value of the resource. Because this is so, any tax changes that affect the value of rents in the future will be immediately capitalized into the current value of the resource. In that sense, current resource owners bear future expected resource taxes.

2.2.2 *Reasons for Taxing Resource Industries*

Given the different types of resource taxes used in practice, it is not surprising that there may be differing motives for taxing them. We present here a non-exhaustive list of some of the reasons for taxing resources in general and for the specific types of taxes sometimes used.

2.2.2.1 Rent Collection

The main justification for taxing resource firms is to obtain a share of the rents for the public sector. From a tax policy point of view, the taxation of rents is an ideal source of revenue since a rent tax is non-distorting (i.e., efficient) if designed properly. By definition, rents are the net value of the resource and do not represent the return to any variable factor of production. Since the objective of a firm will be to maximize the present value of rents, a proportional tax on rents will not affect the choices of the firm. Maximizing pre-tax rents will call for the same behavior as maximizing a given proportion of pre-tax rents.

The equity properties of taxing rents are not as clear-cut. For one thing, the ownerships of rents are not necessarily correlated with a characteristic of taxpayers deemed worthy of special taxation on equity grounds. Furthermore, as mentioned above, taxes on rents can get capitalized into current values and thus effectively be incident on current owners. This is questionable on equity grounds.

2.2.2.2 Capital Income Taxation

It may be desirable to tax resource industries as part of the general taxation of capital income in an economy. In this case, capital income can be thought of as including both the normal return to capital plus rents. The task of taxing capital income falls jointly upon the corporate income tax and personal income taxation. In these systems, capital income on debt tends to be taxed primarily at the personal level. The corporate tax is usually levied on equity capital income, which includes rent. Special measures might be applied to resource industries as a way of ensuring that rents are included properly in the base.

2.2.2.3 Industrial Policy
The design of the tax system as it applies to resources may be chosen so as to achieve certain objectives of industrial policy such as the encouragement of further processing of resources or the maintenance of some minimum level of activity for strategic reasons. This is more often a reason for encouraging the activity through subsidization than the taxing of it to obtain revenues.

2.2.2.4 Risk Pooling and Financing
As mentioned earlier, taxation of resources can be analogous to the public sector becoming a silent partner in the firm. The deductibility of costs combined with the taxation of revenues is like the acquisition of new equity for the firm. This can be advantageous for the firm in a couple of ways if capital markets are imperfect. For one, if the government is better able to pool risks than the firm, the taxation of resource profits can encourage risk-taking and be socially beneficial. Also, the taxation system can serve to improve cash flows in periods of expansion thereby assisting firms which have liquidity problems because of difficulties in obtaining outside finance. The effectiveness of the tax system for these purposes depends upon the firm being able to take full tax advantage of deductible costs. In the absence of full loss offsetting, that will not be the case.

2.2.2.5 The Taxation of Foreigners
If foreigners own resources in the country, the ability to extract tax revenues from them will provide an additional incentive for taxation. There are two sorts of circumstances in which taxes may be obtained from foreigners. The first is when the tax applies on rents, in which case the motivation is exactly as in 2.2.2.1 above. The second is to exploit foreign tax crediting arrangements. If foreign governments offer tax credits on investments made abroad, it is in the interest of host countries to tax the firm up to the limit of the credit. This can significantly affect the design of the tax system and the level of taxation. In the absence of crediting arrangements, any attempt to tax capital income of foreigners will not succeed if the country is a price taker in international capital markets. The tax will simply be shifted back to non-capital factors in the host country.

2.2.2.6   Exercise Monopoly Powers in World Markets
Some countries may be important enough suppliers of a resource on world markets that they are able to influence its price. One way of exploiting this power is to use tax policy. In this case, the appropriate tax would presumably be an export tax. Alternatively, public participation may serve to monopolize the sale of the resource directly.

2.2.2.7   Conservation of Resources
Finally, tax policy may be used as a way of inducing firms to tax account of external factors in their resource management decisions. Production taxes may be used to reduce the rate of exploitation of resources for social reasons. The latter may include environmental costs which depend upon the rate of extraction or equity concerns for future generations.

As mentioned, of all these reasons for taxing resources, that of capturing a share of rents for the public sector is by far the dominant one. The next section is devoted to issues arising from the attempt by the public sector to tax the rents accruing on natural resources.

2.3   Principles of Taxing Resource Rents

There is a large literature in public finance concerned with the design of a tax on pure profits or rents. Indeed, much of the theoretical literature on the corporate tax has addressed precisely that issue. Most of the analysis has concerned economic profits in general without specific reference to the resource industries, that is, without specifying the source of rents. A firm is simply assumed to have a decreasing returns to scale (i.e., strictly concave) production function involving a current input and a depreciable capital input. Part of the purpose of this section is to apply the results of this analysis explicitly to the resource industries where the rents arise because of a given amount of natural resource, renewable, or otherwise. Although the general principles of taxing rents remain intact whatever the source of the rents, some special issues apply in the case of resources which affect the design of revenue-raising mechanisms. It is useful to begin with a discussion of some general issues that arise in the taxation of resource rents before turning to specific mechanisms.

2.3.1 Some General Issues

As we will see below, the principles of designing a proper rent tax in the ideal world often used by economists are fairly straightforward and can take a variety of alternative forms. However, in attempting to apply this in practice to the resource industries, several conceptual problems can arise. It is useful to begin with a list of some of these conceptual problems as a prelude to considering the various mechanisms.

2.3.1.1 Ex Ante Versus Ex Post Rent Taxation

A stock of resources will yield a flow of rents over time. In the case of renewable resources, this flow can go on indefinitely, while for non-renewable resources the flows can only sum up to the given stock. Rent taxation can be designed so as to divert a share of the rents to the public sector from the private sector after they accrue. This is referred to as *ex post* rent taxation. On the other hand, as will be seen later, some rent tax mechanisms take a share of the rents before the rents actually accrue. This is *ex ante* rent taxation. In principle, *ex post* and *ex ante* rent taxation can be designed to yield equivalent revenues in present value terms, and part of the literature is devoted to ensuring that the base of the rent tax is equivalent in present value terms to the flow of accrued rents themselves. Economists have tended to view these taxes as having the same efficiency properties as actual rent taxes and have advocated their use. Some of them are attractive precisely because they are easier to implement than accrued rent taxes. The flip side of this is that whatever the rent tax collected, only its present value counts anyway since future taxes should be capitalized into the value of the resource property.

However, the very fact that the public sector can apparently choose the time pattern of rent tax revenues gives rise to a couple of fundamental problems which are related to one another. The first is that governments can change tax rates at will over time as circumstances change. Thus, there will be some uncertainty about future tax liabilities on this account alone. In a sense, this would argue in favor of a tax base in which tax liabilities are incurred as up front as possible. Then, the consequence of possible tax changes later on will be less since the base will be lower then.

Related to this is the fact that there is a fundamental *time inconsistency* problem inherent in the taxation of natural resources. Once a resource property is acquired either through outright purchase of the rights to a known stock or by incurring exploration and development expenditures, governments have an incentive to tax the stock fully. If they could commit to a predetermined tax policy, they might choose a policy which induces the optimal amount of exploration, development, and renewal. However, such commitment is not possible. Since private operators know that such commitment is impossible, they will adjust their behavior in anticipation of future government tax policies. The result is inefficient behavior. This seems to be an unavoidable problem.

It is one that also applies to foreign investors. If host governments could commit themselves to future policies, both taxation and expropriation, they could choose their policies to attract the most efficient level of foreign investment. However, once the foreign investment is in place, it becomes a fixed factor which is a good target for taxation. Foreign investors will anticipate this and act accordingly. The result will be a sub-optimal level of investment.

2.3.1.2 Problems of Measuring Rents

We have already made some reference to the fact that rents are virtually impossible to measure as they accrue. To do so requires being able to measure accrued real capital costs accurately, including real depreciation, real costs of financing, real capital losses, replacement cost of inventories, the cost of risk-bearing, etc. Special problems arise in the resource industries, both renewable and non-renewable. In the case of renewable resources, there may be costs associated with using the resource property for resource extraction as opposed to some other use (e.g., recreation, farming) and this must be accounted for. The cost associated with current extraction itself is a particularly difficult concept. In principle, the opportunity cost of increased current extraction is postponed future extraction. Given that the dynamics of extraction is itself liable to be rather complicated, this opportunity cost is difficult to measure. Similarly, replenishment or renewal costs are difficult to measure on an accruals basis since they should be imputed to the period at which the resource is eventually extracted.

Similar problems arise with non-renewable resources. The costs of extraction are somewhat simpler to account for since they are simply the value of the resource currently extracted, it being no longer available for use. However, exploration and development costs should be capitalized as should any resource acquisition costs. This gives rise to problems not unlike the measurement of capital depreciation costs.

2.3.1.3 Monitoring and Implementation Problems

All tax systems are subject to enforcement problems, especially those administered on a self-assessment basis. Resource taxes would not be immune to this; in fact, such problems may be more severe in the resource industries if additional taxes are to be imposed. Problems can arise both through outright evasion and through avoidance. Evasion is an illegal activity which involves deliberately under-reporting tax liabilities. Given the fact that firms cannot be perfectly monitored, it is impossible to eliminate evasion entirely. Its incidence can be reduced by increasing resources devoted to auditing and by increasing the penalty for being detected. Of course, if administrative corruption is present, evasion becomes more difficult to control.

Avoidance refers to the reduction of tax liabilities by undertaking measures to divert revenues and costs among activities. Unlike with evasion, under-reporting is not involved. However, the means of reporting certain items may be affected. There are various ways of doing this. One is by the use of *transfer pricing*. Transfer pricing is a phenomenon that occurs primarily in vertically-integrated firms in which sales from one to another are not done at arm's length. Profits are diverted from high- to low-taxed firms or activities by changing the price that is charged in intra-firm transactions. Thus, if a resource firm is also involved in downstream processing, it may be able to avoid part of any special resource tax imposed upon it by arranging to sell its resource output to the processing firm at artificially reduced prices thereby taking more of its profit in the upstream firm. As well as shifting profits through transfer pricing, financial transactions can also be used. For example, if interest is deductible, firms can arrange to do their borrowing through the firm with the highest tax rate thereby reducing their overall tax burden. Again, resource firms may be particularly susceptible to these practices since they may face extra taxation. Finally, firms can rearrange their overhead and administrative costs by changes in marketing, head offices, research and development, and so on.

A final technique for avoiding taxes is to masquerade profits as costs. This is a particular problem with cash flow types of taxes. Closely-held firms can arrange to take some of their profits as salary payments thereby making their cash flows appear smaller and reducing taxes based on cash flows. This can be an issue in the design of taxes based on cash flows.

2.3.1.4 Relation with Other Taxes

Resource taxes will typically be part of a more general business tax system which includes corporation income taxation as well as personal taxation. The issue then arises as to whether one type of tax liability should be deducted against the tax base or credited against the tax liabilities of another. In the case of corporate and personal taxes, there is a strong argument in favor of integrating the two systems by giving some sort of credit at the personal level for taxes having been paid by corporations. This is usually done by means of a dividend tax credit administered at the personal level. This reflects the fact that the corporate tax is intended essentially as a withholding device against domestic tax liabilities for personal taxation. However, resource taxation is intended to be an additional source of tax burden over and above income taxation. Thus, crediting it, or even allowing a deduction for it against corporate taxes, is not desired. Indeed, the opposite is the case. It can be argued that corporate taxes should be deducted against the resource tax base. If so, the rent tax would impose no further distortions over and above those already imposed by the corporate tax. Failing to allow a deduction would imply that the resource tax further compounds the distortion of the corporate tax. In fact, an efficient system would allow a full tax credit of the corporate tax against the resource tax. This would undo the distorting effect of the former. However, it would also undo the effect the corporate tax has on taxing capital income thereby defeating its purpose. Furthermore, it may undo the advantages of obtaining a foreign tax credit, in the case of foreign firms.

2.3.1.5 Absence of Loss Offsetting and Uncertainty

Most tax systems, resource taxes included, do not offer full loss offsetting. At best, they offer partial loss offsetting by allowing firms in a loss position to carry forward or backward losses for a limited number of years. Firms can be in a loss position for a number of reasons. They may be young, growing firms who are involved heavily in investment but whose revenues are expected to accrue only in the future. They may be firms who are temporarily in a loss position because of depressed output prices. Or, they may be declining firms. The absence of full loss offsetting is particularly harmful for the first two types of firms. These can be firms which are stretched for financing or which are in uncertain environments. Imperfect loss offsetting can exacerbate both problems.

Resource firms are typically relatively more likely to experience periods of loss. Since they are highly capital-intensive, they are typically in a loss position when young and growing. As well, their fortunes are likely to be much more uncertain since resource prices are known to fluctuate more than for other products.

2.3.1.6 Treatment of Foreign Income

A final relevant general consideration is the fact that the resource business is typically an international one. That is, resource firms often operate in more than one country. This can have several implications for their tax treatment. For one thing, international operations open up opportunities for avoidance of the sort discussed earlier. This means that if one country's tax rates are out of line with those in others, it may be difficult to monitor and enforce tax collections. Also, international tax conventions will have a bearing on the tax treatment of resource firms. Capital importing countries will need to take account of the home country tax treatment of foreign firms. For example, if home countries credit taxes paid abroad, which is often the case for business taxes, it is in the interest of the host country to take advantage of the credit by mimicking the home country's tax system. If such credits are not available, or if a deduction system is used, attempts to tax capital income of foreign firms will be frustrated. Because of the mobility of capital, the tax will end up being shifted back to other factors of production in the host country. On the other hand, if the tax is on the rent component of equity income, it need not be shifted. Indeed, it will not be, except by the use of avoidance techniques. Typically, resource tax systems will not be eligible for foreign tax crediting so will constitute an additional tax burden on foreign corporations. This will provide some incentive for tax avoidance measures.

Resource taxes, unlike income taxes, are generally levied using the *source* principle rather than on a worldwide or residency basis. Each country treats as its own property rights some share of the resource rents accruing within their boundaries. This is probably a necessary feature of resource tax regimes rather than being an abstract principle of the division of international property rights. It would be very difficult to monitor rents earned abroad by domestic firms.

Given this background of general issues, let us now turn to a consideration of some of the means by which resource rents can be taxed. In principle, the resource tax base could be defined as economic profits or rents of resource firms and a tax applied to that. However, as mentioned earlier, such a tax base would be virtually impossible to implement. It would involve imputing costs to the firm which are not directly observable, including depreciation, depletion of non-renewable resources, the cost of current uses of renewable resources, risk, and the real cost of finance. Thus, from a practical point of view, it is not feasible to tax rents as they accrue.

Fortunately, there are other ways of devising a tax base which are equivalent in present value terms. We begin with an outline of alternative equivalent measures of economic rents.

2.3.2 Some Equivalent Ways of Measuring Rents

It is useful to begin by recalling precisely what is included in the definition of economic rents in principle before turning to alternative equivalent measures.

2.3.2.1 Economic Rents

Current rents are defined to be the value of current output sold by the firm in the current period less the full opportunity cost incurred by the firm during the period to produce those outputs. The costs can be subdivided into two categories—the costs of current inputs and the costs of capital inputs. Current inputs are those which are used in the period in which they are purchased. Capital inputs are those which produce services over several periods. Their contribution to each period must be appropriately capitalized. All costs must be measured in terms of a common numeraire, typically either current dollars or constant dollars. The fact that prices are changing over time gives rise to two complications. One concerns the price of capital goods and the other concerns the discount factor to use. These will be discussed below.

Current Inputs. Current inputs are typically taken to include such things as wages and salaries, materials, fuels, rents, and so forth. The classification of inputs as current is not without ambiguity. Some inputs which may appear as current may actually have a capital component to them. One example concerns labor costs. In many cases, labor once hired can be viewed as a *quasi-fixed factor*. Typically, there is a period of training involved early in the tenure of the worker. To the extent that the firm bears the cost of that training (e.g., if the training yields skills which are specific to the firm), part of the wage payment reflects not a payment for the production of current input, but for the production of future input. In this case, part of the wage represents a capital cost and should be capitalized. Also, the wage pattern may not follow the productivity pattern of the worker over the employment tenure of the worker. For example, the firm may use the wage profile to increase attachment to the firm essentially by postponing wage payments. Alternatively, the firm may act as a sort of financial intermediary to the worker by providing more funds in the form of higher wages earlier in the work life. Finally, labor of the firm might also be used to produce and/or install tangible capital for the firm, such as buildings, machinery, and inventory. That part of the wage bill ought to be treated as a capital input, though it is difficult to distinguish the amount of the age bill that goes for these purposes. For all these reasons, wage payments may not properly reflect current output. A true measure of profits would require wages to be appropriately adjusted. Of course, that would be very difficult to do, and to that extent, rents will be incorrectly measured.

For closely-hold businesses in which owners are also managers, another difficulty arises. The reward that the owner-manager receives for operating the business will be partly a return to capital and partly a return to labor. In practice, the two will be difficult to distinguish. This will be important if capital income and labor incomes are treated differently for tax purposes.

Another example concerns the acquisition of intangible capital by the firm including goodwill and knowledge. Often this is a result of particular types of expenditures such as advertising and marketing. These costs should, in principle, also be capitalized, but are typically treated as if they were current costs. Again, to capitalize the costs of using intangible capital would be extremely difficult, if not impossible. This will be another source of inaccuracy in the measurement of rents.

These sorts of examples can occur in the resource industries as well. In non-renewable resources, exploration expenditures help to create information about the location and size of deposits. This is a form of intangible capital which ought, in principle, to be treated as such.

Capital Inputs. Even more difficult conceptual issues arise in the treatment of capital inputs. They yield productive output over more than the period in which they are acquired. The problem is to attribute to a period the full cost of using the capital. In principle, there are three sorts of costs associated with the use of capital for a period:

i. Depreciation. We will use the term depreciation in a general sense to include all forms of using up capital including wear and tear of machinery and buildings, depletion of a stock of non-renewable resource, the use of an item from inventory, and the use of the existing stock of renewable resource. Some of these are more readily measured than others. For non-renewable resources and inventories, the current usage should simply be costed at the full value of the amount taken. These may be readily measurable using market values. In the case of depreciable capital, the reduction in the value of the capital due to depreciation through use should be treated as depreciation. Since full markets for depreciating capital typically do not exist, this is virtually impossible to measure precisely. For renewable resources, as we have mentioned earlier, the opportunity cost of taking some resource now is the change in the amount that may be taken in the future. This requires that the optimal path of future extraction be known. In all cases, depreciation should be costed at its replacement value.

ii. Financing Costs. Holding a stock of capital of any kind for a period of time involves financing costs, either payments such as interest that must be made to creditors, or compensation for the use of one's own capital. The latter is the cost of equity capital and is the rate of return that is just required to compensate the owner for using his funds in this firm instead of placing them elsewhere. Thus, it is an opportunity cost which partly takes the form of a forgone return. The cost of equity financing for a given firm will consist of two components: the market rate of return that could have been earned elsewhere and the risk premium associated with this firm. The latter is difficult to measure. The financing cost should be based on the full replacement value of capital of all forms held by the firm. This includes the net value of accounts payable (i.e., accounts payable less accounts receivable). Furthermore, the cost of finance should be the real cost rather than the nominal cost. For example, the nominal interest rate will include a component which compensates creditors for the fall in the value of their asset due to inflation. As such, it represents a change in the principal rather than an interest cost. The nominal interest should be reduced by the rate of inflation, unless, of course, the asset is indexed for inflation.

iii. Capital Losses. Finally, if the relative price of a capital good falls over the period, that should also be treated as a cost of holding the capital. Of course, this term could either be positive or negative. If the price of a non-renewable resource in the ground rises, this reduces the cost of holding it and vice versa. Indeed, in the theory of resource extraction, expected changes in price are a key determinant of the decision as to how much to extract.

Capital costs should include each of the three items as appropriate for all forms of capital whether depreciable capital, land, inventories, non-renewable resource stocks, or renewable resources. There should be no other deductions for these items. In particular, costs of acquiring the capital, including leases and property rights to resources, should not be deducted. To do so would involve double-counting.

Present Value and Discounting. The above discussion concerns rents in the current period. Firms will typically operate for several periods and will take decisions from a long-term perspective. At a given point of time, what will be relevant is the present value of future rents rather than just current period rents. This should be what a profit-maximizing firm is interested in maximizing. There are several issues involved in measuring the present value of future rents. One concerns the time horizon itself. The typical practice is to take the time horizon as being the indefinite future (i.e., infinity) if there is no reason to expect the firm to terminate operations before then. Even though the current owners will no longer be owners at some time in the future, they still have an interest in the subsequent operations of the firm since that determines the value for which they (or their estate) can sell the firm. A finite-time horizon will be relevant if, for some reason, the firm expects to cease operations. In the resource industry, a firm may expect the resources it holds to run out. Or, it may have acquired property rights for a fixed length of time only. Another reason for ceasing operations is the possibility of bankruptcy. In any case, in the event of ceasing operations, there must be an accounting of the disposal (scrap) value of assets on hand at the time. There may also be certain costs associated with shutting down, such as responsibility for disposing of hazardous waste in the case of mines.

Another issue is the choice of a discount factor. Assuming well-functioning capital markets, this should be the rate at which the shareholders of the firm are able to convert present into future consumption. Presumably, this is some variant of the market interest rate. Note that there is no need to incorporate into the discount factor a risk component. This is already included as part of the cost of earning income in each period.

A final issue in discounting is the treatment of inflation. We have already noted that in accounting for depreciation, the replacement value for capital ought to be used, and the same applies for all forms of capital from inventory to non-renewable resources. That correction is intended to correct for changes in the relative value of capital. There is, in addition, the issue of how to treat changes in the general price level or inflation. There are two alternative but equivalent procedures that can be used. One is to measure all revenues and costs in current dollars and to discount using a nominal interest rate. The other is to deflate all future prices to some constant dollar value and discount them using a real discount rate. Note that this is quite separate from the use of a real interest rate for measuring the cost of finance. The latter should be done in any case.

We can summarize succinctly the present value of future rents (economic profits) for a representative special case in the following expression which ignores taxes:

$$V = \sum_{t=0}^{\infty} (1+R)^{-t} \left(P_t Y_t - W_t L_t - Q_t \left(\delta + R - \frac{\Delta Q_t}{Q_t} \right) K_t \right) \quad (2.1)$$

where R is the discount rate of the firm, P_t is the price of output in period t, Y_t is the quantity of output sold, W_t is the price of the current input L_t, Q_t is the price of the capital good, δ is the depreciation rate, and K_t is the stock of capital. Note that all prices and rates of return are in nominal terms. It is assumed for illustrative purposes that the firm produces a single output using one current input and one current output. It is also assumed that depreciation is a fixed proportion of the existing stock (i.e., exponential or declining balance), and that the nominal discount factor is fixed. Assume further that the inflation rate is constant at the rate π. Then (2.1) can be rewritten in the following equivalent form:

$$V = \sum_{t=0}^{\infty} (1+r)^{-t} \left(p_t Y_t - w_t L_t - q_t \left(\delta + r - \frac{\Delta q_t}{q_t} \right) K_t \right) \quad (2.2)$$

where r, p_t, w_t, and q_t are real equivalents of their associated nominal values and are defined as: $(1+r)(1+\pi) = (1+R)$, $(1+p_t)(1+\pi)^t = (1+P_t)$, $(1+w_t)(1+\pi)^t = (1+W_t)$, and $(1+q_t)(1+\pi)^t = (1+Q_t)$. This illustrates the equivalence of using nominal prices and discounting by a nominal discount rate, and using real prices with a real discount rate.

2.3.3 Cash Flow

The above description of economic rents confirms that it is very difficult to measure rents. However, there are alternatives which have the same present value as rents but which are much easier to measure. As we have mentioned, one of these is the cash flow of the firm, which is simply the net value of all real transactions of the firm during a period. More specifically, the cash flow of the firm would include the cash receipts from sales of output less the full cost of purchases of all inputs, both capital and current. Revenues and current costs would all be accounted for on a cash basis rather than an accrual basis, so would all capital costs. The cost of capital installation would be deducted fully as the investment occurred. There is no need to account separately for depreciation, cost of finance, and capital gains. The cost of inventory use would be deducted when the inventory was acquired rather than when it is used, and at the actual price of acquisition. There is thus no need to impute replacement costs or to worry about the cost of financing and capital gains. As well, the cost of acquiring resource properties including exploration, development, property rights, etc., would all be deducted up front as would the cost of intangibles. Thus, there would be generally no need to worry about either imputing costs which did not go through the market or to index capital costs. Furthermore, there is no need to include the cost of risk-taking as a separate cost.

That is not to say that there would be no problems at all in measuring cash flows. There are still a couple of difficulties. One concerns owner-managed firms. These firms could arbitrarily reduce the values of their cash flows by paying profits out as salaries. As well, international companies could change their cash flows in various jurisdictions by means of transfer pricing. However, these difficulties already exist in the rent tax.

The present value of cash flow would be obtained by simply discounting rents at the shareholders' discount rate. (Of course, there may be some ambiguity here as well since different shareholders may have different discount rates, say, due to different tax rates. Again, a similar problem also arises with discounting rents.) The important feature of the present value of cash flow for our purposes is that it should be exactly the same as the present value of rents. This can be illustrated using the same example as above.

The present value of cash flow is defined as:

$$C = \sum_{t=0}^{\infty} (1+r)^{-t}(p_t Y_t - w_t L_t - q_t I_t) \qquad (2.3)$$

where I_t is investment expenditures. To see the equivalence between (2.3) and (2.2), note first that the terms involving revenues and current costs are identical so we can concentrate on the capital costs. To make things as simple as possible to explain, suppose that the rate of increase in capital goods prices is constant at $\rho = \Delta q_t / q_t$. Then, the price of capital goods at time s is related to that at time $t < s$ as follows:

$$q_s = (1+\rho)^{s-t} q_t \qquad (2.4)$$

Consider the total amount of investment undertaken at time t, I_t. Given the depreciation rate δ, it gives rise to a stream of capital at each time s in the future equal to $(1+\delta)^{-(s-t)} I_t$. Using (2.4), the value of this stream of capital is given by:

$$q_s K_s^t = \left(\frac{(1+\rho)}{(1+\delta)} \right)^{s-t} q_t I_t$$

where K_s^t is the amount of capital at time s that resulted from investment at time t. The total capital at time s is given by:

$$q_s K_s = \sum_{t=0}^{s} \left(\frac{(1+\rho)}{(1+\delta)} \right)^{s-t} q_t I_t \qquad (2.5)$$

Substitution of (2.5) into (2.2) and simplification yields (2.3). Intuitively, the present value of the future stream of accrued costs resulting from $1 of investment is just $1.

Thus, the present value of cash flow is equivalent to the present value of economic profits. Naturally, the time profile of the two will differ. It should be obvious that the net cash flow is typically lower than rents in early periods and higher later on. This may cause difficulties for governments in attempting to tax cash flows, and it would be useful to seek ways of avoiding the problem. Fortunately, there exists an alternative to cash flows which has the same present value, which is almost as easy to implement, and whose net value can take on any arbitrary time profile. We turn to that next.

2.3.3.1 Cash Flow Equivalent

A very general tax base can be defined which has the same present value as rents and cash flows, and for which rents and cash flows are special cases. First of all, define an accounting stock of capital A_t implicitly in the following way:

$$\Delta A_t = Q_t I_t - \alpha_t A_t \tag{2.6}$$

where α_t is the proportion of the existing accounting stock of capital that is written off in period t. We will refer to α_t as the tax depreciation rate at time t. Note that it can vary over time. This idea is that all new investment increases the accounting stock of capital, while any tax depreciation reduces it. Thus, the accounting stock of capital is simply the aggregate of past undepreciated investment evaluated at historic cost (i.e., there is no inflation indexing imposed). The cash flow equivalent income base is defined as:

$$P_t Y_t - W_t L_t - (R + \alpha_t) A_t$$

The present value of the cash flow equivalent income base is therefore:

$$E = \sum_{t=0}^{\infty} (1 + R)^{-t} (P_t Y_t - W_t L_t - (R + \alpha_t) A_t) \tag{2.7}$$

Several observations can be made about the cash flow equivalent tax base. First, by a technique analogous to that used for cash flows, it can be shown that the value of E is equivalent both to R and to C. The form of the cash flow equivalent base is similar to that of the rent base except that capital costs are based on the accounting capital stock. Nominal deductions are given for the cost of finance of RA_t, and depreciation is also based on A_t. The rate of depreciation α_t is quite arbitrary. It can vary by size and over time as well. The higher is the depreciation rate, the lower will be the accounting stock of capital and the lower will be the cost of finance write-off. The cash flow equivalent base would replicate rents if the depreciation rate were set equal to the true economic depreciation rate, that is, if $\alpha_t = \delta - \Delta Q_t / Q_t$. That can be seen directly. Of course, it is difficult to do so exactly since true depreciation cannot be observed. At the same time, the cash flow equivalent base approaches cash flow as the depreciation rate approaches infinity.

In principle, the depreciation rate can be arbitrarily chosen. It can even be chosen by the firm. However, it might be natural to constrain the choice of α_t by the firm. For example, the firm might be tempted to choose α_t as high as possible to postpone tax liabilities. The government might then constrain the firm never to have a negative cash flow. If this constraint was imposed, the system would be exactly like a cash flow system with loss carry forward at the interest rate R.

As mentioned, a useful property of the cash flow equivalent tax base is its ease of implementation relative to true rents. There is no need to observe true depreciation. Nor is there any need to index for inflation. The depreciation rate used is completely arbitrary. It can be at different rates for different types of capital. It is even possible to treat current inputs as capital ones for this purpose. Similarly, all expenditures on resources can be included as forms of accounting capital and have a book value associated with them. It is always possible to lump together various types of expenditures into a single composite stock of capital for accounting purposes as long as they have the same tax depreciation rate.

In short, the theoretical literature tells us that it is relatively easy to devise an income measure which is equivalent in present value terms to rents. With this background, let us consider the sorts of mechanisms that have been used for taxing resources and compare them against the rent benchmark.

2.3.4 Rent-Maximizing Decision Rules

Equations (2.2), (2.3), and (2.7) all yield the same value. Any of them could be viewed as being the objective function for a profit-maximizing firm in the absence of taxes. Maximizing them will give rise to a stream of demands for current and capital inputs by the firm. It is worth at this point indicating the conditions that characterize the optimal choice of current and capital inputs in the absence of taxes so we can indicate later how taxes impinge on these decisions by the firm. To do so, we suppose that the quantity of output is determined by a production function $Y_t = F(L_t, K_t)$, and that the firm is a price taker in all markets. Under these assumptions, the marginal conditions determining the choice of current inputs L_t and capital inputs K_t in each period are given by:

$$P_t F_{Lt} = W_t \qquad (2.8)$$

$$P_t F_{Kt} = Q_t\left(\delta + R - \frac{\Delta Q_t}{Q_t}\right) \qquad (2.9)$$

These equations state that inputs should be used up to the point at which marginal benefits equal marginal costs. The marginal benefit is the value of the marginal product given by the left-hand side of the two equations. The marginal cost of using the current input is simply its price per unit, W_t. For the capital input, the marginal cost is the right-hand side of (2.9) and is referred to as the *user cost of capital*. It consists of the three costs of holding capital: real depreciation, the cost of finance, and the capital loss. Note that equations (2.8) and (2.9) can be rewritten in terms of real prices as follows:

$$P_t F_{Lt} = w_t \qquad (2.8')$$

$$p_t F_{Kt} = q_t\left(\delta + r - \frac{\Delta q_t}{q_t}\right) \qquad (2.9')$$

The above equation for capital costs is a general one that can be applied to all sorts of capital, although it is most directly applicable to depreciable capital. It is useful to recast it to apply to other types of capital specifically used in the resource industries. Three cases are considered—non-renewable resources, renewable resources, and inventories.

2.3.4.1 Non-Renewable Resources

Consider the case in which a firm has a stock of non-renewable resource and has to choose the rate of extraction. Let the real price of a unit of the resource be p and the real marginal cost of extracting a unit of the resource be c'. The stream of prices is given to the firm, but the marginal cost rises with the quantity extracted in each period. Then, the optimality condition which determines the rate of extraction is given by the so-called

Hotelling Rule which states:

$$\frac{\Delta(p-c')}{(p-c')} = r$$

The right-hand side gives the opportunity cost of holding the resource in the ground. The left-hand side gives the net rate of return from holding it. If the left-hand side is less than the right-hand side, the firm will want to increase its rate of extraction, causing its marginal cost to rise until the two sides come into equality and vice versa. Of course, this is a very stylized way of looking at the extraction decision, but it does capture the fundamental forces at work.

2.3.4.2 Renewable Resources

As an illustration of a renewable resource, consider a stand of trees which is harvested using clear-cut techniques. Slightly different expressions will be obtained for other types of renewable resources, such as a fishing ground. However, the basic principles involved will be similar. Let $F(T)$ be the output of a forest whose trees are all of age T, and $R(F(T))$ is the net revenue from the cutting and sale of the trees. At the beginning of the planning period, suppose that a crop of trees is planted at a cost of c. Suppose the revenue function and the planting costs are unchanging over time for simplicity. The only decision that the forester must take is the age T at which to clear cut the forest and replant. This is referred to as the *rotation period*. The future operation of the forest consists of an indefinite number of cycles of planting and clearcutting each of length T. Thus, he incurs an initial cost of c and then receives a sequence of net revenues of $R(F(T)) - c$ at T, $2T$, $3T$, and so on. Thus, the present value of the cash flows from the operation is:

$$V = \frac{(R(F(T)) - c)(1+r)^{-T}}{1 - (1+r)^{-T}}$$

Choosing T to maximize this yields the following optimality condition:

$$\frac{\Delta R(F(T))}{R(F(T)) - c} = \frac{\ln(1+r)}{1 + (1+r)^{-T}}$$

This equation has basically the same form as that for the non-renewable resource. The left-hand side is the marginal value from increasing the rotation period while the right-hand side is the financial cost associated from the postponement in harvesting. The complicating feature is the fact that increasing the rotation period affects each and every rotation into the indefinite future.

2.3.4.3 Inventories

Suppose a firm has to decide how much of some good to hold as inventory. The good can be a final product or an intermediate one. Suppose the price of the good at time t is P_t. There may also be a storage cost of c, per unit of inventory held. Then, the user cost of holding a unit of inventory consists of the cost of financing the inventory, any capital loss from holding it, and the storage cost. The holding of inventories presumably gives rise to some benefit to the firm. The benefit could involve cost reductions from production smoothing, or reductions in risk. Let us simply denote the marginal benefits from holding inventories as MB_t, without specifying their source. Then, the optimal stock of inventory holdings will be that at which:

$$MB_t = P_t \left(R + c_t - \frac{\Delta P_t}{P_t} \right)$$

This equation has basically the same form as that for the non-renewable resource. The left-hand side is the marginal value from increasing the rotation period while the right-hand side is the financial cost associated from the postponement in harvesting. The complicating feature is the fact that increasing the rotation period affects each and every rotation into the indefinite future.

Note that in this expression the user cost of inventories is evaluated at replacement cost rather than the cost at which any inventory holdings were originally acquired.

2.3.5 Mechanisms for Taxation of Resource Rents

The theoretical concept of rent, which is the primary basis of most resource taxes, is relatively clear. But the design and implementation of mechanisms for its taxation tend to be less than straightforward. In this section, we deal with general types of such mechanisms and with some of the theoretical and practical difficulties involved in their implementation.

Mechanisms for rent collection differ in many respects. One of the fundamental distinctions is between *ex ante* and *ex post* rent taxation. *Ex ante* collection is based on the sale of the rights to the expected rents from a resource or a site, in the form of some sort of lease or concession arrangement. *Ex post* collection is some form of taxation that is based on the actual rents that are derived as the resource is exploited. One interesting question concerns the appropriate mix between *ex ante* and *ex post* taxation of resource rents. In the following subsection, we deal with the principal form of *ex ante* rent taxation: the sale of leases for the exploitation of a resource. The remaining sections deal with various means of taxing *ex post* resource rents.

2.3.5.1 Auctions
One way to capture the rent from a resource is to auction the rights to its exploitation. In competitive bidding for the right to extract and sell a given resource, a government should expect to be able to collect the full amount of the *ex ante* rent from that resource. This would include the present value of all revenues less all costs, including risk and a normal return to all investments, from its extraction; in other words, what we have called V, C, and E above suitably corrected for expected tax payments. This assumes, of course, that the government is willing to lease the resource-producing property for perpetuity, or at least for as long as the resource has any economic value.

There is a considerable literature on the properties of different types of auctions including sealed-bid first-price, sealed-bid second-price, Dutch, English, etc. While there are many important lessons from this literature, some of the most basic messages are quite simple. The first stresses the importance of competition in the bidding process. Without competition, there can be no assurance that the government will succeed in capturing a significant share of the rents. Competition might be difficult to achieve in many cases because of asymmetries in information about the size, quality, or other characteristics of the resource in question. This makes it even more important that the government not restrict participation in other ways.

The second is that under a set of reasonable assumptions, the above-mentioned four types of auctions all yield the same price on average. The assumptions include1 risk-neutral and symmetric bidders, the value of the item being bid for depending upon the characteristics of the bidders, and payment being a function of bids alone. As the number of bidders increases, the average revenue of the seller increases. As the number of bidders becomes indefinitely large (i.e., the competitive case), the price takes on its highest value. As mentioned, in the case of resources this would be the present value of rents. Of course, if the assumptions do not apply, the different types of auctions will not be equivalent. It would take us too far afield to consider the optimal types of auctions for different circumstances.

For our purposes, the important consideration is that given sufficient competition in the bidding process, the government should be able to capture virtually all of the *ex ante* or expected rents from the exploitation of any resource deposit. A perceived advantage to many governments from this way of collecting the rents is that the payment would be made up front, at the beginning of the extraction process. Only if the private sector and the government had different discount rates would this be of any real significance. If the government had a higher discount rate than the resource extracting firm (which might be the case with large transnational firms working in developing countries), then the pro-payment feature of an auction system might be of some benefit to the host government. If the opposite were the case, then any disadvantage to the government of the pre-payment feature (due to a lower bid price by potential developers) could be eliminated by an arrangement for postponement of payments.

One particular form of postponed payment system is an annual land rent for the use of the site on which the resource is located. Any once-and-for-all payment for the right to exploit a resource has an annual land rental fee to which it is equal in present value. Apart from the time pattern of payments, there are some other differences between land rental fees and pre-payment arrangements. First, the risk to the government is greater under the former arrangement. In the event that the resource turns out to be much less valuable than had been anticipated at the time of the rental agreement, the lessee would find it relatively easy to renege on the agreement by simply ceasing to pay the rent. There is little the government could do to prevent this. Second, under an annual rental arrangement, the lessee would have an incentive to exploit the resource more quickly than under a pre-payment system, given the positive marginal cost of exploiting the deposit for one additional year. This might also lead to under-exploitation of the resource since marginally economical deposits might not be financially attractive to extract if this requires extra time and hence additional rental payments at the end of a lease.

Ex ante rents, of course, are not the same as *ex post* rents. An auction system, as opposed to most other systems discussed below, captures the former. Therefore, auction systems differ from most other forms of resource taxation in that they shift the burden of risk from resource exploitation onto the developers. To the extent that social risk is less than the private risk of the developers, this lends some inefficiency to auction systems as a means of collecting resource rents for the public sector.

A lease auction could be transformed into a partial or full *ex post* payment system by making the bids somehow contingent on the value or quantity of the resource actually extracted. For instance, lease payments could be of the form $R = a + bY$ where Y is the value or volume of resources retrieved and sold from the deposit. The standard pure *ex ante* auction system is one in which b is set equal to zero and competitors bid on a. An alternative, however, would be for a to be set equal to zero and to have potential leaseholders bid on b. This would be a pure royalty system in which the royalty rate is set by a competitive bidding process. A mixed system would be one in which the government entertained bids on both a and b, or in which it set a fixed positive value of one of these parameters and asked for bids on the other.

Auction systems also expose private resource developers to another potentially important form of political risk arising from time inconsistent behavior on the part of the government. Having conducted an auction and collected substantial if not complete pre-payment of the negotiated lease price, the government might be tempted at some later date to alter the terms of the original lease. Such changes might range from breaking the lease altogether (i.e., confiscating the previously negotiated exploitation rights) to the imposition of windfall income taxes when *ex post* rents turn out to be greater than *ex ante* rents. In a world of fluctuating resource prices, the imposition of such windfall taxes based on short-term rents would turn out to be a one-sided bet in favor of the government. Anticipation of this sort of political risk would reduce the ability of the government to collect *ex ante* rents. Of course, an anticipated willingness of the government to entertain short-term rent-based arguments made by leaseholders in times of low resource prices would work in the opposite direction.

In order for the government to maximize the proportion of the rents it is able to collect from an auction system, it is important that all the terms of the lease be specified as clearly and irrevocably as possible at the beginning. This applies especially to the conditions under which the lease might be altered or terminated, the nature of tax and other obligations expected of the developer throughout the term of the lease, and the means through which any future disputes over these matters might be settled. Regardless of the tightness of all such arrangements, reputation effects, based on actual behavior of the host and possibly of other governments, will be important in determining their effectiveness. It is probably because of this moral hazard problem, together with the general unwillingness of governments to enter into long-term lease arrangements with private resource developers, that auctions and other forms of *ex ante* rent collection agreements are seldom observed as methods of taxing economic rents in developing countries.

Resource rents can also be lost through inappropriate provision in long-term leases for external effects of the exploitation activity, such as environmental pollution. These external effects might be an ongoing byproduct of the developer's extraction activities and/or they might be long-term costs that are imposed and felt primarily after the conclusion of the project. The latter might be especially important in conjunction with leases whose lives do not match the economic lives of the resource deposits, particularly in the case of renewable resources. Mine sites might be left in a hazardous state after the expiry of a mining operation.

Forest reserves might be "mined" with inadequate investment into replenishment and/or replanting. A short-run revenue-maximizing view would be to collect all the "rents" that are possible without taking these costs into account. Such an approach might be favored by both short-run revenue-maximizing governments and profit-maximizing resource operators. Permitting mineral operators to mine a site without any restrictions on its condition at the conclusion of the operation would permit the government to maximize the bid it would receive for the rental of that site. But after account had been taken of the costs of site cleanup after the operator's departure, the net rents received by the government would almost certainly be less than those that would have been collected if the bids had been made with the understanding (and the incentive) that the operator would be responsible for the appropriate environmental management of the site. Similarly, a lease for a forest concession might bring in much more money to the government if there were no incentives or requirements for the concessionaire to invest in the long-run management of the reserve. But, once again, this would not be equivalent to maximizing the rent from the resource. Government revenues would have been maximized at the sacrifice of long-term rents and efficient resource utilization.

2.3.5.2 Cash Flow and Cash Flow Equivalent Taxation

In our review of the concept of resource rent and its measurement, we showed how the present value of the net cash flows of a resource firm is equivalent to the present value of the rents from its activities. From this, it follows that a tax equal to x% of a resource developer's cash flow would be equivalent to an x% rent tax and, in the absence of capital market imperfections, would not distort the efficient allocation of resources in the market. Furthermore, if the tax rate were 100%, it would be equivalent to the outcome of a competitive bidding process for resource extraction rights except for the fact that the cash flow tax would be an efficient collector of *ex post* rents, while an auction system would do the same for *ex ante* rents. A cash flow tax shifts all the risks over actual rent manifestations to the government, whereas a lease auction places these risks on the resource developer.

The equivalence between a 100% cash flow tax and a competitive lease auction depends as well on several critical details of implementation. The most important of these is the treatment of tax losses. Most resource ventures have the characteristic that cash flows are negative in early years and positive later. In order for a cash flow tax to be equivalent to a pure rent tax, negative cash flows must be (a) subject to immediate refundable tax credits, (b) permitted to be written off against current taxable income from other sources, or (c) allowed to be carried forward with interest at prevailing nominal market rates. Without such provisions, the base of a cash flow tax would exceed, in present value terms, that of a pure rent tax for a loss firm. Such a tax (i.e., without these provisions for tax losses) would no longer be non-distortionary; it would discriminate against investments with relatively long gestation periods and those undertaken in periods of relatively high nominal interest rates. It would also discriminate against young, growing firms at the expense of older established ones. And, it would discriminate against risky investments and in favor of safe ones. Solving this problem with alternative (b), i.e., write-offs of tax losses against other current income sources, would bias the tax system in favor of large established firms and against new ventures without other income sources.

Cash flow taxes are relatively uncommon. Instead, many governments impose taxes on bases which are, in principle, intended to be equivalent (again in present value terms) to that of a cash flow tax. As demonstrated in the previous section, a tax on current net revenues less capital cost allowances equal to the sum of economic depreciation, interest costs on current capital stock and capital losses during the current period would be equivalent to the same tax levied on current cash flows. The difference between this and a cash flow tax base is in the treatment of capital costs. Instead of being written off at the time of their expenditure, capital costs are amortized and deducted from revenues according to their current user cost. This method tends to smooth out the time path of taxable income for the firm and, in particular, to make it more likely that there will be current revenues against which to write off tax-deductible costs that occur in any time period. However, to the extent that discrepancies still do arise between current revenues and allowable costs, appropriate methods must still be found for carrying forward or backward costs which are in excess of taxable revenues in any time period.

The principal problem that arises with cash flow equivalent taxes is in devising rules for defining the user cost of capital. This is especially so in the case of resource taxation. There are not only the standard difficulties of knowing appropriate economic depreciation rates and rules for deducibility of interest expenses, but also those of determining the appropriate treatment of exploration expenses, "depletion" allowances, and expenses incurred in the maintenance and management of renewable resources. The principal danger in the case of non-renewable resources is that of dissipating the tax base by allowing excessive deductions for exploration and depletion (as is often the case with the use of generous depreciation allowances and/or investment tax credits with the normal corporate tax). For example, firms are often allowed a separate deduction for depletion over and above being able to write off many of the costs of acquiring a resource property up front. This obviously involves double-counting. In the case of renewable resources, such as forests, the more prevalent problem is that of overestimating rents by not allowing proper deductions for replenishment costs. Deviations such as these from a pure rent tax will not only affect the tax base but also distort investment decisions in resource exploration, management, and extraction. We have outlined a general method above for designing a tax system which will have the property that it is equivalent to rent taxation. To date, no countries have taken advantage of it.

Another type of cash flow equivalent tax that is sometimes used in resource industries is a "rate-of-return" tax or a tax on "added value."

The purpose of this sort of tax is to avoid many of the ambiguities and arbitrariness of attempting to measure the user cost of capital by a more certain and uniform measure. The measure employed for this purpose is simply the replacement cost of the current capital stock of the firm times the current market rate of interest. While this avoids some of the arbitrary distinctions that might occur because of differences in debt-equity ratios and differences in historical values of investments combined with the effects of ad hoc depreciation rules, it still faces important problems in the measurement of the replacement value of the current capital stock. The problems of determining economic depreciation and of valuing the firm's investments remain, albeit in a slightly different form.

2.3.5.3 Royalties

Another very commonly used form of tax for diverting rents to the public sector is a royalty or severance tax levied on resource extractions. A system of royalty payments could be equivalent to a pure *ex post* rent tax if the royalty were designed in such a way that it was equal or otherwise proportional to the economic rents associated with the amounts extracted. This would require that it is based on the value of the extractions less all the economic costs associated with them. Very few royalty systems meet this requirement. A per unit royalty system takes account of neither the value of the resources sold nor the cost of their extraction. A per unit royalty where the size of the payment depends on the grade or quality of the resource extracted as well as its quality goes part of the way toward the solution of the first of these problems, but does not deal with the second. An ad valorem system based on the gross market value of resource production deals more satisfactorily with the first problem, but still does not help with the second. Ad valorem systems based on net revenues generally consider, at best, only current costs of resource extraction and hence still overestimate true economic rents in the tax base. The extent of the bias depends on the importance of capital costs in total costs.

Some royalties discriminate on the basis of the final use to which the resource is being put. The most common levy of this sort is an export tax on resource products. Such export taxes differ from pure rent taxes not only by generally ignoring extraction costs in defining the base, but also by exempting resources which are sold in the domestic market. The usual reason for this form of tax is to subsidize domestic users of the resource product. As mentioned earlier, this practice usually is associated with industrial policy goals of promoting downstream processing industries. A commonly used tax structure in this regard is one in which the export tax rate is negatively related to the extent of domestic value-added in processing activities. Whatever the justification for this sort of tax, it is clear that it diverges considerably from a tax on economic rents.

Royalties, therefore, tend to be very imperfect mechanisms for the taxation of resource rents. We postpone to the following section a discussion of some of the adverse incentive effects arising from the use of imperfect rent taxes such as these.

2.3.5.4 Production Sharing and Public Sector Equity Participation

Many governments attempt to tax resource rents through some form of more direct participation in resource exploitation. Two of the most common methods are production sharing and equity participation.

The simplest form of production sharing arrangement is one in which the government receives a certain proportion of the output or of the sales revenue from a resource deposit that they have leased to an operator. This is just like a type of sharecropping which is commonly observed in agricultural production. It is formally identical to a crude (ad valorem) royalty described in the previous subsection and is a very imperfect rent tax. As with royalties, more complex production sharing agreements can be devised in order to correct for the obvious distortions of the crude form. For instance, a fixed amount of the initial production might be reserved for the developer in order to compensate for capital and exploration costs. Only after that initial amount would production sharing with the government begin. Of course, the extent to which this actually covered or exceeded capital costs would depend on the price of the resource at the time it was extracted. And the extent to which the production shares corresponded to economic rents (after capital costs) would depend on the value of the developer's share relative to current extraction costs. In order to properly reflect economic rents, the production shares would have to vary with the price of the resource and the actual value of current extraction costs. The latter would vary across resource deposits and over time with any given deposit. Production sharing agreements, therefore, will be generally a very poor substitute for taxes on resource rents.

Another form of direct government participation is through the purchase or granting of equity in a resource extraction operation. The extent to which such arrangements substitute for a tax on resource rents will depend on the terms under which the equity is acquired. Suppose the equity is acquired through governments contributing to the operation's capital investment in return for an equal share of the flow of net current revenues from the resource extraction operation. Then, the returns that will accrue to the government could be thought of as comprising two parts: (a) its share of the returns to capital investment and (b) an equal share of the resource rents. 100% government ownership would correspond to a 100% rent tax, 50% ownership would be equivalent to a 50% rent tax, and so on. The coexistence of other forms of income and resource taxes on such joint venture firms would complicate this simple relationship.

Of course, if the price the government pays for equity participation exceeds its share of the capital investment of the firm, then it will end up collecting a smaller proportion of the rents by this method. In particular, if the equity price were the same as what would be paid by a new private investor, and hence included the capitalized value of expected rents, then no *(ex ante)* rents at all would accrue to the government through its equity ownership. Any rents that were collected would arise only because of differences between actual and expected rents. These could be positive or negative.

Suppose, as is sometimes the case, that the government equity is obtained free of charge, i.e., without any contribution to the firm's capital. This free equity could be thought of as payment by the firm for the rights to resource extraction. If the equity share were on the same terms as if the government had invested, i.e., it gave the rights to a certain proportion of the flow of net current revenues of the firm, then this would be equivalent to a tax on both resource rents and private returns to capital. The only way to convert this into a pure rent tax would be to deduct from the government's revenue rights an imputed return to the firm's capital investment.

There are several other differences worth noting between such equity or joint venturing schemes and pure rent taxes. First, government participation is sometimes seen to have additional advantages to other forms of rent taxes by giving the government some voting power and hence direct control over the firm's activities and by giving the government "a window" which provides valuable information pertinent to both taxation and other forms of regulation of the resource sector. Second, it cannot be automatically assumed that revenues accruing to government resource companies are equivalent to tax revenues paid directly to the state treasury. Because of their greater independence from traditional government budgetary agencies, resource-rich state companies are notorious for the many ways in which their spending patterns differ from those of these other agencies. In many circumstances, it is most realistic to treat state resource firms' profits just like those of other private companies. Then, state ownership makes no contribution to the government's efforts at rent taxation. In fact, the taxation of state companies is often more problematic than it is for private companies.

2.4 The Costs of Imperfect Rent Taxes

2.4.1 Introduction

Most taxes are levied on proxies or imperfect substitutes for the bases at which they really are directed. This certainly tends to be true of those on economic rents from the exploitation and sale of resources. This has implications both for government revenues and for the allocation of a country's scarce resources (natural and other). A pure rent tax can be levied at rates of up to 100% without reducing the efficiency of resource allocation. However, if the tax base diverges from true economic rent, then any tax on that base will affect investment and other allocation decisions of private agents and cause inefficiencies in these decisions when viewed from the vantage point of aggregate economic welfare. The nature and extent of these inefficiencies will depend on the form of the divergence of the tax base from true economic rent. But in general, the size of the efficiency cost will depend, among other things, on the rate of tax, or, more precisely, on the square of the tax rate. As long as tax revenues are increasing in the tax rate, there will then be a trade-off between government revenues and efficiency of resource allocation. This is not true of a pure rent tax. In a world of imperfect taxes, therefore, it is important to understand the nature and the costs of inefficiencies arising from different methods of taxing resource rents. This will facilitate the design of tax systems that will best promote the government's revenue goals while minimizing the efficiency costs imposed on the economy. The ideal tax system from this viewpoint might be expected to differ across countries and even within countries depending on the mix of resource products and the specific circumstances of their exploitation.

Most countries do use a wide variety of mechanisms for taxing resource rents. Royalty formulas might differ considerably across resource products. Partially or completely prepaid leasing arrangements might be used in some sectors and not in others. The same is true of government participation through production sharing and/or equity ownership. Arrangements sometimes differ across firms within the same industry. Furthermore, it is the norm rather than the exception for the same activity to be subject to a number of different types of taxes and royalties. Many of these differences in and mixtures of taxes are due to historical accidents and other reasons that have little or nothing to do with the design of an efficient or otherwise appropriate tax system. Nevertheless, the number of varieties and combinations of taxes that are possible for the collection of economic rents suggests the importance of understanding some of these incentive effects and the determinants of their significance as a guide to the design of resource taxation systems. The purpose of this section is to provide some insights into these questions. An exhaustive treatment of all these possibilities would be almost impossible and not particularly useful. The alternative that we attempt here is to provide some general principles for the understanding of these issues and some illustrations of some interesting types of cases.

Our perspective is generally that of looking at divergences from neutrality in taxation. In the absence of other distortions from economic efficiency, a neutral tax system will also be efficient. When a particular resource extraction activity involves significant externalities, then offsetting non-neutralities in the tax treatment of that activity might be appropriate. Of course, other types of regulation or institutional innovation might be much more effective and less costly means of achieving the same goals. In these cases, it is still important to understand the nature of the distortions that result from different types of taxes. Without such knowledge, the design of appropriate non-neutral tax treatment of that activity would not be possible, nor would be a comparison of this with other forms of regulation.

2.4.2 Decisions Affected by Rent Taxes

Resource exploitation involves a number of different types of activities. In the case of non-renewable resources, these range from exploration to extraction to processing to marketing. Renewable resources involve all of these types of activities as well as those related to the long-term management and replenishment of the resource. Taxes and other regulations might even determine whether a resource is renewable or non-renewable. The tax system can affect decisions at all points in the production process. The decisions which are affected at any stage might involve the level of the activity in question, the input mix and/or the technology utilized, the disposal of the outputs (marketed and non-marketed), and the timing of the activity. In the remaining sections of this chapter, we discuss some of these effects in relation to different types of resource taxes and illustrate a method by which one can measure their quantitative importance.

2.4.3 Royalty Structures

Even the best designed royalty systems are very imperfect proxies for taxes on economic rent. Their basic difficulty is that they ignore all capital costs involved in resource exploitation. In many cases, they also ignore at least some components of current costs and/or imperfectly account for them. This means that royalties generally tend to overestimate economic rents, with the extent of the divergence depending on the importance of the underestimated and/or ignored elements of costs. This will discourage at least some resource-related investments. At low rates of tax, this might not discourage many socially desirable resource exploitation activities. But at higher rates of tax that might be necessary to collect significant public revenues, considerable amounts of such desirable investments might be discouraged.

Consider first the effect of ignoring capital costs in the definition of the tax base. The general effect of this defect in defining the base is to bias the tax system against capital-intensive resource investments. Consider two resource projects, both of which have the same net present value of cash flows over their lifetime, but one of which has a much higher level of capital costs which are offset by higher sales revenues at the time of marketing the product. The more capital-intensive of these projects would be subject to much higher royalty payments over its lifetime than the other. Despite the fact that the projects are equally socially desirable (from the efficiency viewpoint), the more capital-intensive project would be much less likely to be undertaken. The royalty system creates a distortion by driving a wedge between the returns of marginal investments of different capital intensities.

In the case of non-renewable resources, for instance, this would discourage projects with relatively high exploration costs. With renewable resources, this would create a distortion against projects with high replenishment costs. It would bias forest activities in favor of mining of the natural forest and against cutting programs involving significant silvicultural management or the development of plantation forests.

The non-deductibility of capital costs is especially harmful when the effects of the royalty system are considered in conjunction with those of corporate taxes. The treatment of capital costs in royalty systems means that royalties are taxes not only on economic rents, but also on capital income derived from resource exploitation activities. Corporate taxes are also levies on capital income. The combined effect of these two different taxes, therefore, is double taxation of capital income. Relative to other sectors, therefore, the imposition of royalty payments discourages investment in resource projects.

The effects of the mismeasurement of elements of current costs in a royalty system can be thought of in a similar fashion. First, the exclusion of current costs, as is done in the crudest form of royalty system, also overestimates rents and, at least at high rates of tax, discourages socially desirable resource exploitation projects. Second, such systems create a distortion against projects which are relatively intensive in the use of current inputs which are excluded from consideration in the base. Consider two projects or activities of the same pre-tax net present value and which are similar in every other respect except that one is more intensive in some current input whose costs are not taken into account in calculating the base of the royalty. Because the costs of that input cannot be deducted from the tax base, this project or activity will be subject to higher royalty payments and hence will be disfavored by the tax system.

The most common manifestation of this sort of distortion is the phenomenon known as "high grading" of a resource deposit. In the presence of a royalty system which provides a fixed (possibly zero) allowance for current costs in determination of royalty payments, developers will extract only those resources with relatively high values and/or low costs and ignore high-cost and/or low-value deposits or parts of deposits that still have positive social value. Despite their positive social value, the royalty system discourages their extraction by charging a tax in excess of the net current revenues from their extraction. Such systems encourage forest concessionaires to cut only the high-value stems in a stand and leave behind and often even damage or destroy smaller stems of significant social value. Similarly, mining operators are encouraged to close down mines before all socially valuable deposits have been extracted.

2.4.4 Export Taxes

An export tax bears a close resemblance to a crude royalty and has all of the same efficiency costs. In addition, it discriminates between resources marketed domestically and those sold in the world market. If a resource has no outlet in the domestic market, there is no additional efficiency cost due to this form of discrimination. However, this is seldom the case. When resources can be sold in the local market, an export tax induces them to be sold at a lower tax there than in export markets. Rents become dissipated by selling the resources at below world market prices to domestic users. The loss of government revenues arising from the use of an export tax rather than an equivalent royalty on all sales, export and domestic, is proportional to the size of the domestic market. The efficiency cost depends on the size of the tax and on the elasticity of domestic demand.

The only case in which this efficiency argument against export taxes might not apply is when the country is sufficiently large in the world market for the resource in question that it has some monopoly power in that market. In this case, there is an optimal export tax which is inversely related to the elasticity of world excess demand for the product. A general observation that is relevant here is that world markets for most resource products generally tend to be much more elastic than is claimed by the proponents of optimal export taxes. This is especially true in the longer run when other sources of supply become available and users are able to adapt to higher prices through various forms of substitution. The second observation is that an optimal export tax is not a substitute for other taxes to collect economic rents. An ideal export tax facilitates the collection only of the rents arising from a country's monopoly position in world markets. The rents arising from differences between the competitive price of a resource and the costs of its extraction are left untouched by an optimal export tax.

A common reason for using an export tax rather than a uniform royalty is to promote the development of downstream processing industries. An export tax gives domestic processors access to the raw material at a price that is less than that faced by foreign processors, with the gap equal to not only the cost of transporting the resource to the foreign plant, but also the size of the export tax. The amount of the subsidy provided to domestic users depends on the rate of the export tax and on the importance of the resource in total processing costs. This form of subsidy gives rise to several types of inefficiencies. First, to the extent that this effective protection is actually necessary to encourage domestic processing by marginal firms, it substitutes high-cost ways of earning or saving foreign exchange (exporting locally processed raw materials) for lower cost ways of doing the same thing (exporting the unprocessed resource). Resource rents and government revenues, in effect, are dissipated in the subsidization of inefficient marginal domestic producers. Second, by artificially lowering the domestic cost of natural resource inputs, export taxes induce local producers to be wasteful in the use of these raw materials. Plywood and saw mills in countries with significant export restrictions on logs, for instance, tend to have much lower log recovery rates than do mills in log importing countries.

2.4.5 Concessions and Leasing Arrangements

The leasing of concessions to a natural resource deposit can yield revenues which are identical to the *ex ante* rent from that resource. As mentioned earlier, however, it is important that the length of the lease corresponds to the useful life of the deposit. Most governments are reluctant to enter into sufficiently long-term leases for this purpose. This generally means that lease revenues will be less than what could have been collected otherwise.

In the case of non-renewable resources, the short term of the lease makes it difficult for operators to extract all the usable resources from the project. This leads them to offer a lower bid for the concession. It also induces them to engage in inefficient mining practices aimed at speeding up the extraction process. This generally reduces the value of the deposit to potential future operators. Even if the current operators have a right of first refusal on future leases, political and other uncertainties will cause them to discount this possibility and to shorten their time horizons in planning current activities. Therefore, short-term leases, even when offered consecutively, will generally yield less revenues than long-term leases for non-renewable resource deposits.

The same will generally be true in the case of renewable resources. However, in this case, short-term leases might yield much greater revenues over the early years of exploitation than would be obtained from perpetual leases. The reason is that short-term leases give the operator very little incentive to engage in investments in replenishment or renewal of the resource. In these circumstances, the operator would simply mine the first generation or rotation of the resource stock (assuming that this was the length of the lease) without regard for the consequences for future generations or rotations. Therefore, short-term "rents" might be much greater than with a long-term leaseholder, and renters might be willing to pay quite high prices for short-term leases. But, of course, what appear as rents to the short-term leaseholder are largely postponed investments in replenishment and/or the destruction of much of the potential for longer-term rents. The present value of the future stream of all rents that could be received would certainly be less with a succession of short-term leases than with one perpetual lease in the presence of these sorts of incentives. The burden on other types of regulation of leaseholder behavior is very great when leases for renewable resources are relatively short.

2.4.6 Measuring the Distorting Effect of Taxes

Up to now, our discussion of the effect of resource taxes has been largely qualitative in nature. For some purposes, it may be desired to obtain quantitative measures of the extent to which different tax instruments distort decisions. A conventional tool for doing so is the use of *marginal effective tax rates*. These were initially devised as ways of measuring the size of the distortion imposed by capital income taxes on the decision to invest in depreciable capital. However, they can be used to measure the tax wedge imposed on virtually any capital decision and have been applied to such things as inventory holding and non-renewable resource exploitation. Since the methodology for calculating marginal effective tax rates is somewhat technical, we have relegated it to an Appendix. It can be omitted without loss of continuity. In the Appendix, we illustrate the use of marginal effective tax rates in the non-renewable resource context, concentrating on capital investment and extraction decisions. We do so for fairly simple examples, ignoring such important complications as risk and the absence of full loss offsetting.

2.5 POLICY IMPLICATIONS

2.5.1 Introduction

We conclude this review by summarizing some of its implications for resource tax policy. It is useful to begin by briefly recalling the role of resource taxes and their place in the system of taxes. Resource taxes are part of the overall system of taxes which impinge upon the incomes of businesses. The system usually includes direct taxes of a general nature such as the corporation income tax and taxes on personal and unincorporated business income, indirect taxes of various sorts including sales and excise taxes as well as export and import duties, and taxes specifically designed for resource industries.

The system of income taxes is intended to tax capital and personal income of residents and where possible of non-residents earning income in the country of taxation. Such systems typically include both a personal tax system and a corporate tax system. The corporate tax system ought to be viewed as supplementary to the personal tax, that is, as a withholding tax on capital income earned in corporations. It essentially ensures that equity income earned in the corporation is taxed as it is earned, whether or not it is distributed. Many countries recognize this withholding role by integrating the corporate tax with the personal tax system through the use of measures such as dividend tax credits or dividend paid deductions from the corporate tax base. This essentially ensures that double taxation of equity income is mitigated. In the case of foreign corporations, the corporate tax also facilitates a tax transfer from foreign treasuries in cases in which foreign governments offer foreign tax credits. Host country tax systems are often (or should be) designed with this in mind. Interest income tends to be taxed at the personal level since withholding is not necessary here. Income taxes, if designed properly, tax all capital income on a uniform basis, including both the normal return to capital and any rents. Of course, the design of many tax systems is imperfect in the sense that this uniformity is not achieved.

One of the ways in which non-uniformity is evident is in the treatment of resource industries. In most countries, the capital income tax system treats resource industries quite favorably relative to other industries. This occurs mainly because of the favorable treatment afforded various capital expenses which are specific to the resource industries. For example, some items of a capital nature are given rapid write-offs in tax systems which are meant to be abiding by the accrual method of accounting. These include the costs of acquiring resource properties and exploration and development expenditures. Furthermore, double write-offs are often given by virtue of deletion allowances for resources used up. And many developing countries have traditionally given generous incentives in the form of tax holidays, investment tax credits, duty exemptions on imported equipment, and valuable loss carry forward provisions. The consequence is that equity income in the resource industries is often undertaxed relative to other industries. Some corporate tax systems also allow deductions for resource taxes paid. To the extent that this is the case, it vitiates the effect of the resource tax. To the extent that it is desirable to supplement general income taxes with resource taxes, this is undesirable.

The case for special resource taxes is precisely to tax resource rents over and above the levies that are implicit in general income taxes. There are two sorts of arguments for this. One is the efficiency-based argument that resource rents are non-distorting and therefore are an ideal source of revenue from an efficiency point of view. The other one, which is complementary, is that the property rights to resources ought to accrue to the public at large rather than to private citizens since they represent the bounties nature has bestowed on the economy rather than a reward for economic effort of some sort. This can be viewed as a sort of equity argument. However, one must be careful in applying it. In an economy with no resource taxes, the value of known stocks of resources will be capitalized into existing property values at least to some extent. If a government then imposes a new resource tax, the incidence of the tax will fall on the existing property owners or leaseholders. Thus, there will be redistributive effects to be accounted for. If the government is the principal owner of the resource properties, this will be much less of an issue, except to the extent that they have leased the resources on a long-term basis at a predetermined price that reflects the pre-resource-tax value of the rents.

In our view, the main reason for taxing resources over and above that of other general tax measures is precisely to acquire for the public sector a share of the rents generated from resources. In principle, special rent taxes could be imposed on other sectors. However, the argument is strongest for resource industries since those are where economic rents are most likely to reside.

Given that the main purpose of resource taxation is to capture rents, the appropriate form of taxation is one whose base is economic rents. We reiterate below the form that might take. For now, we simply note that actual resource taxes seem to differ from rent taxes in significant ways.

Unlike with the general income tax which includes provisions which allow the resource industries to understate capital income, resource taxes often overstate rents. This is because they frequently do not offer full deductions for all costs, particularly capital costs. Some systems tax revenues without giving any deduction for costs; others allow current costs to be deducted. As a consequence, they discourage investment activity in the resource industries, encourage the exploitation of high grades of resources at the expense of low grades, and make it difficult to impose high tax rates for fear of making the marginal tax rate greater than 100%.

2.5.2   *Policies for Capturing Resource Rents*

As we have discussed earlier, there are three alternative ways for the government to divert a share of rents to the public sector. They are as follows:

2.5.2.1   *Cash Flow or Cash Flow Equivalent Taxes*

The ideal sort of rent tax is a tax on the real cash flows of resource firms. For non-renewable resource firms, the base would include all revenues on a cash basis less all current and capital costs including costs of acquiring resource properties, exploration expenses, development expenses, and any processing expenses incurred by the resource firm. For renewable resource firms, similar costs would be deducted including costs of property rights, harvesting costs, any renewal costs such as replanting or restocking, as well as any processing costs done by the firm. There should be no deductions for other taxes paid. Of course, cash flow accounting should be done from a social point of view so any external costs should be included as costs on a cash basis. It may also be necessary to require the firm to cover the external cost associated with shutting down, though that may be done by forcing firms to post bonds and/or through other forms of regulation. Both corporations and unincorporated firms should be subject to the tax. This is a relatively straightforward type of tax to administer, though there are likely to be incentives to evade. For example, there is an incentive to engage in transfer pricing for vertically-integrated firms as a way of passing rents forward to non-resource firms. (Note, however, there is no disadvantage to extending the base as far forward as is necessary for a vertically-integrated firm since if there are no rents downstream, there will be no tax collected.) As well, there is an incentive to have capital income masquerading as wage and salary payments to avoid the tax. These are inevitable consequences of a tax which applies differentially to some activities and not to others. In principle, the cash flow tax rate could be extremely high, approaching 100%.

The public sector may balk at a full-fledged cash flow tax since it generally implies that tax liabilities will be negative for growing firms. Although the cash flow implications of these may be beneficial for the firms, governments can raise tax revenues only with some welfare cost and they may prefer a system which smooths tax receipts into the future. Such a compromise is easily achieved with a modified cash flow tax base in which the firm can capitalize cost deductions in a straightforward way. In particular, any costs which are capitalized receive a full nominal interest deduction based on the full book value of the capitalized cost. The rate of depreciation used for capitalization purposes is arbitrary. It may well be chosen by the firm subject to the constraint that tax liabilities cannot be negative. Such a system is equivalent to one in which negative tax liabilities are carried forward at full interest. It is therefore equivalent to a straight cash flow tax base.

2.5.2.2 *Auctioning of Leases or Property Rights*

Rents may be transferred to the public sector by requiring firms to bid for the rights to exploit resources. In the case of non-renewable resources, this would occur prior to the exploration stage. For renewable resources, the bid would be for a known stock of resources. As long as the bidding system was competitive and all bidders were equally well informed, the value of the bid would be equal to expected future net rents (net of future expected taxes) corrected for a risk factor. Furthermore, to ensure that optimal rents were obtained, the property rights obtained must be perpetual. If they were for a fixed term, there would be an incentive for the operator to extract the resource inefficiently.

Even with a well-functioning auction, the consequences can differ from that under a rent tax. For one thing, the auction will yield 100% of the expected value of the rents to the bidder, whereas the tax rate may be less than that. Under an auction, the cash flow consequences are much different as well. Net rents must be entirely paid up front, whereas with taxes they are spread out into the future. If there are any capital market constraints, this will be reflected in the size of the bid. Also, the risk effects can be different. Under the auction system, the firm is forced to bear the risk associated with resource exploitation whereas with the cash flow tax the public sector shares the risk. To the extent that the public sector is better able to pool or spread risk, the outcome may be more efficient. Of course, one important reason why the public sector may be better at dealing with risk is that some of the risk facing the operator is the risk of higher taxes in the future. The time inconsistency which gives rise to this will be more severe under a system, such as an auction, which captures rents up front. Thus, while this risk makes it more appropriate to use an auction system, it also reduces the price that bidders will be willing to pay for a long-term lease.

The auction may be inefficient for various reasons. If bidding is not competitive, it will not be efficient. Also, if the auction requires firms to bid not only on a once-and-for-all payment but also on a future royalty payment, the outcome will not be efficient since the firm will be induced to behave inefficiently in the future.

2.5.2.3 Public Sector Equity Participation

Finally, the public sector may obtain a share of the rents by taking on a share of equity in the firm in particular ways. One way of doing so is for the government to contribute to a share of the costs of exploiting a resource and claim an equivalent share of the equity of the firm. This would be financial exactly the same as a cash flow tax, though perhaps more difficult to implement. The public sector would have to identify both the cash costs and the revenues accruing on the relevant operation of the firm. On the other hand, unlike with a cash flow tax, if the public sector actually does become a full partner in the ownership of the firm, it presumably has a say in the decision-making responsibilities that come with share ownership. As well, it may be privy to information that it would otherwise not obtain. This is in contrast to cash flow taxation where the government is a silent partner.

The above method involves the government providing cash up front and obtaining revenues in the future. The government could become an equity participant while avoiding these cash flow consequences for itself. Instead of providing money up front, it could deduct its share of the costs later on against dividends. This is referred to as acquiring *free equity*. As long as the costs were appropriately deducted with interest, the scheme would be financially equivalent to the cash flow equivalent schemes outlined earlier.

As with taxation but in contrast to auctions, equity participation schemes will divert less than 100% of the rents to the public sector. Furthermore, there may be an issue in the case of foreign firms of the extent to which foreign tax credits can be claimed against home country governments. Of course, that may be an issue with resource taxes as well.

2.5.3 How Actual Policies Differ from True Rent Collection Devices

Revenue-raising policies actually used differ from those outlined above in their design. This implies that they are not pure rent-collecting devices, but distort decision-making as well. There may be various reasons for this, some of which involve other policy objectives by the government (e.g., capital income taxation, protection, etc.). However, it is also possible that policy makers are ill-informed about the proper design of rent-collecting devices, or that purely political factors are at work. Rather than second guessing the reasons, we simply discuss the ways in which actual measures deviate from optimal rent-collecting instruments. We concentrate largely on measures specific to the resource industries.

2.5.3.1 Tax Measures

Historically, it has been the exception rather than the rule that rent taxes have been used in the resource industries. Indeed, there are very few examples of cash flow type taxes. We consider the various taxes in turn.

i. *Royalties/Stumpage Fees/Severance Taxes.*
   Perhaps the most common form of resource charge has been a levy based on the quantity extracted, variously referred to as a royalty or severance tax in non-renewable resources and a stumpage fee in forestry. It is difficult to understand the attraction of this type of charge apart from simplicity. Sometimes these levies have been viewed less as a form of tax than as a fee charged by the public sector for removing resources from public or Crown lands. However, from an economic point of view, they are equivalent to a production tax. In their simple form, they tax revenues with no accounting for costs. As such, they act as a disincentive for investment and extraction of resources and coincidentally generate less revenue for the public sector than could be obtained by a rent tax. Furthermore, since no account is taken of costs, they discriminate against high-cost revenue sources at the expense of low-cost ones. This effect of crude royalty systems is generally known as high grading of the resource. In the case of mines, socially valuable but high extraction cost deposits are left in the ground. In selective logging operations,

lower value stems are left unharvested and are often damaged and left to rot in the forest. Also, since costs are not deducted, they do not serve as risk-sharing devices by the public sector, nor do they provide any assistance with the cash flow of firms as is the case with other measures. Against this must be set the fact that production taxes may have a role in correcting for externalities associated with resource production. However, this would not justify their use as primary revenue collection devices.

The effect of production taxes can differ according to whether the tax rate is based on quantity produced (per unit tax) or upon the selling price *(ad valorem)*. In principle, an *ad valorem* rate can always be chosen such that it is equivalent to a given per unit rate. However, when prices are changing, maintaining that equivalence would require constantly changing the tax rate. If the tax rates remain fixed while prices change, the two will have different effects. In particular, when prices rise, the *ad valorem* tax rate rises relative to the per unit and vice versa. This implies that the *ad valorem* tax has some risk-sharing effect that the per unit does not have, and in periods of rising resource taxes, it discourages investment more. Similarly, when the quality of a resource varies within a given deposit (e.g., less rich ore seams in a mine and different tree species within any part of a forest concession), maintaining equivalence between an *ad valorem* and specific tax rate would require different per unit rates for different parts of the deposit which is extracted.

Several countries have moved away from simple per unit royalty systems and export taxes in recent decades. These include Bolivia and Indonesia for hard minerals, Colombia for oil, and Jamaica for bauxite. Sabah and Indonesia have also moved in a similar direction in the case of tropical timber by varying the royalty rate by type of tree species.

Increasingly, royalty schemes have been designed to be more sophisticated than simple production taxes. There are two main ways in which this has been done. For one, some royalty bases have been defined to be revenues net of current costs. Sabah has refined its tropical timber royalties by allowing a deduction meant to represent presumptive logging costs. This goes part way toward making royalties reflect rents. The other method is to make the royalty rate itself a sliding scale based on either resource prices (an *excess price tax)* or the quality of the resource. These are sometimes referred to as *windfall taxes* reflecting the fact that purpose has been seen as a way of creaming off resource rents generated by price increases. Such sliding royalty systems have been used for oil (Peru and Malaysia for example), tropical timber (Sabah), coal (Indonesia), and tin (Malaysia). Again, this is an imperfect way of taxing resource rents in general, although the procedure of basing royalties on price can succeed in obtaining changes in rents from existing resource firms who have benefited from an unexpected increase in price. However, this is done at the expense of discouraging incremental investments. The latter can be mitigated in some instances by basing the royalty rate differentially on new and existing resource properties. Such a procedure will work only once.

ii. *Income-Based Taxes*

Resource properties are usually subject to general income taxes. However, in some instances, taxes specific to the resource industries are also based on some measure of income. In such cases, the tax is often designed in similar ways to the general income tax and has built into it some of the same biases. That is, it affords rapid write-offs for acquisition costs, exploration, and development and often gives a depletion allowance. Although this generates some revenues, it also has the effect of providing a subsidy to marginal projects. That is, average tax rates are positive while marginal tax rates are negative. Furthermore, the way such taxes have been implemented in most developing countries (e.g., for coal in Colombia and hard minerals in Indonesia) the rate of return to equity at which they become effective has tended to be extremely high. Thus, they have not been very effective collectors of excess profits or rents.

We have outlined earlier how income-based taxes could be designed to reflect economic rents, using a modified cash flow approach. However, such systems have not been used. Elements of cash flow taxation have appeared in some developed countries. For example, the mining tax regime in Alberta, Canada, has the following features. It is basically a cash flow tax except that a royalty is also applied until capital and start-up costs have all been deducted. A similar system is used by the Canadian government to tax oil and gas on federal Crown lands. Thus, the principle of cash flow taxation has not been completely ruled out. However, these systems are not fully efficient since they deny the full tax advantages of expensing all capital costs.

iii. *Property Taxes and Leasing Fees*

Some tax regimes impose an annual rental fee or charge for the use of resource properties. This is often done in the case of timber concessions and plantations in states of Malaysia. If their rates were such as to reflect the true capital value of the properties being used, they would be like a rent tax. However, they are typically set at arbitrary and more or less nominal rates. It would be difficult to administer such a tax based on the true economic value of the resource property in question since market values do not exist. Thus, some administrative discretion would be required. If an annual rent tax is to be charged, it seems preferable to use a proper rent tax.

iv. *Export Taxes*

Export taxes are frequently used in developing countries as a source of revenue from primary resources. In primary product exporting countries, they have been a major source of government revenue. In the case in which the country is a price taker on international markets, an export tax has exactly the same effect as a production tax from the point of view of the producers. However, consumers pay a lower price under the export tax. There may therefore be some distributive reasons for preferring an export tax, though it may be more for reasons of administrative simplicity. However, countries have found that export taxes on many resource products (e.g., rubber in Malaysia) have been quite regressive and have tended to eliminate these taxes in favor of other more general taxes on spending and income. In many cases, domestic consumption is a small proportion of production, and so, the differences in the revenue implications of production and export taxes may not be great. However, the efficiency costs arising from diverting high-value resources to lower value domestic uses depend not on the absolute value of domestic use relative to exports, but rather on the responsive of domestic demand to price changes caused by the export tax. Taxes on exports to induce local downstream processing industries can also be a very costly way of dissipating resource rents. Even in cases where the resource-exporting country might have a long-term comparative advantage in further processing, the use of export taxes to speed up the process can be very costly.

The same shortcomings of production taxes as rent collectors apply to export taxes. On the other hand, export taxes may be justified if the country has some monopoly power in world markets by the usual optimal tariff arguments. If so, that would be a separate justification for export taxes over and above rent collection devices.

2.5.3.2 Auction Systems

We have listed auction systems earlier as one of the ways in which rents can be extracted from resource producers up front. However, they tend not to be used much, especially in developing countries. Presumably, one reason is that the conditions do not lend themselves to competitive bidding procedures. Many resource projects are large and may not involve more than one different investor at the same time. For whatever reasons, individual deals are struck with resource producers involving different types of public participation. These can take various forms as discussed next. One feature of such contracts which distinguishes them from other arrangements is that they tend to involve a major element of administrative discretion. That may be viewed as a drawback from an economic point of view when compared with schemes for which eligibility and conditions are non-discretionary.

2.5.3.3 Production Sharing

There are various non-tax ways in which governments acquire shares of the proceeds of resource projects. Two common methods are by sharing of the output of production and government acquisition of equity shares in resource firms. Variants of the first of these are considered here.

The simplest case is that in which the government simply takes a given share of the product. The analogy would be a system of sharecropping in agriculture in which a landowner allows a tenant to farm a plot of land in return for a share of the crop produced. The basic scheme is identical to an *ad valorem* production tax at the same rate. It differs from a tax on pure rent since no costs are deducted. Since it is *ad valorem*, some risk-sharing is implicit in the scheme.

Since production sharing schemes are subject to negotiation, the proportion of sharing could vary from project to project. In this way, some account can be taken of different potential rents. However, as long as costs are not explicitly deducted, such schemes will not reflect pure rents.

Some schemes account for costs partially by having the production sharing cut in only after some minimum guarantee level of revenues for the firm (e.g., oil in Indonesia). As well as allowing the firm to cover some part of initial costs before sharing its output, this provides an additional measure of risk-sharing. However, even if the minimum were set such that total costs were covered, there would still be a marginal disincentive involved in such schemes once the production sharing begins to apply.

A variant on production sharing is a requirement that a certain proportion of production be "made available" to the domestic market. If such local market sales are at the prevailing world price, this does not transfer any rents. If the price is less, then some rents will be transferred, and it will be similar to a simple production sharing arrangement. Of course, if the sales at subsidized prices are to private traders, the rents will not accrue to the public sector. Lack of clear specification of the terms of such sales in the local market (including the price and the eligible buyer) can be a source of contention with resource investors (e.g., aluminum in Indonesia).

2.5.3.4 Equity Participation

Finally, governments may negotiate to adopt equity positions in resource firms. Again, this can take various forms, and the ability to obtain rents depends upon the form taken. At one extreme, the government could simply purchase shares of a resource firm on the open market. Divestiture of a given proportion of shares to local investors within a specified time period is a standard condition of foreign hard mineral investments in Indonesia. The government has often put forward as an obvious investor in such circumstances. Since the market value of the firm should capitalize all expected future net rents of the firm, this would not be expected to yield any net revenues to the government. All it would do is to provide the government with whatever decision-making authority goes along with share ownership. To facilitate rent transfer to the government, the government must succeed in obtaining shareholding privileges at below the market value of the shares.

At the other extreme, the government may simply take "free equity" in the firm, thereby entitling itself to a share of future dividends of the firm. This will differ from a rent tax regime by the fact that no implicit deduction is given for the initial equity put in by the firm. This may approximate the initial capital costs incurred by the firm. It would then be similar to a royalty system with current costs deducted. There are many instances of such free equity arrangements, especially in hard minerals (copper in Panama, copper and nickel in Botswana, and uranium in Gabon).

Instead of taking free equity, the government may pay some price for it. As mentioned, to obtain some share of the rents, the price would have to be less than the market price of the shares taken. This could be done up front or it could be made later by reducing future dividends. Equity sharing schemes of this form will be equivalent to rent taxes if the payment made by the government is equal in present value terms to an equivalent share of the cash costs of the project. If this payment is made up front, it would have the identical financial effect as a cash flow tax. The only real difference is that the government obtains voting rights. If the payment is spread out into the future (e.g., taken out of future dividends), it should be carried forward with interest. In either case, the government will obtain only a share of the rents rather than the entire rents under an ideal auction system.

2.5.4 Other Design Issues

There are a number of other design issues involved in resource taxation which may cause them to differ from ideal rent taxes. Some of them are as follows:

i. *The Time Horizon*
As mentioned, arrangements with the private sector for sharing rents may be viewed as being for a limited period of time. This may be because of conscious design, as in the case of forestry concession. Or, it may be because of the inevitable inability of governments to commit to fixed policies for long periods of time. In any case, the result is an inefficiency which is hard to avoid.

iii. *Shut-Down Costs*
Many non-renewable resource operations face costs of shut down such as cleanup costs to avoid environmental damage. Simply requiring firms to meet such costs may be unenforceable since they may be able to avoid them by just abandoning the site. Cleanup could be enforced by requiring the firm to post bonds against the cost of cleanup, or, equivalently, by imposing a withholding tax in respect of resource management which is refundable once the cleanup is completed.

iii. *Discretionary Policies*

Some sorts of policies may involve administrative discretion. Economists generally view these sorts of policies with some suspicion and prefer those for which the terms of eligibility are automatic. Discretionary policies lend themselves to costly rent-seeking activities as well as to possibilities for dishonest behavior.

iv. *Jurisdictional Issues*

In many countries, jurisdiction over resources is decentralized at least partly to lower levels of government. Examples include Malaysia and Canada. This can give rise to problems of tax coordination among various levels of government as well as to different fiscal capacities among lower levels of government. As the literature on fiscal federalism makes clear, the latter can cause inequities across the federation and inefficiency in the allocation of mobile factors of production in favor of the wealthier states. Many countries have instituted mechanisms to enable at least some share of resource rent to be shared among states.

v. *International Aspects*

Many of the firms that operate in less developed countries are foreign firms. This gives rise to various other issues. For one, certain tax measures may be preferred to others to the extent that foreign tax crediting is facilitated. Use of the income tax system rather than free equity or production sharing arrangements may have that property. As well, the ability of foreign companies to shift profits through transfer pricing and other means will limit the extent to which such types of taxes on resource rents will be effective. This may help to account for the growing use of other measures such as royalties, equity participation, and leasing of property rights.

2.5.5 *Conclusion*

Developing country governments have become increasingly conscious of the desirability of levying taxes on economic rents arising from natural resources occurring within their boundaries. At the same time, they have shown increasing sophistication in modifying the crude fiscal instruments that have been traditionally used for this purpose in order to both decrease the efficiency costs arising from the use of imperfect rent taxes and increase the proportion of the rents that they are able to attach for public purposes. The time has now been reached in many countries at which the gains from further refinement of what are basically very crude taxes such as royalties and export levies might be far exceeded by replacing them with much simpler forms of pure rent taxes.

Appendix: Measuring Marginal Effective Tax Rates in Resource Industries

The marginal effective tax rate measures the difference between the pre-tax rate of return on the marginal investment and the after-tax return to savers. The latter can be inferred from observed market rates of return. The former is more problematic because the marginal investment project cannot be identified. Instead, the return on the marginal investment project is inferred from the user cost of capital. Consider, for example, the case of depreciable capital discussed above. The value of the marginal product of one unit of capital in real terms is given by (2.9′). To convert it into a rate-of-return expression, two steps must be taken. First, the entire expression is divided through by q_t so it represents the marginal product per dollar of capital. Then, to make it a rate of return, the economic depreciation rate $(\delta - \Delta q/q)$ is subtracted out. This leaves r as the rate of return on the marginal investment. That is also the rate of return on saving, so the marginal effective tax rate is naturally zero in the absence of taxes.

Suppose now we take a very simple, but representative, corporate tax system. Let the rate of depreciation for tax purposes be σ applied on an historical basis to undepreciated capital. Suppose that interest deductions are allowed on debt, but no deductions are allowed for the costs of equity. Also suppose that there is an investment tax credit in place at the rated ϕ based on gross investment. The tax rate is u. Then, it can be shown that the expression for the value of the marginal product of capital (2.9′) must be amended as follows:

$$p_t F k_t = \frac{q_t\left(\delta + r - \frac{\Delta q_t}{q_t}\right)}{1 - u}\left(1 - \phi - \frac{u\sigma}{r + \sigma}\right) \quad (2.9'')$$

where r is the real cost of funds to the firm. Suppose a proportion β of the firm is financed by debt and the rest by equity, and the nominal costs of debt and equity are i and ρ, respectively. Then, with interest deductibility,

r is given by:

$$r = \beta i(1-u) + (1-\beta)\rho - \pi \tag{2.10}$$

In interpreting Eq. (2.9''), note that $u\sigma/(r+\sigma)$ is the present value of future tax savings due to depreciation. Thus, given the investment tax credit, the second bracketed term on the right-hand side of (2.9'') can be thought of as the effective price of new investment.

The pre-tax rate of return can be constructed as above. It is given by:

$$r_g = \frac{\left(\delta + r + \frac{\Delta q_t}{q_t}\right)}{1-u}\left(1 - \phi - \frac{u\sigma}{r+\sigma}\right) - \delta + \frac{\Delta q_t}{q_t} \tag{2.9''}$$

Given the tax parameters and estimates of the true depreciation rate and the cost of funds to the firm, r_g can be calculated. To obtain the marginal effective tax rate, the after-tax rate-of-return r_n must be subtracted from r_g. The after-tax rate of return is given by $r_n = \beta i + (1-\beta)\rho$.

Next, we want to apply the same methodology to a non-renewable resource firm. We consider a firm which is simultaneously involved in exploration, investment in mining facilities, and extraction. Inventories are excluded so that sales equal extraction; it would be relatively straightforward to add inventories. The taxation of resources is notoriously complex in practice. For illustrative purposes, we consider a relatively simple scheme which incorporates most of the key issues.

In the exploration stage, the firm hires current inputs L at a price W and produces a depletable asset according to the strictly concave function $S(L)$. (We are deleting time subscripts for simplicity.) It then invests in mining capital K at a price Q to make the asset ready for extraction. The production function is $Z(K,F)$ where F is the current use of previously discovered asset. This is the only stage at which depreciable capital is used, though it would be straightforward to allow for it at either of the other two stages. Finally, the firm extracts an amount Y of the resource according to the strictly convex nominal cost function $C(Y)$ and sells it at a price P.

The tax regime facing the firm consists of two taxes—a corporate tax and a simple royalty or severance tax based on total revenues. The corporate tax involves write-off provisions for depreciation and interest costs and an investment tax credit as above, as well as some deduction for the use of the asset itself (a depletion allowance). We assume a royalty tax rate of g based on total revenues. The corporate tax liability will be written:

$$T_c = u[PY - C(Y) - WL - \sigma A - R - iB] + \phi QI$$

where A is the accounting value of the capital stock for tax purposes. Here, R is the depletion allowance and is defined to be:

$$R = t(PY - C(Y) - \sigma A)$$

though most systems are more complicated than that. All other variables are the same as defined earlier.

Given this, the expression for the cash flow of the firm is defined to be:

$$\begin{aligned}CF = {} & PY(1 - u(1-t) - g) - C(Y)(1 - u(1-t)) \\ & - WL(1-u) - Q(1-\phi)I + \sigma A u(1-t)\end{aligned}$$

where the accounting capital stock is defined as in (2.6) and investment is related to the real capital stock as in (2.5).

The firm maximizes the present value of its cash flow discounted by the nominal cost of funds $r + \pi$ defined by (2.10) and subject to the following two resource constraints:

$$\int_0^\infty (Y - Z(F, K))dt \leq 0$$

$$\int_0^\infty (F - S(L))dt \leq 0$$

The first states that the total resource extracted cannot exceed the total developed, while the second states that the total resource developed cannot exceed the total found. (In a more general version of this problem, this constraint would have to hold at each point in time.) The solution to this problem yields the following marginal conditions to be satisfied:

$$\frac{p-c'}{q}Z_K = \left(\frac{\delta + r - \frac{\Delta q_t}{q_t}}{1 - u(1-t) - g\frac{p}{p-c'}}\right)(1-\phi) - \frac{\sigma u(1-t)}{r+\sigma}$$

$$\frac{p-c'}{q}Z_F S_L = \frac{1-u}{1 - u(1-t) - g\frac{p}{p-c'}}$$

$$\frac{\Delta(p-c')}{p-c'} = r - \frac{rg}{(1-u(1-t))\left(1-\frac{c'}{p}\right)}$$

The first of these is simply the pre-tax marginal product of capital. To convert it to r_g simply subtract $\delta - \Delta q/q$ as before. The second equation is the social value of marginal product per unit of the current input L. An effective tax rate can be obtained directly by subtracting unity from it. The final equation is a form of Hotelling's rule. It gives the pre-tax rate of return to society from not extracting the resource. It can be converted to an effective tax wedge by subtracting r_n. These can be used to calculate marginal effective tax rates for a given institutional setting. Notice that the corporate tax and the royalty system interact in each of the decisions of the firm—the current input decision, the depreciable capital input decision, and the extraction decision.

Suggested Readings

For a general treatment of the economics of both renewable and non-renewable natural resources, see:

Hartwick, John M. and Olewiler, Nancy D. (1986). *The Economics of Natural Resource Use* (New York: Harper and Row).

For a survey of the literature on taxes and other instruments for obtaining revenues from, and regulating, various types of natural resources (fisheries, forestry, mining, oil and gas, and hydro-electricity), see:

Heaps, Terry and Helliwell, John F. (1985). "The Taxation of Natural Resources". In Alan J. Auerbach and Martin Feldstein (eds.), *Handbook of Public Economics,* Volume I (Amsterdam: North-Holland), 421–72.
Boadway, Robin and Keen, Michael. (2010). "Theoretical Perspectives on Resource Tax Design". In Philip Daniel, Michael Keen and Charles

McPherson (eds.), *The Taxation of Petroleum and Minerals: Principles, Problems and Practice* (London: Routledge), 13–74.

Boadway, Robin and Keen, Michael. (2015). "Rent Taxes and Royalties in Designing Fiscal Regimes for Nonrenewable Resources". In Robert Halvorsen and David F. Layton (eds.), *Handbook on the Economics of Natural Resources* (Cheltenham, UK: Edward Elgar), 97–139.

A general outline of the special problems of taxing natural resources in developing countries may be found in:

Gillis, Malcolm. (1982). "Evolution of Natural Resource Taxation in Developing Countries". *Natural Resources Journal* 22, July, 620–48.

A survey of the theory and calculation of marginal effective tax rates, including alternative approaches and applications, may be found in:

Boadway, Robin W. (1987). "The Theory and Measurement of Effective Tax Rates". In J.M. Mintz and D.D. Purvis (eds), *The Impact of Taxation on Business Activity* (Kingston, Canada: John Deutsch Institute), 60–98.

An application of the role of rent taxation and the concept of effective tax rates to non-renewable resources is developed in:

Boadway, Robin W., Bruce, N., McKenzie. Kenneth. J., and Mintz, J.M. (1987). "Marginal Effective Tax Rates on Capital in the Canadian Mining Industry". *Canadian Journal of Economics* 20 February, 1–17.

Some general issues of taxation in developing countries are surveyed in:

Newbery, D.M.G. and Stern, N.H. (1987). *The Theory of Taxation for Developing Countries* (Washington: The World Bank).

For a treatment of the effects of taxes on investment in developing countries, see:

Shah, Anwar (ed.). (1995). *Fiscal Incentives for Investment in Developing Countries* (London and New York: Oxford University Press).

CHAPTER 3

Green Taxes and Policies for Environmental Protection

Neil Bruce and Gregory Ellis

3.1 INTRODUCTION

This chapter presents an overview of corrective green taxes and public policies to deal with environmental externalities. The chapter discusses the theoretical rationale for public intervention, highlights tax and regulatory policies to advance public interest in environmental protection, and elaborates the efficiency and equity implications of various policies. It provides welfare analysis of the relative merits of different policy instruments for pollution control. In this context, significant consideration is

---

As both the authors are deceased. any communications regarding this chapter may be submitted to the editor at Email: shah.anwar@gmail.com.

N. Bruce (✉) · G. Ellis
University of Washington, Seattle, WA, USA

© The Author(s), under exclusive license to Springer Nature Switzerland AG 2023
A. Shah (ed.), *Taxing Choices for Managing Natural Resources, the Environment, and Global Climate Change,*
https://doi.org/10.1007/978-3-031-22606-9_3

given to the importance of uncertainty, asymmetric information, enforcement, and fiscal policy objectives. The chapter concludes by offering a few policy recommendations.

3.2 The Foundations of Environmental Policy

In this section, we briefly discuss the rationale for government policies toward environmental control and regulation, and the role that the tax system can play. The criteria of economic efficiency and cost minimization are stressed, although possible conflicts with distributional considerations are noted. It introduces criteria for evaluating alternative policy instruments for achieving environmental improvements associated with pollution control.

3.2.1 Rationale for Public Policies: Missing Markets

A market economy relies on price and profit signals to direct resources into highly valued uses. Firms seeking to maximize profit and consumers seeking to acquire material well-being are all led to achieve their ends at least cost to themselves. While such private cost-minimizing behavior is a social virtue when goods and factors are priced to reflect their costs to society, it results in economic inefficiency and reduced social welfare if markets are missing and externalities are present. In particular, it results in excessive pollution and environmental degradation.

Environmental quality can be considered an economic good and the degradation of the environment caused by other economic activities can be considered as an input or cost into those activities. Unlike most commercial goods, environmental quality is naturally endowed rather than produced. But the production and/or consumption of other commercial goods may reduce the level of environmental quality, therefore, it is in variable supply like other goods. For example, the supply of clean air and water, which are valued for their own sake, are used up by production processes that dump waste products into the environment.

The demand for environmental quality comes from people who wish to enjoy air and water which is clean and safe to breathe and drink. Like any good, the willingness to pay for more of it declines as the amount of it available rises and rises as the ability to pay for it (household income) rises. The supply of environmental quality comes from producers and consumers of pollution-generating activities who supply more of it when they reduce the level of polluting activities or when they purchase equipment that reduces the amount of pollution caused at given levels of production. The cost of providing more environmental quality is the net value of foregone output (that is, value of output less the value of the resources used) or the extra costs of the pollution abatement equipment, respectively. Normally, we expect that the marginal cost of "supplying" an extra unit of environmental quality to rise as the amount supplied rises (i.e., as the amount of pollution abatement rises). The "optimal" level of pollution occurs where the marginal willingness to pay for an increase in environmental quality is just equal to the marginal cost of supplying it.

Environmental degradation arises as an economic policy problem because of a market "failure" or a "missing" market. There is no way for demanders and suppliers to express their relative willingness to pay for, or marginal willingness to accept a reduction in, the quantity of environmental quality. Correspondingly, there is no price to be paid by firms and consumers who degrade environmental quality by their activities. Polluters treat the degrading of environmental quality as practically costless to themselves and ignore the costs they impose on others. When an input is free, a cost-minimizing producer wants to use a lot of it, so excessive environmental degradation results. But the degradation of the environment is not free to the economy as a whole. Rather, high social costs are imposed on the economy in terms of reduced recreational opportunities, health hazards, reduced productivity of workers, general unpleasantness of day-to-day life, etc.

Why does a market for environmental quality not exist like those for other goods? The reasons have to do with the absence of private property rights and the fact that environmental quality is a public (i.e., non-rival) good. In order for something to be priced by the market, it is necessary to have a legal right to control its use. The environment is owned by everyone and hence by no one. A "common property" cannot be priced for its use and therefore there is competitive overuse (Afzal et al., 2015; Pearce, 2014).

Coase (1960) pointed out that such overuse is not an inevitable outcome. In principle, the demanders of higher environmental quality should be willing to find some way to "bribe" polluters to reduce the level of pollution to the efficient level. This doesn't happen because environmental quality is also a non-rival or public good. Clean air purchased for oneself yields benefits to everyone, but there is no way the purchaser can charge for the benefits he provides to others. Moreover, since the marginal cost of an extra consumer of environmental quality is zero, it would not be optimal to charge a price even if it were possible. As a result, no individual has much incentive to pay polluters to reduce their pollution. Collective action is needed to prevent free-riding.

3.2.2   The Role of Government Policies

The above suggests the rationale for government policy. One possibility is to force payments by the people who enjoy the increased environmental quality. In this case, a tax is placed on everyone in the economy and the proceeds are used to pay the polluters to reduce the level of pollution they cause. This is a "consumers pay" policy.

Coase's proposition suggests another possibility. According to Coase, it doesn't matter whether the property rights are given to the polluters or to the consumers of environmental quality. In the latter case, the consumers can demand compensation from the would-be polluters. When environmental quality is very high, the amount polluters would be willing to pay to degrade the environment by some amount is more than the people need to receive (their marginal willingness to accept) in order to tolerate some amount of degradation. Thus polluters are willing to pay for and households are willing to accept some level of environmental degradation. This is the "polluters pay" scenario.

This outcome can be achieved if the government charges an emissions fee to producers for polluting the environment. Alternatively, it could regulate the level of pollution that firms can do. Both of these are "polluters pay" policies. With lump—sum taxation, either a "consumers pay" or a "polluters pay" policy can achieve the economically efficient level of pollution (environmental quality). The two types of policies differ in terms of their distributional impact, their administrative ease, and the revenue implications for the public sector.

3.2.3   Pigouvian Taxes for Environmental Control

As described above, pollution levels are excessive because polluters do not bear the full social cost of their actions. Over seventy years ago, Pigou suggested that the government should impose taxes on activities that involve external social costs and provide subsidies for activities that confer external social benefits. "External" denotes costs and benefits which are not incorporated into the market prices faced by private economic decision-makers.

Consider the act of consuming a gallon of gasoline which entails an external cost. If the gasoline market is well-functioning in other respects, the consumer pays the full marginal cost of production in the purchase price. But when the consumer burns the gasoline in an automobile engine, another social cost is incurred which the consumer does not pay. The consumption of the gasoline contributes, albeit slightly, to the level of air pollution in the area. A small increment in air pollution in a large population can have a finite marginal cost because air pollution is a "public" bad—that is, it is a bad incurred on many people in the community. The consumer of gasoline ignores this part of marginal social cost when deciding whether to consume an extra gallon of gasoline.

Because the consumer does not pay the full social cost of burning gasoline, the activity appears cheaper than it really is, the idea of the Pigouvian tax is to impose a tax on gasoline equal to that part of the marginal social cost which is not included in the production price—the external marginal cost. The tax-inclusive price faced by the consumer is then equal to the marginal social cost of the product. For example, if the production cost of gasoline is a dollar per gallon and its combustion increases the social cost of pollution by 10 cents, then the marginal social cost of a gallon of gasoline is $1.10. The consumer pays only a dollar per gallon in the absence of government policy, but with a 10 cent Pigouvian tax, the consumer will perceive the socially correct price of $1.10.

In the above example, the Pigouvian tax achieves the outcome that would have occurred if, somehow, the consumers of air quality were able to charge the gasoline consumers for the costs of the pollution. It is not necessary, indeed under some circumstances it is undesirable, that the proceeds of the Pigouvian tax be used to compensate consumers of the air quality for their loss. The revenue collected can be added to general revenue and used to make overall reductions in tax rates or to purchase public goods.

The separation of the efficiency and distributional impacts of correcting pollution levels means that there are alternative ways of imposing the Pigouvian tax. For one, a general tax can be imposed on the population with the revenue used to bribe consumers of gasoline to reduce their consumption. In this case, the government offers consumers (say) a ten cent per gallon payment to reduce their gasoline consumption. Again, the consumer of gasoline perceives the cost of consuming gasoline as $1.10, dollar for the gasoline plus the ten-cent foregone payment from the government. Although the incentive to reduce gasoline consumption provided by the two policies are the same, they differ in their distributional impact—in the first case the consumer is worse off, in the second he is better-off.

While the concept of a Pigouvian tax is simple, it may be difficult to implement due to imperfect information and monitoring costs. These difficulties are discussed in detail in subsequent sections. It is useful, however, to discuss what the policy can accomplish in abstraction from these difficulties (Jenkins and Lamech, 1994).

Two efficiency objectives should be achieved in reducing the level of pollution. First, a given amount of pollution abatement should be accomplished at least social cost. This objective is sometimes called "cost-effectiveness". It should be stressed that an improvement in the level of environmental quality will be costly to the economy. Other economic activities will have to be curtailed and their value to society foregone, or more costly methods of production or consumption must be used. An advantage of Pigouvian taxes, and other "market-based incentives", is that they automatically achieve the reduction in pollution at least cost.

Second, it is desirable that the right amount of pollution abatement be done. Too much pollution abatement is undesirable as well as too little. Accomplishing this objective requires that the Pigouvian tax be set at a rate such that the economically efficient level of pollution is attained. This is a more information ally-demanding objective than cost-effectiveness.

A third objective must also be kept in mind—the distribution of income in the country. Different environmental policies have different distributional impacts and this may be an important consideration.

We now examine in more detail the policy of Pigouvian taxes and its ability to achieve the stated objectives.

i. *Cost-Effectiveness*
Continuing the gasoline example, suppose that there are two types of gasoline consumers. Type I, who commutes to work from a location not served by public transport, finds it very costly to reduce gasoline consumption and would be willing to pay a high price to continue using gasoline for commuting purposes. Type II drives mostly for pleasure and would not be willing to pay a lot to continue gasoline for this purpose. Pollution abatement

using achieved by curtailing type I's gasoline consumption has high marginal cost while the abatement achieved by curtailing type II's consumption of gasoline has low marginal cost.

If the government mandates less use of gasoline by all consumers, say in equal amounts or proportions, pollution abatement would not be accomplished at least social cost. At the mandated total gasoline consumption level, letting type I consume an extra gallon and requiring type II to reduce by an extra gallon would leave total consumption unchanged but yields social cost savings. The least cost method of pollution abatement requires that the low abatement cost consumers reduce gasoline consumption more than the high abatement cost consumers. The least cost abatement policy requires that each type of consumer reduces his consumption of gasoline until the marginal cost of abatement, and therefore the marginal willingness to pay for a gallon of gasoline, is equal to that of every other type.

To mandate pollution abatement at the least social cost, the government would have to know the marginal abatement cost for each type of consumer and set mandated consumption levels accordingly. The advantage of the Pigouvian tax is that this is done automatically. The government simply imposes a Pigouvian tax sufficient to reduce total gasoline consumption to the desired level. The consumers with high costs of reducing gasoline consumption will reduce their consumption by a small amount with the bulk of the reduction coming from consumers with low costs of reducing consumption. In this way, the reduction is accomplished at least social cost.

ii. *Economic Efficiency*

To accomplish an efficient level of pollution abatement, the Pigouvian tax must be set so that the tax-inclusive price of the activity is equal to its total marginal social cost. Facing this market price, private economic decision-makers will undertake the activity only if the marginal benefits are at least this high, and are thereby be led to make efficient choices.

There are many difficulties in implementing this policy. First the external marginal social cost element must be determined so that the Pigouvian tax can be set equal to it. Compounding the problem is the fact that this cost element may depend on the level of the activity undertaken. For example, the marginal social cost of pollution resulting from burning an extra gallon of gasoline may be low when gasoline consumption is high. The simplest cases occur when the external cost element is invariant to the amount consumed and the Pigouvian tax can be imposed at a rate which is fixed per unit of the activity, or when the external cost element is a constant fraction of the market price so the Pigouvian tax can be imposed at a rate which is fixed as a percentage of market price. Either case would be fortuitous.

In practice, the best alternative is often to determine the level of the polluting activity which is most desirable and set the tax to achieve that level. This can be done using estimates of the elasticities of market supply and demand. If the supply curve is horizontal, the appropriate tax rate can be determined directly from the price elasticity of market demand.

The Pigouvian tax is imposed over and above any tax that is imposed for revenue purposes. For example, gasoline may already be taxed as part of a country's value—added tax. The Pigouvian tax is added to this "revenue tax", A Pigouvian subsidy to an economic activity with a positive externality implies a tax rate below the going revenue tax rate. The tax rate may be, but is not necessarily, less than zero (i.e., a nominal subsidy).

iii. *The Distributional Impact*

A tax on a particular economic activity raise the price to the consumer and/or lower the price to the producer. These price effects, and subsequent effects in other markets, determine the incidence of the tax. In the simplest cage where the supply curve is horizontal, the price to the consumer rises by the amount of the tax. The relative incidence of this tax across consumers depends on the income elasticity of the taxed good. If the good is income inelastic (elastic), the budget share of the good rises as the income of the consumer fails (rises) implying that the tax burden is distributed regressively (progressively).

An important problem arises if a Pigouvian tax falls on goods which have low income elasticity. To the extent that the tax increases the price to consumers, the burden will be borne disproportionately by lower income groups. A potential conflict exists in this case between environmental and distributional policies. We briefly discuss the importance of this conflict and possible remedies.

One issue is the extent to which low income elasticity of the taxed good in fact implies regressivity. First, as Poterba (1991) shows in the case of gasoline, consumption may be inelastic with respect to income but proportional or even elastic with respect to expenditure. Expenditure increases less than proportionally with income so that, in principle, all goods for present consumption could be income inelastic. What matters is whether a good is more or less income elastic than the average, Poterba argues that the expenditure elasticity of a good may be a better criterion for determining whether a tax is regressive, Also, empirical studies indicate that annual expenditure tends to be proportional to "permanent" or lifetime income, which may be a better indicator of ability to pay than annual income. Income which is saved will be spent by the household in the future and does not escape taxation altogether. The burden of those future taxes is ignored when judgments concerning the progressivity/regressivity of a tax are based on how current spending on the taxed good is related to current income. As an alternative, estimates of lifetime incidence of a commodity can be used, although these are not readily available, particularly for developing countries.

A second consideration in determining the distributional impact of a Pigouvian tax is the distributional impact of the resulting improvement in environmental quality. If low income households benefit disproportionately from environmental improvements, then a Pigouvian tax on an income inelastic good may not be a regressive policy overall. However, the burden of the Pigouvian taxes can exceed the efficiency gains because of the so-called "primary" or revenue burden of the tax. As a result, it is unlikely that the poor will be made better-off from high Pigouvian taxes on goods with very low income (expenditure) elasticities even if they do gain disproportionately from the resulting improvements in environmental quality. Of course, if the rich benefit disproportionately from environmental improvement, the re-distributional impact is even more perverse.

The distribution of the benefits of pollution control and environmental improvement may, in fact, disproportionately favor the poor in developing countries. The poor currently have less access to safe drinking water and adequate sanitation than those with higher incomes (Munasinghe, 1990). The poor are also more likely to suffer from indoor air pollution problems (Krupnick, 1997). To the extent that environmental policies improve these particular pollution conditions, the poor are more likely to benefit. Eskeland and Jimenez (1991) note that the poor are more likely to benefit from pollution control also because they tend to live in poor health and sanitary conditions in polluted urban areas and cannot afford to protect themselves or move. However, Eskeland and Jimenez also point out that some empirical evidence suggests that the willingness to pay for environmental improvement among wealthier individuals may be higher than that among the poor, and such differences could make the wealthy the principal beneficiaries of environmental improvement.

Where Pigouvian taxes do lead to regressivity, some alternatives are available. The first is to make a compensating change in other components of the tax system. For example, if the country levies a personal income tax, it can be made more progressive, perhaps by giving an additional tax credit to low income taxpayers. In developing countries that depend on commodity taxation for revenue purposes, tax rates can be reduced on basic foodstuffs or low-priced clothing. This may increase the costs of administering the commodity tax system, however.

A second alternative is to structure the policy as a subsidy. Rather than taxing a polluting good with low income elasticity, the government should provide a subsidy for conserving it. Thus, instead of taxing home heating fuel, the government can subsidize home insulation. Unfortunately, rather than generating government tax revenue, this type of policy raises the government's revenue needs. These are already severe in many developing countries.

A third alternative is to ration the available consumption of the negative externality good and distribute the rations disproportionately toward the poor. This type of policy is very unlikely to achieve pollution abatement at least cost unless a gray market in ration coupons is permitted. If not, the distributional effects of the tax instrument would have to be quite adverse before the rationing instrument should be considered.

3.3 Possible Environmental Policy Instruments

The government has a number of instruments available for pursuing policies aimed at improving environmental quality. In this section, we describe the set of instruments from which the government may choose and in the next section, we discuss the criteria for selecting one instrument over another and identify the circumstances in which the Pigouvian tax is likely to be the preferred instrument.

3.3.1 Assignment of Relevant Property Rights

In Sect. 3.2, we argued that a major reason why there is excessive pollution in the first place is that, typically, environmental quality is a common property resource. Often, there are no private property rights established for an environmental resource so everyone is free to use it. When the use of the resource by one individual reduces its availability to others, it imposes costs on them but no price reflects this fact to the user since there is free access.

For example, suppose there are many users who have free access to a lake. Each desires clean water but in using the lake, he reduces its cleanliness for others. With many users and free access, no user has the incentive to maintain the cleanliness of the lake. Suppose, however, that private property rights are established to the lake allowing the owner, in effect, to sell the use of the water to the various users. The price for which the water can be sold, and hence the profitability to the owner, depends on its cleanliness. In this case, the private owner has an incentive to maintain the cleanliness of the lake at its economically efficient level.

In many cases, the assignment of property rights is not a very good instrument to accomplish environmental policy objectives. The assignment of property rights could create a monopoly, or it may be impossible for a private owner to monitor the use of the resource by others and therefore charge the appropriate price. Also, many environmental assets have a "public good" quality to them, so it is not to be efficient to charge prices for enjoying the asset in some of its uses. In these cases, other solutions may be possible.

3.3.2 Marketable Pollution Quotas

A related idea is a policy of marketable pollution quotas. This policy may emerge naturally from a policy of regulation. Under a regulatory policy, firms are limited in how much pollution they can cause—for example, how much Sulfur dioxide they may emit into the air or how much effluent they can dump into a watershed. Often these limits or pollution "quotas" are the same for all firms. Typically, different firms have different costs of pollution abatement. As discussed in the previous section, the cost of pollution abatement is not minimized if firms with different marginal abatement costs are required to do the same amount of abatement.

Marketable pollution quotas are a method of ensuring that pollution abatement is done at least cost. Keeping the total amount of pollution permitted constant, the government can allow firms to "sell" their pollution quotas to other firms. Firms with low marginal costs of abatement are willing to sell their quotas and firms with high marginal costs of abatement are willing to buy them at some intermediate price. The pollution quota market is in equilibrium when the price of a pollution quota is just equal to the marginal cost of pollution abatement to all polluters. In this way, the least cost pollution abatement is obtained for a regulatory policy, as in the case of a Pigouvian tax. For this reason, marketable pollution quota and Pigouvian taxes are lumped together as market-based incentives (MBIs). In fact, except for distributional impact of the policies and the fact that Pigouvian tax policy raises revenue, the two policies are equivalent under perfect information.

A policy of marketable pollution quotas can be carried out at different levels of formality. The most informal policy is simply to let pollution quotas be traded within the firm. In this case, a firm can increase the pollution emitted by one of its plants if it makes a compensating or more than compensating reduction in pollution emission by another. Alternatively, the firm may increase emissions in one year if it decreases them in another. These informal trading arrangements are incorporated in policies variously referred to as offsets, bubbles, and banking. An even more market-oriented approach is to allow inter-firm trading. This can range from informal trading among firms to a formally established market in pollution rights like that recently announced by the Chicago Board of Trade for Sulfur dioxide emissions (in the global context, see Larsen and Shah, 1994).

3.3.3 *Indirect Taxes on Inputs and Outputs*

Rather than taxing the pollution-causing activity itself, the government may levy excise taxes on outputs and inputs closely associated with the pollution-causing activity. This approach has the advantage that the government may already have in place an indirect tax system on goods and services. Thus, environmental policy can be accomplished simply by setting the existing tax rates to incorporate a Pigouvian element. Also, taxable outputs and inputs usually are readily monitored as part of raising public revenue. This policy would be as good as taxing the pollution-causing activity if the latter occurs in fixed proportions with the taxable output or input.

One difficulty that taxing an output or input is that it may be too blunt of an instrument. In the gasoline example of Sect. 3.2, an output tax on gasoline is used instead of taxing the pollution-causing activity burning of the gasoline. Since gasoline is generally purchased only to burn it, and if all methods of burning gasoline contribute equally to air pollution, the output tax is almost as good of a policy as taxing the burning of gasoline. But suppose gasoline can be used in ways that do not cause air pollution? A gasoline tax would discourage these socially harmless activities as well as those that cause pollution. In the process of trying to correct one economic inefficiency, another would be created.

In fact, it can be shown that a small tax on gasoline will improve economic efficiency even if gasoline is used for other, harmless purposes. But the existence of the harmless uses limits the amount of welfare improvement that can be attained through this policy. Taxing output is a "second-best" policy, in that there is an additional cost element to the policy—the cost of discouraging socially harmless uses of the output. Whether one has to settle for second best or choose another instrument can only be determined with further analysis and information.

A special example of this problem occurs when pollution is caused by consumption of the output (say, gasoline) but can be mitigated by the purchase of a pollution-abating input (say, catalytic converters for cleaner burning). Again an output tax will not accomplish the least cost method of reducing pollution. The output tax by itself provides no incentive to purchase the pollution-abating inputs even though they may be the least cost method of reducing pollution. Similarly, subsidizing the abating inputs provides no incentive to reduce polluting output, in fact it may increase it.

The least cost method of reducing pollution can be achieved, however, if an output tax is imposed on a polluting firm and some of the revenue is used to subsidize the purchase of the pollution-abating inputs. This policy would leave unchanged the relative cost of reducing pollution by decreasing output or by increasing abatement inputs since the after-tax price of the output and the pollution-abating are reduced in the same proportion. This ensures that an efficient means of reducing pollution is chosen. Alternatively, as an approximation, one could combine a tax on output with regulatory standards (see F and G below) requiring the use of pollution abatement inputs. In the gasoline example, a tax on gasoline could be combined with a regulation that requires cars be equipped with catalytic converters or requires that cars be subject to emissions tests and fuel carburation adjustments.

3.3.4 Effluent and Emissions Charges

In general, in the absence of monitoring and other information costs, the best policy is to tax the activity most directly related to the pollution-causing activity. This requires information about how the pollution comes about and the production technology available to reduce it. It may also require monitoring and taxing activities which are not normally taxed for revenue purposes.

Where the effluent emitted by the polluter can be monitored, the most direct policy is to impose an effluent fee. For example, firms dumping waste water into a watershed may be required to pay an effluent fee per unit dumped. This is perhaps the closest example of a pure Pigouvian tax. Unfortunately, there are perhaps few cases where such a policy is administratively feasible.

3.3.5 Content Taxes

In between excise taxes on inputs and outputs and taxes or charges on the polluting effluents or emissions themselves are what might be called content taxes. With this instrument, a tax is levied on the amount of a particular component in a commodity. The best-known example is the "carbon" tax levied by Finland and some other Scandinavian countries which tax the carbon contained in fossil fuels (see Larsen and Shah, 1995 and Chapter 8, this volume). Other examples could include a "Sulfur" tax which taxes fossil fuels according to their Sulfur content, or the BTU tax (British Thermal Unit) under consideration in the United States which taxes all energy commodities based on their heat content.

The case of the carbon tax is illustrative. The main pollutant here is carbon dioxide, a greenhouse gas, which is thought to be the main contributing factor to global warming. Taxing the actual emissions of carbon dioxide difficult or impossible since they are not readily monitored. On the other hand, an ad valorem (equal percentage) tax on all carbon-bearing fuel is less than ideal since some fuels contain much more carbon per unit of energy than others. For example, a low carbon fuel like natural gas would be taxed at the same rate as a high carbon fuel like coal. Rather, if one desires to reduce carbon dioxide emissions, it is better to target the tax on the carbon contained in the fuel. Thus a $5-ton (say) carbon tax translates into a 13% tax on coal, which contains 0.605 tons of carbon per ton of coal, and a 5% tax on natural gas, which contains 0.207 tons of carbon in a volume which has the same pre-tax value as a ton of coal (see Larsen and Shah, 1995, 1992a, 1992b; Shah and Larsen, 1992).

An important consideration in judging the suitability of content taxes is appropriate targeting. A carbon tax is most appropriate in the case where the policymaker is most concerned with reducing carbon dioxide emissions to slow global warming. Although such a tax may also reduce other pollutants, most notably carbon monoxide, it is not targeted specifically at them and therefore may be a second-best policy if the other pollutants are the main environmental concern. For example, where Sulfur dioxide is a prime concern, the carbon tax provides no incentive to substitute low Sulfur-content coal for high Sulfur content coal, although it will undoubtedly lead to lower Sulfur dioxide emissions overall. Shah and Larsen (1992) have shown that a carbon tax could appreciably reduce emissions of local and regional pollutants such as nitrous oxide, carbon monoxide, particulates, and Sulfur dioxide.

Of course, content taxes can be combined. Thus, a Sulfur tax can be levied along with a carbon tax. The total tax on a unit of fossil fuel will then depend on both its Sulfur and carbon contents.

3.3.6 *Emission (Abatement) Standards*

This is perhaps the most common environmental policy adopted. In effect, it amounts to a non-marketable pollution quota. Strict limits are set on the quantity of emissions that a firm or economic agent can produce during a given period. Alternatively and equivalently, a firm may be required to reduce its emissions by a certain amount relative to what it has done in the past.

While the firm is required to satisfy some standard on the level of its emissions, typically the government does not control or care about how it accomplishes this objective. The firm may reduce the level of its economic activity or install pollution control equipment. It is left up to the firm to choose the least cost method.

3.3.7 Abatement Technology Standards

With this policy, the firm is required to install certain pollution abatement equipment or adopt certain abatement methods. That is, the government specifies the method which the firm must use to reduce its emissions, unlike for the policy of emission standards. While such a policy seems to obstruct the goal of least cost emission reductions, it may be desirable if the level of emissions is difficult or costly to monitor whereas confirming the use of the technology standard is not (see the discussion in the next section). Also, as mentioned, it may be a useful policy in conjunction with output taxes.

3.4 CRITERIA FOR COMPARING POLICY INSTRUMENTS

In this section, we introduce criteria for evaluating alternative policy instruments for achieving environmental improvements associated with pollution control. Policy instruments differ in administrative expense, level of bureaucratic control over the actions of polluters, flexibility afforded polluters in abating emission levels, requirements for monitoring and enforcing compliance, incentives for polluters to engage in the research and development of new pollution abatement technologies. Lastly, policy instruments for pollution control differ in their ability to meet other fiscal policy objectives of government.

3.4.1 Level of Control by Regulators and Flexibility Offered Polluters

Several of the policy instruments previously discussed in Section 3.3 share the property that the environmental regulatory authority directly controls the quantity of pollution generated by firms or consumers (e.g., emission or abatement standards) or, at the very least, the aggregate quantity of pollution generated by an entire industry or set of industries (e.g., marketable pollution permits). In some instances, the regulatory authority even directly controls the method of pollution generation and abatement (e.g., abatement technology standards, mandated input mix for particular production processes). Controls on permissible quantities of pollution or methods of pollution generation and abatement constitute the most direct form of government intervention into markets with environmental externalities. Mandating emission levels and abatement technologies gives polluting firms little flexibility in achieving the abatement targets.

By contrast, several of the policy instruments described in Sect. 3.3 affect the prices firms or consumers face for goods and services (e.g., indirect taxes and content taxes), some policy instruments establish prices for nonmarket goods like pollution (e.g., effluent charges, marketable pollution permits), and others influence the cost or price of pollution abatement (e.g., abatement subsidies). These are the so-called "market-based incentives" for pollution control or the "price-type" policy instruments. They allow polluting firms' flexibility when implementing pollution abatement strategies. Market-based incentives for pollution control establish artificial prices for environmental externalities directly, as in the case of effluent charges, or indirectly, as in the case of well-functioning markets for tradable pollution permits which establish a market price for a unit of pollution. Once prices for pollution are established, firms and consumers determine the quantity of pollution (conversely, abatement) to generate. Profit-maximizing firms facing a per unit effluent charge will abate pollution as long as the marginal cost of abatement is less than the per unit price of generating the effluent. Profit-maximizing firms will purchase permits (and pollute) as long as the market price for a permit (on a per unit basis) is less than the marginal cost of abatement. In this context, the cost of abatement includes reductions in outputs, the cost of altering the input mix, as well as the cost of installing and operating the abatement equipment.

The key feature of market-based incentives or price-type policy instruments is that they delegate control over decisions about the relevant quantities to the self-interested firms and consumers. Much of the subsequent analysis about the relative desirability of market-based incentives or price-type instruments versus command-and-control regulations or quantity-type instruments focuses on the informational settings and circumstances under which such delegation is desirable or undesirable.

3.4.2 *Monitoring and Enforcing Compliance*

Regardless of the form of environmental regulation chosen by the policymaker, regulations will have little success in controlling the generation of pollution and its damaging effects if compliance with the regulations is not adequately monitored and enforced. Many environmental economists view enforcement as the weakest link in the efforts to control pollution. Most environmental policy instruments impose costs on polluters, and these costs can be avoided if polluters do not comply with the intent of the environmental policy. To circumvent polluters' incentives to not comply, regulators must implement monitoring and enforcement procedures and strategies.

Nearly every form of environmental regulation entails an enforcement burden. Regulators must ensure compliance with quantity-type emission standards by monitoring and punishing violators. If effluent charges are the regulation of choice, the regulator must collect the appropriate revenue. If abatement technology standards are employed, the regulator must check that the appropriate equipment is installed and is subsequently operated and maintained.

The best policy instrument in a given environmental and industrial context may depend critically on the associated enforcement considerations. Without considering monitoring and enforcement, effluent charges may seem the most desirable because of their efficiency properties when compared with inflexible policy instruments such as uniform abatement technology standards. If, however, it is nearly impossible to monitor the discharge of effluent accurately, it will be impractical and prohibitively expensive to collect the effluent fees from, and therefore obtain the efficient level of pollution abatement by, the targeted industry. On the other hand, it may be relatively easy to monitor the installation and operation of mandated abatement technologies. Similarly, indirect taxes on polluting inputs and outputs, or content taxes, may be easier to collect than effluent and emissions charges, and properly designed fiscal reforms may be more efficient than environmental regulations.

3.4.3 *Incentives for Innovation in Pollution Abatement Technology*

Choosing one policy instrument over other affects not only the allocation of resources and the associated level of net benefits enjoyed by society in the present, but also in the future by increasing the incentive for polluters to invest in newer and cost-reducing abatement technologies. In the presence of a pollution tax, the polluter incurs two types of costs. First, the costs of abatement and, second, the tax revenue which is paid on the unite of pollution which are not abated. If a polluter can reduce the marginal cost of abatement by R&D investment, he not only enjoys the saving in abatement costs, he also reduces the amount of pollution taxes paid to the government by increasing the level of abatement. Under a pollution tax, the polluter determines the level of abatement by equating the marginal cost of abatement to the pollution tax. Thus, the incentive to invest in abatement cost-saving technology includes the tax savings from any additional abatement which becomes worthwhile. The polluter adopts the new technologies if the present value of future cost savings, including savings of pollution taxes, covers their R&D costs.

Contrast this to the case where the same polluter faces a quantity-type regulation such as an emission standard. The polluter initially faces the cost of complying with the regulation, which is the total cost of abating pollution to the allowable level. If, thereafter, the polluter invests in measures to reduce abatement costs, the only return on the investment is the cost saving of meeting the mandated level of abatement. In particular, the polluter does not have the incentive of cost savings achievable by reducing abatement levels below those mandated.

Thus, we might expect less investment in more efficient methods of pollution abatement by polluters who face quantity-type environmental regulations as compared with those subjects to market-based incentives for pollution reduction. These differences may be even more pronounced once the incentives of the regulators are considered. If the regulator sets a pollution tax to equate the marginal benefit of abatement with its marginal cost, then the best policy response to a reduction in the marginal cost of abatement is to lower the per unit pollution tax. This lowers the tax costs to the polluter even further and, if anticipated, further increases the polluter's incentive to invest in reducing abatement costs. On the other hand, when an emission (or abatement) standard is used and set so as to equate the marginal benefit of abatement to its marginal cost, the best policy response to a reduction in the marginal cost of abatement is to tighten the abatement standard. If this is anticipated, it reduces further the polluter's incentive to invest in reducing abatement costs.

3.4.4 *Fiscal Policy Objectives*

This discussion of the incentive effects (for R&D) of the alternative policy instruments does not mean that regulated polluters prefer market-based incentives like taxes to quantity controls. On the contrary, the increased incentive for investing in new abatement technology under a pollution tax occurs because of the revenue burden associated with the tax policy instrument, a burden which is not welcomed by the polluters.

However, the revenue generated by pollution taxes may be very much welcomed by the government, particularly in revenue-short developing countries. Pollution tax revenue can support programs of environmental improvement or help to achieve other fiscal policy objectives. In addition, to the extent that the revenue from pollution taxes replaces revenue obtained from distortionary taxes, there can be a further efficiency gain over the regulatory instruments.

Environmental regulations seldom can be considered in isolation of other government policies. Industries subject to environmental regulation may also be subject to other tax/subsidy policies designed to promote growth. In the next section, we explore the welfare implications of alternative environmental policy instruments in greater detail, including an examination of the impact of environmental regulations where there are pre-existing fiscal policy distortions.

3.5 Welfare Analysis of Environmental Policy Instruments

In this section, we extend the analysis of the relative merits of different policy instruments for pollution control. In particular, we consider in greater detail the importance of uncertainty, asymmetric information, enforcement, and fiscal policy objectives.

3.5.1 The Equivalence Information of Different Instruments with Perfect Information

In a world of certainty and full information, the choice of one pollution control instrument over another may, in fact, be of little consequence. In this idealized world, quantity instruments such as emissions standards and market-based instruments such as emissions charges achieve the same objectives at the same costs. To achieve efficiency, the regulator can set an effluent tax, an emission standard, or issue pollution permits—it makes no difference, at least in the short run, if the policy instrument effectively equates the marginal benefit of pollution abatement to the marginal cost, the inefficiency due to the pollution externality is eliminated.

As economist Martin Weitzman (1974) forcefully noted, under conditions of full information and perfect certainty, there is a complete equivalence between price-type planning instruments and quantity-types.

The regulator can either mandate the optimum quantity directly or induce indirectly the optimum quantity from the self-interested parties by setting the right prices.

3.5.2 Comparative Advantage of Policy Instruments Under Uncertainty

Weitzman's seminal work illustrates how this equivalence breaks down in the more realistic context of uncertainty and asymmetric information. Uncertainty in the environmental policy context may result from several sources. First, it may result from imprecise estimates of the levels of pollution damage and the benefits of environmental improvement. While much of the environmental economics literature of the last three decades has been devoted to the development and improvement of techniques designed to measure pollution damage or elicit information about the willingness to pay for environmental improvements, benefits measurement is still inexact.

A second kind of uncertainty may confront regulators: asymmetric information. Asymmetric information describes a situation where one party to a transaction possesses relevant private information that the other does not. Environmental regulators typically have less information about the abatement capabilities and costs than the polluting firms themselves do. Firms have a better understanding of the production process and therefore better information about the least cost way of obtaining a particular level of pollution abatement.

Weitzman considers a situation in which a regulator chooses between a price-type instrument and a quantity-type instrument. The objective is to maximize the expected net benefits (the difference between expected gross benefits and expected total costs). The policy instrument must be chosen under conditions where both types of uncertainties described above prevail: that is, general uncertainty about the benefits and asymmetric information about abatement costs. The important question in choosing between these policy instruments is the desirability of delegating the decision about the quantity of abatement to the firm, a delegation which occurs under the price-type instrument. The firm has better information about costs; however, its self-interest does not coincide with the social interest.

In particular, let gross benefits be the willingness to pay for pollution abatement (and the corresponding environmental improvement) and let total costs be the costs of pollution abatement. To simplify, assume that the regulator knows the shape and slope of the marginal benefit and marginal cost curves of abatement (see Figs. 3.1 and 3.2), but not their heights (vertical intercepts). Although the height of the marginal benefit curve is uncertain to all parties, the height of the marginal cost curve is known by the polluting firm but not the regulator. The polluter knows the marginal cost of abatement and, in the case of a pollution tax, chooses the level of abatement which equates the marginal cost of abatement to the pollution tax. On the other hand, when the regulator mandates the quantity of abatement directly, the firm needs to simply comply with the mandate. We consider later the possibility that the firm can choose to comply or not.

Figure 3.1 shows when a pollution abatement standard is preferred to a pollution tax. The regulator, who must set either a tax, t, or a standard,

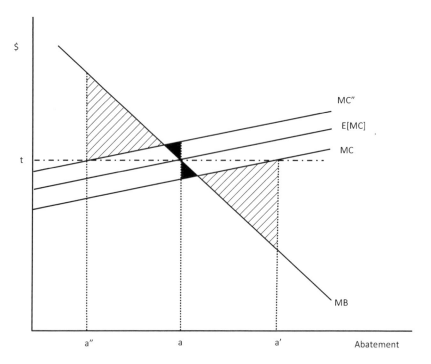

Fig. 3.1 Taxes versus standards—standard preferred

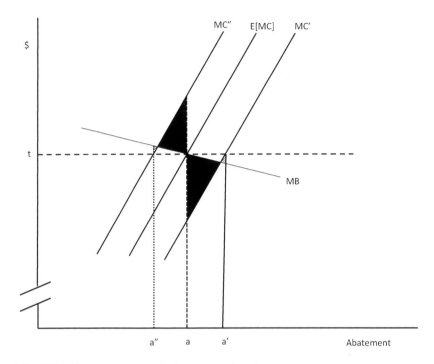

Fig. 3.2 Taxes versus standards—tax preferred

a, in the face of uncertainty about the exact location of the marginal cost of abatement curve, will choose to equate the expected marginal cost, E[MC], with the marginal benefit, MB, of abatement. The optimal level of abatement will, in contrast, be given by the intersection of the true MC of abatement (either MC" or MC') with the MB of abatement. A tax, t, will induce abatement of only a" units from a firm with high costs (MC") and a very large amount, a', from a firm with low costs (MC'). These abatement levels are not close to the optimal levels, and they result in losses of societal net benefits measured by the large-lined triangles. In contrast, an abatement standard will come much closer to the optimal level of abatement when costs are either high or low, and the resulting welfare loss is smaller (measured by the small darkly shaded triangles). The comparative advantage of a standard to a tax emerges because of the relatively steep slope of the MB of abatement curve.

Figure 3.2 presents a case when a pollution tax is preferred to an abatement standard. The regulator, who must issue a tax, t, or a standard, a, in the face of uncertainty about the exact location of the marginal cost of abatement curve, will equate the expected marginal cost, E[MC], with the marginal benefit, MB, of abatement. A tax will induce abatement levels of a" or a' if the true marginal cost is MC" or MC' respectively. These abatement levels (a" and a') are close to the optimal abatement levels (given as the abatement levels which equate the true MC of abatement with the MB. In contrast, if the standard is used, far too much abatement is required of a firm with high abatement costs (MC") and far too little abatement is required of a firm with low abatement costs (MC'). The relatively large errors occur because the marginal cost of abatement curve is steeply sloped, compared to the MB curve.

It turns out that the comparative advantage or disadvantage of a price-type instrument over a quantity-type instrument depends only on the relative slopes of the marginal benefit and marginal cost of abatement curves and on the nature of the asymmetric information about costs. Interestingly, the decision about the policy instrument does not depend on the more general uncertainty about the level of benefits. The desirability of delegating than quantity of abatement choice to the firm (by using the price-type instrument) depends on how much better the firm's information about abatement costs is as compared to the regulator. If the price instrument is used, the level of abatement which results is not known with certainty beforehand by the regulator: the firm resolves the uncertainty when it chooses its abatement level. If the quantity instrument is used, the regulator fixes the quantity of abatement (assuming no compliance and enforcement problems), but the cost of the abatement is uncertain.

A quantity instrument is more likely to be preferred when the marginal cost of abatement rises slowly as the level of abatement rises and when the marginal benefit of abatement falls sharply. Under these conditions, the costs imposed by setting the "wrong" price due to the uncertainty facing the regulator will be large relative to the costs imposed by setting the "wrong" quantity of abatement. Thus, a price-type instrument, where the resulting level of abatement is uncertain, is less desirable than a mandated quantity of abatement which achieves the same expected level of abatement (see Fig. 3.1) but with certainty.

On the other hand, if the marginal cost of abatement rises steeply with the level of abatement and the marginal benefit changes little, the price instrument will be preferred. In this case, setting the "wrong" quantity of abatement due to the limited information facing the regulator is more serious than setting the wrong price. It is better to set the price based on the marginal benefits and marginal costs of abatement, and let the firms, who know the true marginal cost of abatement curve, choose the level of abatement (see Fig. 3.2).

In practical terms, this analysis suggests that in cases such as toxic waste disposal, where a little toxic waste may be all that is needed to have drastic and dire effects on environmental quality, the government is best off relying on quantity restrictions. Using a price instrument and relying on the firms' superior knowledge risks the possibility that the marginal cost of abatement is high to the firms and they will follow their self-interest and produce too little abatement of an activity with a very high social cost.

On the other hand, in the case where firms generate non-toxic air pollutants which degrade the environment by lowering visibility, it may be advantageous for the government to use a price or market-based instrument. The marginal benefit of abating this type of pollution is not likely to change much as the quantity of abatement changes. But if the firms have heterogeneous costs of abatement, it might be wise for regulators to set effluent taxes and let each firm, which possesses better information about its own costs, decide the level of abatement. The cost savings from firms choosing minimum cost abatement levels are likely to outweigh the costs of having the "wrong" overall level of abatement due to regulators' uncertainty.

3.5.3 Enforcement Considerations

Another practical consideration facing the regulators is the fact that the polluting agents may not comply with the regulations. Just because a regulator forbids emissions beyond a particular level does not guarantee that polluters will not continue to exceed those limits. Similarly, just because a regulator establishes a price for pollution by setting an effluent charge does not guarantee that firms will honestly report their discharges and pay the correct fees. And if the firms don't pay the fees, they won't have an incentive to choose the optimal level of pollution abatement.

There is anecdotal evidence of noncompliance with technology-based standards even in the United States where institutions are well-developed. While it is fairly easy to check that mandated abatement equipment has been installed, if operating the equipment is costly, firms may circumvent the regulation by "unhooking" the equipment when regulators are not looking.

We discuss below the implications of costly monitoring and enforcement for the case where regulators set emission or abatement standards. Much of what we have to say applies with equal force to price-type policy instruments. The analysis is based on the (economics-of-crime analysis of Becker [1968]). Becker argued that "rational" criminals will commit crimes as long as the private marginal benefit of the crimes exceeds the expected marginal cost of committing the crimes. Similarly, when deciding whether to comply with an environmental regulation, firms will compare the cost of compliance with the expected value of the consequences of noncompliance.

Denote the probability of detection and prosecution of noncompliance (as determined by the enforcement budget and strategy of the regulator) and suppose further that, if found in violation of an emission standard, the firm must pay a fine F and comply with the standard incurring abatement costs of C. A risk-neutral firm will choose to comply if $C < p(F + C)$ or if $c < (pF/(1-p))$. This simple model of firm behavior suggests that the lower the costs of compliance, the greater the chance that a firm will comply. It also suggests that where firms have heterogeneous cost, some will comply while others, which have the higher abatement costs, will not. Finally, compliance is more likely that the greater is the probability of detection and the larger is the fine.

Interestingly, enforcement considerations affect the standard-setting process. Viscusi and Zeckhauser (1976) point out that tightening standards (i.e., requiring more abatement) may have perverse consequences. Suppose a regulated industry, where firms have different abatement costs, is confronted with a uniform abatement standard per firm. If enforcement is costly, so that the probability of detecting violations is less than one, firms with high abatement costs may choose not to comply with the standard while other firms, with lower abatement costs, do. This determines the aggregate level of pollution abatement obtained. If the standard is tightened by increasing the required amount of abatement per firm, then abatement will rise to the extent that firms who complied with the less stringent standard continue to comply. However, some of the firms who complied with the less stringent standard will find it in their interest not to comply with the more stringent standard. For these firms, who have intermediate levels of abatement cost, the cost of complying with the new standard is too great, so they will now cheat and their abatement will fall. The non-complying firms will continue not to comply. If the increase in pollution from the second group of firms is greater than the decrease from the first, pollution may rise as standards are tightened.

While this perverse outcome from tightening pollution standards is not always the rule, it is a fairly general result that the optimal pollution standards are legs stringent in the presence of imperfect monitoring and costly enforcement than in the idealized world of perfect monitoring and costless enforcement.

An important implication is that the level of environmental quality depends on the level of compliance with environmental regulations in practice, and not on how "tough" the environmental standards are in statutes. Unfortunately, as reported by Eskeland and Jimenez (1991), monitoring, enforcement, and regulatory capacities have been weak in developing countries. They note that in Mexico, for instance, the influence of regulations has been limited by the resources of the enforcement agency and the low level of fines. In Columbia, laws have included formulas for calculating a tax on discharges to water, but no apparatus has been in place to monitor and bill polluters. In India, inefficient legal processes have reduced the disincentive effects of lawsuits against polluters.

Indeed, enforcement efforts have been weak in some developed countries as well. Magat et al. (2013) conducted a study of regulatory effort to enforce safety and environmental regulations in the United States and found that the resources devoted to enforcement were inadequate.

Several economists have emphasized recently the importance of targeting current monitoring and enforcement efforts and resources on particular firms chosen on the basis of their past compliance records. Monitoring firms with a record of previous violation of environmental regulations more frequently than firms with no such record is found to lower the enforcement costs of obtaining a given aggregate level of compliance from the regulated industry. That is costs are lower when past performance is used as a factor in targeting enforcement expenditures than when all firms are monitored with the same frequency regardless of their past performance. This feature of optimal enforcement practice emphasizes the importance of accurate record keeping regarding (non)compliance.

3.5.4 Constraints on Environmental Policy in Developing Countries

There are a number of differences between developing and developed countries that should be considered when setting environmental policy in the former. Among these differences are the facts that in developing countries per capita income is much lower, access to quality medical care lower, the baseline level of pollution control is lower, the capital base is often older and not as well maintained, institutions in charge of administering fiscal reforms and promulgating and enforcing environmental policies are generally weaker (among other things this means the expertise, funds, and technology for data collection and monitoring are often lacking), and the demand on the public sector for the provision of public goods is greater while the revenue base is much narrower.

From this list, we emphasize low per capita income and the lack of medical care for much of the developing world's population. These are important determinants of the level of pollution control or abatement that is beneficial and cost-effective. In developed countries, there is evidence that the demand for environmental quality depends positively on income—that is, environmental quality is a normal good. Although estimates of the income elasticity for environmental quality from developed countries should not be used as precise estimates for their counterparts in developing countries, it is likely to be the case that the demand for environmental quality in developing countries will rise as their economies grow.

In the initial stages of growth, perhaps only modest reductions in optimal level of pollution are warranted on the basis of the income elasticity of environmental quality. However, the existing levels of pollution presently suffered by residents of Mexico City, Calcutta, Cairo, Beijing, or Warsaw are far from optimal. Up until now, very little pollution control measures have been employed in the developing world, and even modest reductions in pollution emissions or modest improvements in the treatment of polluted water could produce large gains in the welfare of the general population. In addition, with so much of the population lacking adequate medical care, modest improvements in environmental quality may represent one of the least cost methods of improving health in developing countries.

The differences between developing and developed economies are also relevant to the question of instrument choice, in particular, the relative desirability of marketable pollution permits versus regulations or pollution taxes. In developed countries, one can argue that marketable pollution permits may be a very attractive alternative to command-and-control regulations or to indirect (and distortionary) taxes designed to curtail pollution. First, the formal market economy is comprehensive and well-established, so there is reason to believe that pollution permits will be traded among firms with heterogeneous pollution abatement costs. Second, by establishing a price for pollution (the price of a permit), profit-maximizing firms with heterogeneous abatement costs should equate the marginal cost of abatement with the price of pollution. Equating the marginal cost of abatement across firms of all cost types is a necessary condition for cost-effectiveness in an environmental regulation. Third, environmental regulatory agencies have existed for over two decades in most developed countries, and monitoring and enforcement practices are well-established, so it is expected that firms in a developed country are more likely to comply with the pollution restrictions inherent in the permit system.

In contrast, in developing countries, the formal market economy is not as comprehensive, so there is reason to question whether a market for pollution quotas would match buyers and sellers efficiently. Furthermore, with a system of pollution quotas, the regulatory authority must monitor the quantity of pollution emitted at each source (and account for the changing pollution rights of individual firms as trades of quotas occur). If meager enforcement budgets and a lack of technological expertise preclude effective monitoring, the supposed efficiency properties of marketable pollution permits will not be realized.

3.5.5 The Efficiency Value of Environmental Taxation in Developing Countries

Other facts about environmental regulation in developing countries (viz. the lack of administrative expertise in environmental policymaking, general weakness of legal institutions, and potentially small enforcement budgets) enhance the desirability of using environmental policy instruments which are similar in structure to existing fiscal instruments. These are instruments with which developing countries have administrative experience, and which minimize enforcement requirements. As a first step toward curbing pollution, developing countries should reduce their sizable energy subsidies to highly polluting industries. Kosmo (1989) notes that governments in many developing countries keep energy prices at levels well below the world prices. These subsidies result in excessive energy-related emission of pollutants in the industries relying on subsidized energy as an input.

A second step in reducing industrial pollution in the energy-consuming sectors is to tax energy inputs, in addition to removing the subsidies. A tax on energy inputs is likely to have lower enforcement costs than alternative regulatory approaches. For example, an environmental tax on polluting inputs requires less monitoring than levying an emissions tax or enforcing an emissions standard. Purchases of important inputs to production, like energy, should be relatively easy to document. Of course, as Krupnick (1997) notes, "such an approach only provides incentive to reduce purchase of the input, not necessarily to find the least-cost means of reducing pollution and, if the input is not chosen with a careful eye towards substitutes, there is no guarantee that emissions will fall. Nevertheless, on balance, this approach seems to be a reasonable second-best policy" (p. 434).

The second-best policy of taxing polluting inputs (e.g., energy, leaded gasoline, pesticides) is particularly appealing in developing countries because it draws upon the administrative capability most of them possess from levying commodity taxes. Most developing countries do not have extensive experience with administering and enforcing other forms of environmental regulation. Assistance from international agencies and academics is changing this, but the change is slow. The lack of government expertise, record keeping capability, and enforcement funds probably rule out, for the present, the more sophisticated regulatory schemes are being tried in parts of the developed world (e.g., tradable pollution permits).

The use of environmental taxes, whether on polluting inputs and outputs, content taxes, or emissions fees, have the added appeal that they generate revenue, unlike emission standards, abatement technology standards, or chemical bans which collect no revenue and are costly to enforce. By the same logic, pollution taxes are superior to abatement subsidies. The former generate and the latter add to the government's revenue requirements. Public sector budgets in many developing countries are severely limited, and raising additional revenue through the existing tax structures can involve large inefficiencies because the tax bases are typically quite narrow and taxation badly distorts resource allocation decisions. To the extent that environmental tax revenues replace those obtained through more distortionary means, environmental taxes have an additional "efficiency value". As Terkla (1984) explains, "it is defined as the reduction in excess burden resulting from the substitution of these revenues for current and future resource distorting tax revenues" (p. 107).

Terkla calculates the efficiency value of potential particulate emissions and Sulfur dioxide taxes on stationary sources in the United States and finds that the estimated values range from a possible value of US$O.63 to US$4.87 billion (in 1982 dollars), depending on the wide range of plausible abatement cost levels and on whether environmental tax revenue replaces labor or corporate income tax revenue. His working assumption is that the taxes would be set to achieve national air quality standards for Sulfur oxides and particulates based on the US Environmental Protection Agency emission standards for new sources. Terkla uses the mid-seventies estimate from the public finance literature of $0.35 for the marginal welfare cost of a dollar of labor income tax revenue. The estimates of marginal welfare loss are consistent with figures generated more recently from a computable general equilibrium model of Ballard (1985).

Terkla's argument is important since so many developing countries currently raise revenue from a very narrow tax base with highly distortionary commodity taxes. The efficiency value of pollution taxes is best viewed as an element in the instrument choice debate. The argument is relevant whether or not the polluting inputs and outputs are already taxed as part of the country's revenue base. Suppose all taxable goods in the economy are initially taxed (for revenue-raising purposes only) so that the marginal excess burden (exclusive of the impact on environmental quality) is the same across taxable commodities. If good X is a particularly polluting commodity, either in its use as an input to production or through its consumption as a final good, its use can be curbed in two ways: a quantity restriction on polluting emissions or an increase in the per unit tax levied on the good. Notwithstanding differences in administrative or enforcement burden, the commodity tax approach is preferable because of its additional value in raising revenue.

One way to demonstrate this is to note that an emissions standard, which limits the quantity of the polluting good X sold, creates an economic rent (the difference between the marginal willingness to pay [price] and the marginal cost (including the baseline commodity tax) of production)—a rent which would be taxed away if the revenue structure is optimized since taxes on rents are lump sum (i.e., non-distortionary) taxes. Increasing the tax on the polluting good so as to achieve the same level of use as under the standard achieves the goal of the environmental policy and captures the economic rent as a low cost source of government revenue.

Even if environmental tax revenues are not used to reduce the revenues raised from other distortionary taxes, they provide an additional efficiency value relative to other forms of environmental regulation if the revenue is used to finance public infrastructure projects which generate large net benefits. Water treatment plants and improved sanitation facilities are good examples of the types of projects that could be financed through environmental taxes and user charges. They hold the promise of generating large environmental benefits, given the current state of water pollution problems in most developing countries.

3.5.6 Environmental Regulation and Pre-Existing Market Distortions

In developing (and developed) economies, environmental policies are formulated in the presence of existing fiscal policies aimed to raise revenue and to encourage targeted industries. In this section, we analyze the relationship between environmental regulations and pre-existing fiscal policies.

We begin by considering an existing tax/subsidy structure which is clearly suboptimal. This is, unfortunately, the case in many developing countries where energy-consuming sectors are heavily subsidized while other sectors face substantial commodity taxes. In this situation, what are the welfare effects of an environmental regulation which curtails output in a heavily subsidized sector? The social costs (benefits) of the regulation are usually less (greater) than they would be if the sector was not subsidized. An example from the United States clarifies this point further.

Lichtenberg and Zilberman (1986) study the desirability of pesticide bans for agricultural crops which are heavily subsidized for political or distributional reasons. The crops under consideration receive substantial subsidies in the form of output price supports. These supports encourage an overproduction of the commodities in question (corn, rice, and cotton). Banning certain harmful pesticides not only generates desirable environmental benefits, but welfare may be further enhanced because the ban reduces the output of the overproduced commodities. This argument is relevant in many developing countries where subsidies have led to an undesirable expansion of some outputs (Kosmo, 1989). On the other hand, stringent environmental regulations which reduce output in sectors that are already hampered by high rates of commodity taxation can be very costly in welfare terms. In this case, stringent environmental regulations are only warranted where the pollutant is extremely damaging.

While the existence of a polluting output or input which is subsidized or taxed too little is a particularly attractive target for a Pigouvian tax, it should be remembered that commodity tax rates can vary for sound fiscal reasons and these variations do not identify an extra efficiency reason for imposing a Pigouvian tax. Conversely, the presence of pre-existing high tax rates on polluting goods does not eliminate the argument for taxing them even more for environmental reasons. If the pre-existing tax rates are chosen as is, optimal for fiscal reasons but without regard to the environmental impacts, they can and should be increased further if the goods cause external environmental costs. That is, the Pigouvian tax on a good should be added to whatever tax rate is appropriate for revenue purposes.

3.6 A Few Specific Policy Recommendations

We have emphasized the usefulness of eliminating subsidies and increasing taxes on polluting inputs and outputs as an important first step in reducing the pollution problems prevalent today. In addition, governments can consider content taxes such as taxes on the Sulfur, carbon, and lead content of fossil fuels. Such taxes are more finely focused on the offending agents of pollution. Beyond this, it is difficult to make specific recommendations about environmental policies which would be appropriate for all countries given differences in the structures of their economies and the pollution problems they face. However, most developing countries are confronted with the emerging problems of air pollution from vehicular sources and the industrial and domestic use of coal as a fuel. Also, nearly all developing countries are plagued with the contamination of water resources with raw sewage, agricultural runoff, and some industrial point-source pollution. For these problems, we offer a few suggestions.

Concerning air pollution problems in developing countries, Krupnick (1997) contends that the efficiency case for market-based incentives is particularly strong and that a limited system of emission fees, along with price reforms on subsidized polluting production inputs, are attractive approaches for the control of urban air pollution from industrial sources.

For health reasons, it is imperative that developing countries phase out the use of leaded gasoline. In countries with state-owned refineries, this could be accomplished by fiat or by setting a high price on leaded gasoline. In countries with private refineries, regulations and hefty taxes on leaded gasoline would accomplish the same thing. More generally, controlling present and future vehicular emissions with increased fuel taxes is likely to be a very sound policy, both as a revenue-raising device and as an effective and least cost means of discouraging this increasingly important source of urban air pollution. In the case of mobile source air pollution, fuel taxes should be supplemented with programs aimed at improving vehicle maintenance. Krupnick continues (p. 443):

> ...buses and trucks are such an important transport mode in developing countries and, with some exceptions, these vehicles are old and poorly maintained, more attention needs to be paid to their emissions than now (where emissions are generally ignored). Performance standards and inspection and maintenance programs specifically directed to these types of vehicles are needed. Such monitoring and enforcement requirements are likely to result in inexpensive carburation and other adjustments that would yield substantial emissions reductions. Automobiles and two-wheeled vehicles also need to come under these programs.

The periodic monitoring of emissions, while costly, may still be economically efficient. A fuel tax is broader than the theoretically desirable tax on individual emissions. Therefore, a fuel tax by itself, with no monitoring of actual emissions, will not induce the optimal abatement of emissions per unit of fuel consumed.

Another important source of air pollution in developing countries is that of indoor heating and cooking. Smith (1992) documents that in several developing countries, coal burning stoves are a major problem and that subsidizing improved cooking stoves for the poor may yield large benefits and be relatively cost-effective (compared to the costs of abating comparable quantities of the same pollutants generated by industrial sources). The emission of Sulfur dioxide, from both industrial and domestic sources burning coal, might best be controlled by taxing high-Sulfur content coal at a much higher rate than low-Sulfur coal. The regulation of Sulfur content in fuels has worked well in many developed countries (OECD, 1991).

In the case of water pollution, a mix of policy instruments designed to curtail pollution is needed, in addition to the publicly financed expansion of treatment and sanitation facilities. To partially finance the expansion of water treatment plants, the water authorities in developing countries should institute a system of user charges for those municipalities and industrial sources whose effluent is to be treated. User charges like these have been employed successfully in Germany. Large industrial polluters whose water effluent can be easily identified and measured could be charged a per unit effluent fee. Most agriculturally based water pollution problems are of a nonpoint nature, and monitoring runoff is impractical. Consequently, input taxes (on manure, chemical fertilizers, pesticides, and irrigation water) are the most practical and reasonably efficient policy instruments available.

We have argued the case for indirect commodity taxation as a useful instrument for controlling many of the pollution problems in developing countries. This case is based on the limited administrative expertise in environmental policymaking, general weakness of legal institutions, and potentially small enforcement budgets found in many developing countries. However, for certain classes of pollutants, strict command-and-control regulations are warranted. Toxic pollutants should not be controlled with market-based incentives like emission fees or indirect commodity taxes on inputs. In these circumstances, where the marginal damage from a pollutant depends critically on the quantity emitted, the quantity should be controlled directly with an outright ban or a vigorously enforced limit.

References

Afzal, A., Scott Barret, Karl-Göran Mäler, and Eric S. Maskin, eds. (2015). *Environment and Development Economics: Essays in Honour of Sir Partha Dasgupta*. London: Taylor and Francis.

Ballard, C. L., Shoven, J. B., and Whalley, J. (1985). General Equilibrium Computations of the Marginal Welfare Costs of Taxes in the United States. *The American Economic Review*, 75(1), 128–138.

Becker, G. S. (1968). Crime and Punishment: An Economic Approach. In *The Economic Dimensions of Crime* (pp. 13–68). London: Palgrave Macmillan.

Coase, R. H. (1960). The Problem of Social Cost. *Journal of Law and Economics*, 3, 1–44.

Eskeland, G., and E. Jimenez. (1991). Curbing Pollution in Developing Countries. *Finance and Development*, 28, 15–18.

Jenkins, G., and Lamech, R. (1994). Green Taxes and Incentive Policies An International Perspective. San Francisco: ICS Press.

Kosmo, Mark. (1989). Economic Incentives Industrial Pollution in Developing Countries, The World Bank, Policy and Research Division Working Paper No. 1989—2, Environment Department.

Krupnick, A. (1997). Urban Air Pollution in Developing Countries: Problems and Policies. *The Environment and Emerging Development Issues*, 2, 425–459.

Larsen, Bjorn, and Shah, Anwar. (1995). Global Climate Change, Energy Subsidies and National Carbon Taxes" In Lans Bovenberg and Sijbren Cnossen (eds.), *Public Economics and the Environment in An Imperfect World*. Boston/London/Dordrecht: Kluwer Academic Publishers.

Larsen, Bjorn, and Shah, Anwar. (1994). Global Tradable Carbon Permits, Participation Incentives, and Transfers. *Oxford Economic Papers*, 46, 841–856.

Larsen, Bjorn and Shah, Anwar. (1992a). World Energy Subsidies and Global Carbon Emissions. Policy Research Working Paper Series, *World Development Report*, WPS1002. The World Bank, Washington, DC.

Larsen, Bjorn and Shah, Anwar. (1992b). Combating the 'Greenhouse Effect. *Finance and Development*, 29(4), 20–23.

Lichtenberg, E., and Zilberman, D. (1986). The Welfare Economics of Price Supports in US Agriculture. *The American Economic Review*, 76(5), 1135–1141.

Magat, W., Krupnick, A. J., and Harrington, W. (2013). *Rules in the Making: A Statistical Analysis of Regulatory Agency Behavior*. RFF Press.

Munasinghe, M. (1990), Managing Water Resources to Avoid Environmental degradation: Policy Analysis and Application. The World Bank, Environment Working Paper No. 41, Environment Department.

Organization for Economic Cooperation and Development. (1991). *The State of the Environment*. OECD: Paris.

Pearce, D. (2014). *Blueprint 3: Measuring Sustainable Development*. London: Routledge.

Poterba, James M. (1991). Tax Policy to Combat Global Warming: On Designing a Carbon Tax. National Bureau of Economic Research Working Paper #3649.

Shah, Anwar and Larsen, Bjorn. (1992). Carbon Taxes, the Greenhouse Effect and Developing Countries. Policy Research Working Paper Series. *World Development Report*, WPS 957, World Bank, Washington, DC.

Smith, S. (1992). Taxation and the Environment: A Survey. *Fiscal Studies*, 13(4), 21–57.

Terkla, D. (1984). The Efficiency Value of Effluent Tax Revenues. *Journal of Environmental Economics and Management*, 11(2), 107–123.

Viscusi, W. K., and Zeckhauser, R. (1976). Environmental Policy Choice Under Uncertainty. *Journal of Environmental Economics and Management*, 3(2), 97–112.

Weitzman, M. L. (1974). Prices vs. Quantities. *The Review of Economic Studies*, 41(4), 477–491.

CHAPTER 4

Revenue Sharing from Natural Resources: Principles and Practices

Baoyun Qiao and Anwar Shah

4.1 INTRODUCTION

Management of resource revenue represents a major challenge for most resource-rich developing countries as evidenced by their disappointing growth performance (Sachs and Warner 1995). The key issue is related to the distribution of resource revenues. In particular, the revenue sharing of natural resources is a critical question for countries whose government revenues depend significantly on natural resources. Fiscal systems play an important role in determining the extent to which a country can benefit from its resources, and amicable revenue sharing of natural

---

B. Qiao (✉)
Central University of Finance and Economics, Beijing, China
e-mail: baoyun.qiao@gmail.com

A. Shah
Brookings Institution, Washington, DC, USA

© The Author(s), under exclusive license to Springer Nature Switzerland AG 2023
A. Shah (ed.), *Taxing Choices for Managing Natural Resources, the Environment, and Global Climate Change*,
https://doi.org/10.1007/978-3-031-22606-9_4

resource revenues contributes to goals of sustainable development in both industrial and developing countries.

The critical component in the design of revenue sharing arrangement is about the balance of centralization and decentralization. Arguments for both centralization and decentralization for resource revenues have existed for a long time. In general, the literature on this issue suggests that revenues from natural resource exploitation should generally flow to the federal/central budget. In particular, the uneven distribution of natural resources argues for a federal role in exploiting oil, gas, and mineral deposits. In fact, the central government is assigned responsibility for natural resources in several countries such as India, Indonesia, Russia, China, Brazil, and Nigeria. However, due to various assignments of ownership and responsibilities for natural resources management and development, the subnational government sharing of natural resource revenues is widely practiced (Boadway and Shah 2009; Bahl and Tumennasan 2002). Varied institutional settings across the countries such as political environment, the property rights in natural resources, the dependency on natural resources, geographical distribution of natural resources, division of powers explain diverse approaches followed by various countries.

While fully recognizing that there would not be a uniform model for the design of revenue sharing that would suit all countries, it is important to base their design on basic economic principles to avoid economic inefficiencies and inequities under a multi-order governance system (Boadway and Shah 2009). This chapter presents these general economic principles in the design of resource revenue sharing and evaluates international practices using these principles, and also draws lessons for future reform. The remainder of the chapter is organized as follows. Section 4.2 discusses the principles in resource revenue sharing. Section 4.3 introduces practices of resource revenue sharing in selected countries. Section 4.4 evaluates these practices and Sect. 4.6 distills lessons.

4.2 Resource Revenue Assignment and Sharing—Basic Considerations

Natural resources as well as the activities involved with exploration and extraction are location-specific and taxing natural resource can be regarded as a way to secure the public sector share of the rents generated from these resources. Although special rent taxes could be imposed on other sectors, the argument is strongest for resource industries since those are where economic rents are most likely to reside (Boadway and Flatters 1993, Chapter 2 this volume). Meanwhile, natural resources tend to be highly unevenly distributed (Shah 1994). Consequently, resource revenues show significant regional inequality for almost all countries. For example, in Indonesia "Own fiscal capacity among the regions varies widely and much of this variation is due to revenues from natural resource sharing" (Hofman and Kaiser 2002). In Russia, "natural resource endowments vary greatly across regions, so that resource-poor regions would find it hard to sustain themselves on natural resource revenues" (Benoit Bosquet 2002).

For non-renewable natural resources, resource revenues could in part be used for compensating for the negative externalities suffered due to resource exploitation. The negative externalities not only arise from the degradation of the environment from resource exploitation, but also may include the cost of economic adjustment when resources are exhausted (Shah 1994).

Resource revenues are volatile and marred with instability and unpredictability over time (Shah 1994). Natural resources are typically traded on international markets, and resource prices fluctuate in world markets generating uncertainty that different levels of government may deal with differently (Boadway and Shah 2009).

Clearly, the resource revenue assignment and sharing needs to fit specific institutions in the economy. For example, resource revenue sharing will require a trade-off and national consensus among various objectives such as fiscal equity, economic efficiency and horizontal balance, and the rightful compensation to subnational governments for environmental damage and the exploitation of natural endowments. However, a few economic considerations apply in all circumstances.

4.2.1   Resource Tax Assignment

(a) *In assigning tax functions resource revenues that play a predominant role in government revenues are a good candidate for centralization.*

Resources are immobile and therefore could be easily taxed by subnational jurisdictions. However, since resources tend to be distributed unevenly across a nation, taxation by subnational jurisdictions perpetuates both regional inequalities and inefficient allocation of resources.

(a) Equity
Because of uneven distribution of tax bases geographically, resource rent taxes are not suitable for assignment to subnational governments. More importantly, the issue of fiscal inequities could be more serious if resource rents for those resources are of significant size such as oil and gas properties and significant mineral deposits. On the other hand, although it is not a sufficient condition, centralization of resource revenues provides the central government fiscal capacity to improve the equity in allocation of resources. Particularly, centralization becomes more important the more significant the value of the resource and the more unequally is it distributed among states (Shah 1994).

(b) Efficiency
There are potential efficiency losses when resource revenues are assigned to subnational governments because of the externality of resource industry and the instability of resource revenues. In general, decentralization of taxing powers is desired mainly to induce political accountability. Local jurisdictions may have preferences for certain features of the tax system, such as the degree of progressivity or the set of tax preferences to use. The decentralization of expenditure responsibilities also implies an argument for decentralizing tax responsibilities. Therefore, decentralization of revenue arrangement would have to be guided also by the assignment of expenditure functions.

The decentralization of tax responsibilities of natural resources, nevertheless, could result in inefficiencies. First, it may encourage fiscally induced migration to resource-rich regions, resulting in differential net fiscal benefits (imputed benefits from public services minus tax burden) being realized by citizens depending on the fiscal capacities of their place of residence. A richer jurisdiction can provide a higher level of public services at a lower tax rate. It is argued that such differential net benefits (NFBs) would encourage people to move to a resource-rich area, although appropriate economic opportunities may not exist. Thus, people in their relocation decisions would compare gross income (private income plus net fiscal benefits minus cost of moving) at new locations, whereas economic efficiency considerations warrant comparing private income minus moving cost. It is argued that the national government should have a role in correcting such a fiscal inefficiency. Second, the resource rent taxes would be highly volatile given the instability of resource revenues. When resource revenues form a large component of the total revenue base, the destabilizing effects can be quite large. The central government, however, has a stronger ability to stabilize revenues in response to uncertainty such as oil price volatility (Boadway and Shah 1993; Boadway et al. 1994; Shah 1994).

(b) *Centralizing resource revenues strengthens central government's ability for economic stabilization.*

(c) *Decentralization of taxing powers improves local political accountability and it can also reduce incentives for "backdoor" approaches to revenue raising.*

If there is no formal system of sharing natural resources with the regions or districts, various types of transit taxes, fees, licenses, and a whole range of nuisance levies are likely to be enacted (Bahl and Tumennasan 2002). Decentralization of resource revenues may also discourage secession movements. Some degree of decentralization of taxing powers accompanied by fiscal equalization and/or revenue sharing would mitigate accountability and secession concerns (Shah 1994). Therefore, the case for centralization of resource revenues must be evaluated primarily from a prudent fiscal management perspective (McLure 1994).

(d) *Tax harmonization and equalization transfers play a critical role in a decentralized resource-dependent economy.*

To the extent that resource revenues are decentralized to the states, some harmonization might be beneficial. Of course, for federations where state revenue requirements are limited, tax sources for which harmonization is relatively less important can be assigned to the states.

If some or all resource taxes are assigned to the states, it is important that the federal government implement a system of overarching equalizing transfers. If the equalization payments are based on a relative measure of fiscal capacity, they should have a stabilization effect on state revenues. The level of payments will move in a direction opposite to that in which the states' own revenue-raising capacity moves. Maximum stabilization of state-local revenues will occur when the payments are based on all revenue sources, when a national average standard of equalization is used, when cycle fluctuations in provincial economies are small, and when the time lag in calculating the grants is relatively short. Some sort of averaging formula should be employed to ease difficulties associated with provincial budgeting in the face of uncertainty (Shah 1994).

4.2.2 Resource Revenue Sharing

Resource revenue sharing should be designed to match revenue means with expenditure needs as closely as possible. Also the regime must not create disincentives for own tax effort in resource-rich jurisdictions.

Although centralization of resource revenues can be regarded as a better arrangement, the degree and form of centralization of resource revenues depends upon the political, economic, and institutional characteristics of the country in question, as well as the role that governments actually assume. Of course, the central governments may loathe to give up their natural advantage (Bahl and Tumennasan 2002), particularly if the constitutional assignment of the ownership of natural resources vests with the central government. From an equity point of view, one can argue that property rights to the bounty of natural resource endowments ought to rest with the national government to be shared among all citizens (Shah 1994). The reasons for assigning resource revenues to subnational jurisdictions are mainly political and turn on either constitutional stipulations that delineate the regional or local ownership of the natural resources or the right to levy taxes on certain bases or sources of income (IMF 2005). Often the property rights in natural resources have been widely used as arguments for claiming the ownership of resource revenues. Taxing resources over and above that of other general tax measures is to acquire a share of resource rents generated for the government. Subnational governments have argued strongly that they may have the right to tax natural resources located within their boundaries, to convert resource wealth (their "heritage") into financial capital—turn "oil in the ground into money in the bank" (McLure 1994).

While it may be desirable to have national ownership of resources, a case can still be made for retaining state control over the collection of production taxes or royalties. Also, small mines and quarries might be good sources of revenues for states. In any case, shared resource revenues should match a properly designed assignment of government responsibilities. In general, revenue means should be matched as closely as possible to revenue needs. Thus, tax instruments to further policy objectives should be assigned to the level of government having the responsibility for such a service.

Resource revenue sharing should be directly related to the roles of the governments in managing, developing (including providing infrastructure), and conserving the resource. These are often the functions whose primary benefits accrue to state residents. To the extent that state tax and royalty systems are useful for these regulatory purposes, decentralizing responsibility for them would be a good thing. If needed, the federal government could always provide general incentives over resource usage through its spending or regulatory power. The upshot is that resource tax assignment must be considered on a case-by-case basis. From general principles follows a preferred assignment of expenditure responsibilities. The states would be responsible for the delivery of such public services as resource management (including local land management and environmental issues). Consequently, "some resource taxes designed to cover costs of local service provision, such as property taxes, royalties and fees, and severance taxes on production and output could be assigned to local governments. In addition, subnational governments could also impose taxes to discourage local environmental degradation. This explains why in Australia, Canada and the United States, intermediate level governments (and some local governments in the United States) impose such taxes on natural resources" (Shah 1994).

4.2.2.1 *Sharing of Resource Revenues—A Summary of Basic Considerations*

In summary, sharing of resource revenues should be structured to:

- Ensure efficiency within the economic union, that is:
  - Dissuade subnational governments from "backdoor" taxation of resource rents.
  - Accommodate regional expenditures on sector-specific public services.
  - Minimize additional differences in net fiscal benefits between jurisdictions.
  - Provide insurance against volatile resource revenues.
  - Promote efficient rate of taxation across all tax bases.

- Furnish revenues adequate for subnational governments to meet their expenditure responsibilities.
- Provide adequate reinvestment for regional adjustments for shared revenues.
- Preserve the federation through an appropriate revenue sharing bargain with resource-rich regions.
- Avoid multiple or "special" agreements with different regions on revenue sharing.
- Minimize the stress on the federal fiscal equalization program.

Table 4.1 presents a comparative perspective on resource tax assignment in Brazil vs. Canada and compares it to a representative tax system. One observes that Canada has a relatively decentralized tax regime that creates fiscal inefficiency through fiscally induced migration and fiscal inequity by accentuating provincial fiscal disparities. It places great strains on the federal fiscal equalization program. Further, it creates political divisions through federal bilateral bargains with resource-rich provinces.

Table 4.1 Resource Tax Assignment: Brazil vs Canada—A comparative perspective

Resource tax	Ideal (representative)	Brazil	Canada
Resource rent taxes (profits and rents)			
Base	F	F	F,S
Rate	F	F	F,S
Tax collection	F	F (tax sharing)	F,S
Royalties, fees, severance, production, output and property taxes			
Base	S,L	F	S
Rate	S,L	F	S,L
Tax collection	S,L	F (tax sharing)	S,L
Conservation charges			
Base	S,L	F	S,L
Rate	S,L	F	S,L
Tax collection	S,L	F	S,L
Carbon taxes	F,S	Not applicable	S
Energy taxes	F,S,L	F	S,L
Motor Fuel taxes	F,S,L	L	S

Notes F-Federal; S-State; L-local
Source Shah (2014)

Brazil, on the other hand has a relatively centralized resource tax regime complemented by tax-by-tax sharing as well as general revenue sharing (shares vary directly with population and inversely with per capita income). Note that the recognition of local government is much stronger in Brazil than in Canada. Brazilian tax and revenue sharing regime (see Table 4.2) creates perverse incentives for own tax effort in resource-rich jurisdictions.

Table 4.2 Brazil—tax sharing of oil and gas royalties and special participation in federating units—2009

Unit	Basic allocation		Supplementary allocation		Special participation	% of total
	Inland	Offshore	Inland	Offshore		
Union	0%	20%	25%	40%	50%	39.4%
States	52.5%	24.5%	52.5%	22.5%	40%	33.9%
Producing/Bordering states	70%	30%	52.5%	22.5%	40%	35.2%
Redistribution through VAT	17.5%	7.5%				
Special Fund		2%		1.5%		0.8%
Municipalities	47.3%	55.5%	22.5%	36%	10%	26.7%
Producing	20%		15%			0.9%
Bordering				22.5%	10%	10.0%
Bordering/producing zone (PZ)		21%				4.6%
PZ Neighborhood		9%				2.0%
Des(embarkation) locals DEL	10%	10%	3%	3%		3.1%
Affected by DEL			4.5%	4.5%		1.1%
From producing states through VAT	17.5%	7.5%				2.1%
All special Funds		8%		6%		3.0%
Grand Total	100%	100%	100%	100%	100%	100%

Source Gobetti et al. (2020)

4.3 Resource Revenue Sharing—Practices

Practices in resource revenue sharing differ significantly across countries. We categorize these practices based on the following dimensions: (a) the authority on revenue raising which can be subdivided into centralized, joint, and decentralized; (b) the arrangements of revenue sharing which can be subdivided into centralized, revenue-shared, base-shared, and decentralized (see also Bishop and Shah 2011). In general, these practices can be categorized into five groups with respect to the order of centralization level as follows:

(1) Full centralization: No revenue sharing—full central control of resource revenues usually under a unitary form of government.
(2) Tax-by-tax Sharing: The national government returns a fraction or all of revenues from one or more resource tax instruments to the originating subnational government on a derivation basis.
(3) Revenue Sharing: The national government either unilaterally or in consultation/agreement with subnational units disperses revenues by formula without or in partial consideration of derivation.
(4) Mostly, provincial/state control. Resource revenues are collected and retained by provinces/states but the federal government shares the tax base for corporate income.
(5) Full provincial/state control with bottom-up revenue sharing. Decentralized arrangement under which resource revenues are controlled by the state/local governments. Resource revenues solely accrue to subnational government and the national government has no role in raising revenues from resources.

Table 4.3 presents a bird's eye view of resource revenue sharing systems practiced in selected countries. These country examples are further elaborated in the following paragraphs.

4.3.1 No Revenue Sharing (Centralized Arrangements or Full Central Government Control over Resource Revenues)

(a) Azerbaijan

Resource industries play a predominant role in the economy. Azerbaijan has a substantial endowment of oil and gas deposits, and the energy sector

Table 4.3 A bird's eye view of resource revenue sharing systems

Arrangement type	Country	Country type	Ownership of natural resources	Authority on revenue bases and rates	Authority on revenue collection	Authority on sharing arrangement	Arrangement related to origin of revenue	% Share of mining and quarrying in GDP	Subnational share of total government expenditure
Full Centralization	Azerbaijan	Unitary	Central	Central	Central	Central	No	24.7	24.1
	Norway	Unitary	Central	Central	Central	Central	No	12.9	32.4
Centralized tax-by-tax sharing	Brazil	Federal	Federal/State	Federal	Central	Central	60%	NA	NA
	Indonesia	Unitary—decentralized	Central	Central	Central	Central	15% (oil)	10.1	12.2
	Pakistan	Federal	Provincial	Federal	Federal	Federal/Provincial	98%	0.5	45.0
	Russia	Federal	Federal	Federal	Federal/State	Federal	60%	30	NA
General revenue sharing	China	Unitary—decentralized	Central	Central	Central	Central/Provincial	100%, inland resources	NA	69.9

(continued)

Table 4.3 (continued)

Arrangement type	Country	Country type	Ownership of natural resources	Authority on revenue bases and rates	Authority on revenue collection	Authority on sharing arrangement	Arrangement related to origin of revenue	% Share of mining and quarrying in GDP	Subnational share of total government expenditure
Mostly provincial control	Nigeria	Federal	Federal	Federal	Federal	Federal	13%	35.7	28.9
	Canada	Federal	Provincial	Provincial	Provincial	Federal and provincial separately	100%	3.5	57.4
	United States	Federal	State	Federal and State separately	Federal and State separately	Federal and State separately	100%	1.6	46.3
Full decentralization	United Arab Emirates	Federal	Emirate	Emirate	Emirate	Federal/Emirate	100%	46.7	NA

Notes Share of mining and quarrying in GDP excludes oil and gas and Subnational share of total government expenditure are from Bahl and Tumennasan (2002) except for Russia, which is from National Account Statistics, United Nations. NA: Not available

Source Authors, Bahl and Tumennasan (2002), and National Account Statistics, United Nations

(and the oil and gas subsector) represents the most promising source of exports and economic growth. The oil and gas sector accounts for 32% of the total GDP in 2001(IMF Country Report 2005).

Natural resources such as oil and gas are the sole and exclusive property of the government of Azerbaijan. This is defined by the Subsoil Law, which governs the exploration, use, protection, and supervision of subsoil reserves located both within Azerbaijan and its sector of the Caspian Sea and serves as the basic legal framework in combination with the Energy Law for natural resources in Azerbaijan.

Practically, Soviet-era oil and gas fields are operated and managed by the State Oil Company. The State Oil Company of Azerbaijan Republic (SOCAR) and its many subsidiaries are responsible for the production of oil and natural gas in Azerbaijan, for the operation of the country's two refineries, for running the country's pipeline system, and for managing the country's oil and natural gas imports and exports. While government ministries handle exploration and production agreements with foreign companies, SOCAR is party to all the international consortia developing new oil and gas projects in Azerbaijan. New fields are operated and managed under the leadership of international partners (PSAs), from which the incomes are shared with government by predetermined production sharing agreements. Azerbaijan allows investors to work through Production Sharing Contracts (PSCs) as well as traditional joint ventures (JVs) for the development of onshore deposits. In a JV, a foreign company can have a maximum ownership stake of 49% and must pay eight separate taxes, while a PSC allows an investor to hold even a greater interest than SOCAR and is subject only to the profit tax. Like a PSC, the Revenue Sharing Agreement (RSA), created by SOCAR to attract foreign investment aims to create the unique fixed term legal and fiscal framework for the existence of an alliance and its cooperation with the Azeri industry. It would be sanctioned by the Azeri Government and incorporated as a national law as an alternate structure to the JV structure. It is modeled after the PSC form with all its benefits and principles that allow an alliance to operate like a foreign subcontractor, avoiding the adverse tax consequences of the joint venture. This is a flexible, robust alliance structure that is familiar to the Azerbaijan authorities and international contracting community.

In 1998–1999, oil-related revenue brought in nearly 50% of budget revenues, including 57% of total indirect taxes. Income tax from both State Oil Company and new oil and gas fields goes to State Budget, and the government share of new oil and gas field goes to State Oil Fund. The State Oil Fund of Azerbaijan was established in December 1999 by presidential decree as an extra-budgetary institution, accountable and responsible to the president. The main objective of State Oil Fund is to ensure collection and effective management of foreign currency and other assets generated from the activities in oil and gas exploration and development, and from the Oil Fund's itself. In detail, assets of the fund are generated from the following sources: the government's profit share of PSAs; rental fees under the contracts with foreign companies; bonus payments starting from year 2000 onward; acreage fees; and income earned on the Oil Fund's assets.

The ultimate authority over all the aspects of the Oil Fund's activities rests with the president, who is empowered to liquidate and reestablish the fund, approve the Fund's regulations, and identify its management structure. An annual budget will be drafted and submitted to the parliament as part of a consolidated annual budget procedure ensuring public scrutiny of the use of the Oil Fund. Consolidation with the state budget is ensured by close coordination with the Ministry of Finance through the joint Budget Coordination Committee co-chaired by the executive director of the Oil Fund and the Minister of Finance and based on a memorandum of understanding signed between the fund and the ministry. The Oil Fund's spending is carried out through the state's treasury system, and the use of funds is subject to the State Procurement Law (SPL) and the provisions therein, which govern other budgetary expenditures.

(b) Norway

Norway is an advanced, natural resources-rich country. The oil and gas sector comprises a significant portion of the total economy in Norway, accounting for 17% of total GDP in 2002. The central government of Norway controls the natural resources. The control mechanisms over petroleum resources are a combination of state ownership in major operators in the Norwegian fields and the fully state-owned Petoro, a company managing Norwegian offshore oil and natural gas properties, and State's Direct Financial Interest (SDFI), on behalf of the government. Meanwhile, the government controls licensing of exploration and production of fields.

Norway has a unique system to capture rent from oil and gas production. In contrast to the Canadian jurisdictions and Alaska, which collect most of the oil and gas revenues from royalties, Norway does not auction off licenses. In Norway, licensees, which give companies the right to explore, drill, produce, and sell oil and gas in the country for a certain period, are awarded to oil and gas companies. The rationale is that it disperses exploration risks over many wells and companies, rather than just those companies willing to place significant bids in an auction. Consequently, revenue is obtained mainly through a system of taxes and direct ownership of resources through royalties, area fees, and a carbon dioxide tax, among other taxes. In addition, the special tax on profits allows the Norwegian government to capture significant revenues and a high portion of economic rent. Meanwhile, the carbon dioxide tax is part of the country's long-term plan to reduce greenhouse gas emissions and is an important part of the country's rent capture regime.

A couple of factors are cited as the major factors permitting resource revenue to be managed transparently as part of an integrated fiscal management system, such as the well-established institutional framework, its long tradition of transparency for both fiscal policy and central bank operations, and its broad revenue base (with oil revenue accounting for typically less than 15% of total fiscal revenues) of Norway. Particularly, Norway has a well-formulated and transparent asset management strategy for its Government Pension Fund. The Ministry of Finance bears overall responsibility for the Pension (Petroleum) Fund's asset management but has delegated the task of the operational asset management to the central bank (Norges Bank) based on a management agreement. The Ministry of Finance defines the strategy for investment by identifying a benchmark portfolio against which Norges Bank seeks to achieve the highest possible return. Meanwhile, the Ministry of Finance also controls exposure to risk so that the actual return should remain within a range around there turn on the benchmark portfolio.[1]

Norway's fiscal policy drives Petroleum Fund operations rather than vice versa. The Petroleum Fund accumulates all oil revenue and returns on financial investments, and transfers from the Petroleum Fund to

---

[1] See http://www.norgesbank.no/english/petroleum_fund/management/strategy.html.

the budget are only made to the extent necessary to finance the non-oil deficit, with the size of the non-oil deficit determined by annual, medium-term, and long-term fiscal policy.

4.3.2 Central and Central/Provincial Government Tax-by-Tax Sharing Programs—Top-Down Tax Sharing (Centralized or Negotiated Sharing Arrangements)

(a) Indonesia

Indonesia is rich in natural resources, and the role of oil, gas, and other mineral sectors is of great significance in the Indonesian economy. The contribution of these natural resources sector to the Indonesian economy constituted 24% of GDP in 2005. By constitution, natural resources are owned by the unitary state in Indonesia. Natural resources' contribution to government revenue is significant. In 2005, about 22% of government revenue came from non-tax natural resources revenue, the second most important revenue source after income tax from non-oil and gas.

Indonesia experienced a significant political and economic decentralization in recent years.[2] Particularly, decentralization brings to the provinces (a) a revenue allocation by the national government through the IRA funds and (b) resource rent revenue sharing, particularly with the resource-rich provinces (Eckardt and Shah 2006). Meanwhile, fiscal decentralization also grants extensive responsibility to all of Indonesia's provinces in all matters except defense, foreign judicial, fiscal, monetary, and religious affairs and matters that are deemed "strategic".

The revised law 33/2004 introduced a new type of shared revenue which slightly increased the subnational share of oil and natural gas revenues. Starting in 2009, 84.5% of oil revenues and 69.5% of gas revenues will accrue to the central budget and the rest to subnational governments. Subnational governments also receive an extra 0.5% of both oil and gas revenues to increase local expenditures on primary education. Table 4.4 summarizes the resource revenue sharing according to law 33/2004 in Indonesia.

---

[2] See the Law on Regional Autonomy and the Law on Intergovernmental Fiscal Relations passed on April 23 and April 25, 1999 respectively.

Table 4.4 Natural resource tax sharing in Indonesia—2012 (% share)

Revenue source	Central government	Originating provincial government	Originating local government	All local governments in the originating province	All local governments—equal share
Income taxes	80.0	8.0	12.0		
Property taxes	9.0	16.2	64.8		10.0
Property transfer tax		16.0	64.0		20.0
Tobacco excise tax	98.0	0.6	0.8	0.6	
Mining land rent (Kabupaten/Kota producers)	20.0	16.0	64.0		
Mining land rent (provincial mines)	20.0	80.0			
Mining royalty (Kab./Kota producers)	20.0	16.0	32.0	32.0	
Mining royalty provincial	20.0	26.0	0.0	54.0	
Forestry license	20.0	16.0	64.0		
Forestry royalty	20.0	36.0	32.0	32.0	
Reforestation	60.0	0.0	40.0		
Fishery royalty	20.0				80.0
Geothermal mining	20.0	16.0	32.0	32.0	

(continued)

Table 4.4 (continued)

Revenue source	Central government	Originating provincial government	Originating local government	All local governments in the originating province	All local governments—equal share
Oil					
Base rate (kab./kota)	84.5	3.0	6.0	6.0	
Conditional rate (kab./kota)—education	0.0	0.1	0.2	0.2	
Base rate—province	84.5	5.0	0.0	10.0	
Conditional rate—province	0.0	0.17	0.0	0.33	
Natural gas					
Base rate	69.5	6.0	12.0	12.0	
Conditional rate—education	0.0	0.1	0.2	0.2	

Source Shah (2012) and the Indonesian Ministry of Finance

(b) Pakistan

Based on articles 141 and 142 of the Constitution of the Islamic Republic of Pakistan, 1973 as amended states that minerals other than oil, gas, and nuclear minerals and those occurring in special areas under the control of the national government (e.g., the federally administered tribal areas and the national area surrounding the capital city) are governed by the laws and regulations of the provinces in which they are located. Particularly, the Federal Government of Pakistan has exclusive authority over the development of petroleum resources, natural gas, and "mineral resources necessary for the generation of nuclear energy", and all other mineral resources occurring in federally administered areas. Meanwhile, the respective provincial governments of the four provinces of Pakistan have exclusive authority over the development of all mineral resources occurring within their respective borders other than those resources under the exclusive control of the Federal Government.

Pakistan is not regarded as a natural resource-rich country. Under current system, all central excise duty on natural gas, surcharge on natural gas, royalty on natural gas, royalty on crude oil will go through straight transfer, i.e., distributed entirely to the provinces on the basis of collection.[3]

(c) Russia

Article 72 of the Constitution of the Russian Federation provides for the joint jurisdiction of the federal government and its 89 subjects (members of the Federation) over "issues of the possession, utilization, and management of land and of sub-surface, water, and other natural resources and administrative, administrative-procedural, labor, family, housing, land, water and forest legislation; legislation on subsurface resources and on environmental protection".

Russia has substantial oil and gas revenues. At the federal level, oil and gas related revenues account for about 50% of total revenues. Natural resource taxes include petroleum production royalties, charges for use of

---

[3] Transfers of funds through the NFC Awards are the dominant source of revenue for the provincial governments, and account for about 80% of provincial revenue receipts.

mineral deposits,[4] and are mainly collected by subnational governments. Natural resources are highly unevenly distributed and concentrated in some small regions. For example, most oil and gas production is located in two autonomous regions (Khanty-Mansi and Yamalo-Nenets), about three-fourths of all metal production originate from 10 regions. Consequently, resource revenues are highly concentrated in a small number of regions. In 1997, the three regions best endowed in natural resources collected 47% of the sum of regional revenues from taxes, fees, and charges on natural resources (Klotsvog and Kushnikova 1998; USGTA 1998).

Natural resource tax revenue assignment is a politically contentious issue, and the federal and subnational governments often disagree as to how to share the fiscal revenues from natural resources (McLure 1994). For example, in 1997–1998, the governors of the regions of Karelia and Khabarovsk challenged the federal ownership of forests before the Supreme Court but lost. Currently, natural resource tax revenues are shared among the three levels of government in Russia, Table 4.5 shows how the revenues collected from some natural resource and environmental taxes are shared among the various levels of government (Table 4.3).

4.3.3 General Revenue Sharing

(a) China

The ownership of natural resources in China vests in the state. According to Article 9 of the 1982 Constitution, "Mineral resources, waters, forests, mountains, grassland, unreclaim land, beaches, and other natural resources are owned by the state, that is, by the whole people, except for the forests, mountains, grassland, unreclaim land and beaches that are owned by collectives in accordance with the law. The state ensures the rational use of natural resources and protects rare animals and plants. The appropriation or damage of natural resources by any organization or individual by whatever means is prohibited". The general framework of

---

[4] Exploration fees are mainly collected by subnational governments, and are important sources of own revenue in the oil-producing regions.

Table 4.5 Statutory assignment of selected natural resource revenues in the Russian Federation (%)

User fee type	Federal share	Regional share	Local share
Subsoil user fees (oil and gas)	40	30	30
Subsoil user fees (other than oil and gas)	25	25	50
Subsoil water user fees	40	60	0
Oil excises	100	0	0
Gas excises	100	0	0
Oil replacement fee	20(100)	20(0)	0
Forest user fees	40	60	0(100)
Surface water user fees	40	60	0
Aquatic biological resource user fees	100	0	0
Pollution charges	19	27	54
Water supply, sewage and treatment	0	0	100
Land taxes and rentals	30	20	50

Source Adapted from Benoit Bosquet (2002)

resource revenue assignment sharing and resource revenue sharing structure between the central government and the subnational government is defined by the Tax Sharing Reform of 1994.

In general, the central and subnational governments have separate revenue bases. The central government has access to resources tax on offshore oil exploitation, and the subnational governments derive revenues from taxation of inland resources. In terms of tax assignments, 100% of inland resource revenues are assigned to the provinces and 100% of offshore resource revenues are assigned to the central government. In practice, almost all resource tax revenues are collected by the central government and transferred to subnational governments through the tax sharing and transfer system administered by the central government. Resource revenues account for less 1% of total government revenues.

It is at the provincial governments' discretion to decide the resource revenue sharing between the provincial and prefecture/county governments. It implies that there are varied resource revenue sharing mechanisms between the provincial and sub-provincial governments. In general, about two-thirds of resource tax revenues (70%) are retained by the provincial government for own use and the rest transferred to local governments in their jurisdiction.

(b) Nigeria

The Federal Government of the Republic of Nigeria owns natural resources. Oil is the main source of revenue in Nigeria and unevenly distributed. 82% of the total revenue of the general government, or 40% of GDP came from oil in 2000,[5] and the revenue is concentrated in only a few of the 36 states, which have onshore or offshore oil production.

Oil revenue is shared between the federal government and the state and local governments in Nigeria. The 1999 constitution provides a common pool of financial resources (the Federation Account) under which all monies are to be distributed among the federal, state, and local government councils in each state. National Assembly defines the Federation grants to a State to supplement the revenue of that State. The share of the local governments is from the Federation Account through the states and is paid into the State Joint Local Government Account. Basically, federal government controls and collects oil revenue, but attributes at least 13% of the net oil revenue to the oil producing states. In addition, about half of the net proceeds (after deduction of first charges) are redistributed to state and local governments according to a formula decided by parliament every five years. Excess proceeds over the budgeted revenue are also redistributed in the same way, after assigning 13% to oil-producing states.

Under Nigeria's current system, only federally collected revenue is subjected to revenue sharing arrangement. All revenue collected by federal government goes to the Federation Account, and the Federation Account is divided into federal government, State Joint, Local Joint, and Special Funds accounts. Both State Joint and Local Joint accounts are shared based on the same criterion for state and local governments respectively, and Special Funds are designed for special purposes.

The states and local governments keep whatever internal revenue they are able to raise themselves and also receive their share of federally collected revenue. It is worth noting that all the major taxes are federal taxes, and the federal revenues constitute about 90% of the total revenue of all the three tiers of government in Nigeria. Table 4.2 shows the overall share, collection of resource revenues and distribution of federally collected revenues.

---

[5] Oil revenues consist of (i) crude export earnings of the NNPC; (ii) profit taxes and royalties of oil-producing companies (usually joint-venture companies with a government majority ownership); and (iii) domestic crude sales and upstream gas sales.

Table 4.6 Resource revenue sharing in Nigeria

Tiers of govt./components	Overall share (annual average %)	Collection (annual average %)	Distribution of federally collected revenues %	Remarks
Federal Government	43.65	90	48.5	
State Govt./State Govt. Joint Account	29.6	8	24	Criterion: • Equality 40% • Population 30% • Social Development 10% • Land Mass and Terrain 10% • Internal Revenue Effect 10%
Local Govt./Local Govt. Joint Account	20	2	20	
Special Funds	6.75		7.5	• FCT 1 • Ecology 2 • OMPADEC 3 • Derivation 1 • Stabilization 0.5
Total	100	100	100	

Source Udeh (2002)

Consequently, States and local governments are highly dependent on revenue sharing arrangements with the federal government. In 1999, 75% of state revenue came from redistributed revenue from the federal government, including their share of the VAT (85% of total proceeds), and 94% for local government revenue. Most of this revenue was oil-related (Table 4.6).

4.3.4 Mostly, Provincial Control with No Revenue Sharing (Resource Revenues Primarily Accrue to States/Provinces but the Federal Governments Share the Tax Base for Corporate Income)

(a) Canada

In Canada, provinces own and tax resources and keep all resource revenues for own use. The Federal Government shares the tax base for corporate income (profit) tax with the provinces and levies its own rate on this base whereas the provinces levy supplementary rates. Canada has a federal fiscal equalization program to address fiscal disparities among provinces. The purpose of the program was enshrined in the Canadian Constitution in 1982: "Parliament and the Government of Canada are committed to the principle of making equalization payments to ensure that provincial governments have sufficient revenues to provide reasonably comparable levels of public services at reasonably comparable levels of taxation" (Subsection 36(2) of the Constitution Act, 1982). Equalization is financed by the Government of Canada from general revenues, which are largely raised through federal taxes. Provincial governments make no contributions to the Equalization program. Equalization payments are unconditional—receiving provinces are free to spend the3 funds according to their own priorities (Department of Finance, Canada 2022).

For natural resources, consistent with the 2006 Expert Panel's recommendation, fiscal capacity is assessed based on partial inclusion of actual revenues collected by the province. However, a province's equalization payment cannot raise its fiscal capacity above that of a non-receiving province when all resource revenues are taken into account. In the pre-2004 formula, 100% of natural resource revenues were included in Equalization calculations, but Alberta's resources were kept out of the standard against which entitlement to Equalization payments was determined. Since 2007, Alberta's energy resources have been included in the standard, and eligible provinces receive an Equalization payment based on a calculation that either includes 50% of natural resource revenues or excludes those revenues entirely. Eligible provinces automatically receive payments according to the option that yields the larger per capita Equalization payment.

The decision to have two options in relation to natural resource revenues is the result of a political compromise. On one hand, the federal government accepted the recommendations of the Expert Panel on Equalization and Territorial Formula Financing, which—in 2006—called for 50% inclusion of resource revenues in the Equalization formula.[6] On

---

[6] Expert Panel on Equalization and Territorial Formula Financing, *Achieving a National Purpose: Putting Equalization Back on Track*, Ottawa, May 2006. Implementing the Expert Panel's recommendations accounted for the vast majority of the reforms to the Equalization program in 2007.

the other hand, the federal government considered itself bound by a pre-2006 election commitment to exclude natural resource revenues from the formula (Roy-Cesar 2013; Boessenkool 2002; Feehan 2002).

(b) United States

In the United States, states are sovereign under the constitution and own the resources (except on federally owned land), and resource revenue bases are assigned to the states (McLure 1994). Alaska—where oil revenue represents about four-fifth of own revenue—presents an interesting example of the methods used for resource revenue management by the state. On oil, the State of Alaska levies (i) a property tax (at 20 mills or 2% on appraised value), a severance tax ranging from 12 1/4% to 15% subject to a minimum tax per barrel of oil; (ii) royalties; (iii) a production tax surcharge for hazardous spill; and (iii) a state corporate income tax. The corporate income tax is based on corporation worldwide net income apportioned to Alaska under a three-factor formula involving (i) the percentage of corporate sales and tariffs from Alaskan operations; (ii) the percentage of production from Alaska; and (iii) the percentage of property represented by Alaska holdings—at a maximum marginal rate of 9.4%. All state taxes and royalties are deductible for federal income tax purposes. The state is fully accountable for the fiscal policy choices related to oil revenues and their possible uses for spending or savings. Alaska has created a fund to save part of the oil revenue for future generations.

4.3.5 Full Provincial Control over Resource Revenues Accompanied by Bottom-Up Revenue Sharing (Provinces/States Contribute a Bilaterally Negotiated Share to the Federal Government)

United Arab Emirates
The United Arab Emirates (UAE) is an important oil producer with the fifth largest proven oil reserves in the Middle East. The UAE has been a member of the Organization of the Petroleum Exporting Countries (OPEC) since 1967.

The UAE has seven autonomous sheikdoms—Abu Dhabi, Dubai, Sharjah, Ajman, Umm al-Qaiwain, Ras al-Khaimah, and Fujairah. Under the country's constitution, each emirate maintains principal control of its own oil revenues and other natural resources. They turn over a set percentage of their revenues to the federal government, which takes care of defense, diplomacy, education, public health, banking, and several other concerns for everyone. The constitution gives each emirate explicit and exclusive control over any matter not specifically assigned to the federation. Oil and gas revenues include royalties, company profit transfers, and income tax. The federal government is mostly financed by cash and in-kind contribution, which is negotiated every year between the federal government and each of oil-producing emirates.

The United Arab Emirates is perhaps the only country practicing fully decentralized arrangement of resource revenues. Oil is the major revenue source for the United Arab Emirates. The emirate of Abu Dhabi is the center of the oil and gas industry, followed by Dubai, Sharjah, and Ras al-Khaimah. Hydrocarbon revenues account for around one-third of the UAE's GDP, and oil and gas revenues accounted for about half of the total government consolidated revenue. The non-oil sectors have grown at a rapid pace in recent years but the country remains dependent on oil revenues. The UAE revenue sharing arrangements are summarized in Table 4.4.

4.4 A Brief Discussion of Resource Revenue Sharing Practices

The central government control over resource revenues provides the possibility for the central/federal government to internalize the externality of resource extraction, fiscal stabilization, overcoming regional disparities as well as ensuring the stability of regional revenues. For example, the Norwegian Fund is an integral part of a coherent fiscal policy strategy that aims at smoothing public spending over time and decoupling it from volatile oil revenues. It also seeks to replace oil wealth with financial assets, which are expected to grow in value over time to be able to deal with the expected rise in public spending associated with an aging population (Scancke 2003). The central control also provides the possibility for Azerbaijan to use its oil wealth to reduce its dependency on potentially volatile and short-lived oil revenues and facilitate development of non-oil sector. The above-mentioned merits of central control are typically not achieved in most developing countries. On the contrary, it often results in corrupt practices and regional discontent over the use of resources and unaccountable and unresponsive governance.

Central government top-down tax or revenue sharing programs have the potential merits of realizing efficiency gains from a harmonized and centralized tax administration as in China. It can achieve simple and transparent division of fiscal pie and matching revenue means with expenditure needs. It can preserve state and local autonomy while offering the potential to achieve national objectives such as national minimum standards of public services across the country and an internal common market. Nevertheless, such programs in practice are beset with major challenges. They can weaken political and fiscal accountability. Resource revenue sharing system in Nigeria may have contributed to heightened regional fiscal disparities and a lack of fiscal discipline in oil producing regions (Davis et al. 2003). Revenue sharing in Nigeria also did not protect the state and local governments from the adverse impact of volatility in oil prices on state and local revenues and fiscal equity became a serious concern as non-oil producing states were left with few resources to discharge their public service provision responsibilities.

Resource revenues returned by origin facilitates tax harmonization and efficiency in tax administration, but it does not deal with fiscal disparities arising from uneven geographical distribution of resource endowments and revenue bases. In Russia, sharing arrangement caused large disparities in revenue between resource-rich regions and the others. The five richest resources regions, which have 5.5% of total population, collected 53% of all subnational government revenues from taxes, fees, and charges on natural resources in 1997 (Martinez-Vazquez and Boex 2000).

Mostly, provincial control of resource revenues as in Canada and the United States offers greater local accountability but accentuates fiscal disparities. This system can work well if it is accompanied by a federal or inter-state fiscal equalization program.

Full provincial control is important for local accountability but weakens the federal government. Federal-provincial negotiations for the federal share can represent an important challenge if the national politics is divisive.

4.5 Some Conclusions

The theoretical literature argues for decentralization of resource revenues to improve local accountability. However, international practice in resource-rich countries places greater emphasis on volatility of resource revenues and negative externalities of resource extraction in designing revenue sharing systems. Our earlier discussion suggested the following considerations in designing natural resources revenue sharing systems.

First, the key objective in designing resource revenue sharing system should be the sustainable development of the economy. Centralization of resource revenues offer the potential of achieving efficiency, equity, and economic stabilization objectives. However, this potential would not be achieved if the federal government does not manage resources prudently, efficiently, and equitably.

Second, under federal collection, sharing of resource revenues with subnational governments requires striking a fine balance between the interests of the producing regions and regional fiscal equity considerations.

Third, if resource ownership and resource revenue administration is decentralized, a political compact on inter-state revenue sharing system would be helpful in mitigating regional fiscal disparities.

Fourth, management of resource revenue funds under any arrangements requires careful consideration as to the proper institutional design. In this context, Norway and Alaska funds offer useful examples of well thought-out practices. However, each country has to design and develop institutions to suit local circumstances.

4.6 Lessons

The following general lessons can be drawn from the principles and practices of resource revenue sharing worldwide.

- Ideally, natural resource ownership should be vested in all citizens of the nation regardless of the place of residence. And all resource revenues should be deposited in a trust fund. The assets of this fund would be held in perpetuity and could not be drawn but capital income of the fund will be available for current use and distributed partially to citizens on an equal per capita basis and to governments based on a political compact which provides funding to federal government to facilitate its spending power and to constituent units through a fiscal equalization program.
- However, what is ideal may not be feasible. This is because of the competing and conflicted federal-provincial goals. Economic and social union objectives requiring equitable sharing of national

wealth, but political cohesion and environmental protection considerations require preferential access of resource revenues to producing provinces (states). Only second-best solutions are possible in federal countries. Such solutions should aim to limit adverse incentives. No "role model" is available to follow in federal countries but some countries do better than others.
• Second-best solutions in federal countries include centralization of resource rent taxes (but with decentralization of royalties and charges) and redistribute through a federal fiscal equalization program. Alternately, decentralization of all resource revenues are accompanied by a fraternal inter-state (net) equalization program to share national wealth more equitably.

References

Bahl, Roy and Bayar Tumennasan (2002). How Should Revenues from Natural Resources Be Shared? Working Paper 0214, International Studies Program, Georgia State University.

Benoit Bosquet (2002). The Role of Natural Resources in Fundamental Tax Reform in the Russian Federation. World Bank Working Paper Series. WPS 2907.

Bishop, Grant and Anwar Shah (2011). Sharing Petroleum Revenues in Iraq: Obstacle of Foundation to Decentralization? (with Grant Bihop). In *Decentralization in Developing Countries*, edited by Jorge Martinez and Francois Vaillancourt, chapter 16, 549–594. New York: Edward Elgar Press.

Boadway, Robin and Anwar Shah (1993). Fiscal Federalism in Developing/Transition Economies: Some Lessons from Industrialized Countries. Proceedings of the Annual Conference on Taxation Held Under the Auspices of the National Tax Association-Tax Institute of America. National Tax Association, America.

Boadway, Robin and Anwar Shah (2009). *Fiscal Federalism: Principles and Practice of Multiorder Governance*. New York: Cambridge University Press.

Boadway, Robin and Frank R. Flatters (1993). *The Taxation of Natural Resources*. Washington, DC: World Bank.

Boadway, Robin, Sandra Roberts and Anwar Shah (1994). The Reform of Fiscal Systems in Developing Countries: A Federalism Perspective. Policy Research Working Paper 1259. Washington, DC: The World Bank. http://imagebank.worldbank.org/servlet/WDSContentServer/IW3P/IB/1994/02/01/000 009265_3961006021533/Rendered/PDF/multi0page.pdf.

Boessenkool, Kenneth J. (2002). Ten Reasons to Remove Nonrenewable Resources from Equalization. Atlantic Institute for Market Studies.

Davis, J., R. Ossowski and A. Fedelino, eds. (2003). *Fiscal Policy Formation and Implementation in Oil-Producing Countries*. Washington, DC: International Monetary Fund.

Department of Finance, Canada (2022). Equalization Program. https://www.canada.ca/en/department-finance/programs/federal-transfers/equalization.html.

Eckardt, Sebastian and Anwar Shah (2006). Local Government Organization and Finance in Indonesia. In *Local Governance in Developing Countries*, edited by Anwar Shah, chapter 7, 233–274. Washington, DC: World Bank.

Feehan, James (2002). Equality and the Provinces' Natural Resource Revenues: Partial Equalization Can Work Better. TS, Memorial University of Newfoundland.

Gobetti, S. W., Orair, R. O., Serra, R. V., & Silveira, F. G. (2020). A polêmica mudança na partilha das receitas petrolíferas. Discussion Paper IPEA, Brazil.

Hofman, Bert and Kai Kaiser (2002). The Making of the Big Bang and Its Aftermath: A Political Economy Perspective. Paper Presented at the Conference: Can Decentralization Help to Rebuild Indonesia? Georgia State University, Atlanta.

International Monetary Fund (2005). IMF Country Report No. 03/130.

Klotsvog, F.N. and I.A. Kushnikova (1998). A Macroeconomic Appraisal of the Regional Resource Potential of Russia. *Studies on Russian Economic Development* 9(2): 182–188.

Martinez-Vazquez, Jorge and Jameson Boex (2000). *Russia's Transition to a New Federalism*. Washington, DC: World Bank.

McLure, Charles E. Jr. (1994). The Sharing of Taxes on Natural Resources and the Future of the Russian Federation. In *Russia and the Challenge of Fiscal Federalism*, edited by C.I. Wallich. Washington, DC: The World Bank.

Roy-Cesar, Edison (2013). Canada's Equalization Formula. Library of the Parliament of Canada, Publication no. 2008-20-E Revised 2013-09-04.

Sachs, Jeffrey D. and Andrew M. Warner (1995). Natural Resources and Economic Growth. Harvard Institute for International Development, Discussion Paper No. 517a.

Scancke, Martin (2003). Fiscal Policy and Petroleum Fund Management in Norway. In *Fiscal Policy Formulation and Implementation in Oil-Producing Countries*, Davis et al., op.cit.

Shah, Anwar (1994). *The Reform of Intergovernmental Fiscal Relations in Developing and Emerging Market Economies*. Policy and Research Series Number 23. Washington, DC: World Bank.

Shah, Anwar (2012). Options for Financing Sub-national Governments in Indonesia. In *Fiscal Decentralization in Indonesia: A Decade After Big Bang*, edited by the Ministry of Finance, chapter 13, 222–254. Jakarta, Indonesia: University of Indonesia Press.

Shah, Anwar (2014). Raising and Sharing Resource Revenues. PowerPoint Presentation to the South Sudan Conference of the Parties. Addis Ababa, Ethiopia, December 19.

Udeh, John (2002). Petroleum Revenue Management: The Nigerian Perspective.

USGTA (1998). *The Choice Between the VAT and the Retail Sales Tax in the Russian Federation*. Atlanta, GA: Georgia State University, Andrew Young School of Policy Studies.

CHAPTER 5

Non-renewable Resource Revenue Funds: Critical Issues in Design and Management

Anwar Shah

5.1 Introduction

Non-renewable natural resource revenues comprise over 25% of government revenues in 50 countries and an even larger share of revenues in most petroleum producing countries (Venables, 2010). Abundance of non-renewable natural resources and associated flow of revenues represents a great challenge for policymakers in resource-rich countries. This is because of the unique features of such revenues. These revenue streams are highly volatile, uncertain, and unstable due to large fluctuations in prices and demand. Such volatility creates difficulties for planning, budgeting and in pursuit of prudent fiscal and economic policies. Such economic shocks are often persistent. Furthermore, the exhaustible nature of these resources with expected depletion over a defined time horizon, has important implications for fiscal sustainability and intergenerational

---

A. Shah (✉)
Brookings Institution, Washington, DC, USA
e-mail: shah.anwar@gmail.com

equity. The level of revenue stream available today may be transitory and current use of such revenues may deprive future generations the degree of economic prosperity available to current generations.

Exploitation of natural resources involves a complex value chain with potential for grand corruption and with high probability of corruption being undetected and when detected even more difficult to prosecute. Large increases in revenues invites rent-seeking behaviors and one observes that resource-rich countries typically have higher incidence of corruption.

Large inflow of revenues also leads to exchange rate appreciation with investments in non-resource sectors becoming less competitive. This may result in contraction of the traded sector and expansion of the non-traded sector, leading to de-industrialization (van der Ploeg, 2010). This has adverse implications for domestic industrial base and economic growth—the so-called 'Dutch disease' or 'oil curse' (Sachs and Warner, 1995; Frankel, 2010; Mehrara et al., 2012; Scott, 2015; Saeed, 2021). This term was coined in the late 1970s to describe the decline in manufacturing experienced by the Netherlands after the discovery of a large gas deposit. Dependence on resource revenues may also lead to excessive resource extraction, investment in white elephants, short-sighted economic policies, weak institutions of governance, higher incidence of corruption and armed conflict. Adrian Gonzales (2016) notes that due to high incidence of corruption and waste, half of OPEC countries were poorer in 2005 than they were in 1971. Overall, this has the potential of resources being seen as a curse rather than a blessing for country's political stability, long-term growth, and development prospects.

Empirical evidence on this issue is mixed. Among developing countries, Nigeria represents the worst example of the Dutch disease with 70% of its population below $1 a day poverty level. In contrast, the United Arab Emirates, Chile, and Botswana have used resource revenues to achieve remarkable progress. Most industrial countries, for example, Norway, Canada, and the USA have used these revenues well to achieve higher levels of economic prosperity. What distinguished these two contrasting experiences is the quality of institutions entrusted with the management and use of resource rents.

To overcome, Dutch disease requires a long-term vision with an intergenerational time horizon of using resource rents to produce long-lived productive assets. An important institution that has been advanced to promote such an approach is to create an autonomous trust fund to deposit all or a fraction of such revenues as done by several countries including Canada, Norway, State of Alaska, USA, Chile, and others (see Table 5.1). There are several reasons that have been frequently advanced in an advocacy of such funds. First, such funds help to deal with the uncertainty, and instability of resource revenues associated with price volatility and enable governments in long-term budget planning. Second, they help to reduce exchange rate volatility. Third, they could be helpful in achieving fiscal sustainability and intergenerational equity. Fourth, they can soften the impact of exhaustibility of resources by providing income when this occurs. Fifth, they can facilitate fiscal discipline. In formerly centrally controlled economies, these funds could be helpful in expenditure rationalization by enabling a closer look at public subsidies for inefficient and uncompetitive state enterprises. Finally, they could reduce opportunities for corruption.

Davis et al. (2001) have argued that all of the above fiscal prudence objectives could be achieved without having an oil fund. However, political economy consideration in resource abundant countries would prevent such prudent measures. This is because politicians have typically short-time horizon. Schick (2000, p. 79) argues that current political and economic pressures determine annual budgetary allocations. This would dictate diverting oil revenues to overcome these pressures in the short run whereas an oil fund would insulate a fraction of these revenues from short-term political exigencies and would preserve them for long-term future use. Oil funds also improve coordination of monetary and fiscal policies by sterilization of some of oil revenues. Oil funds also help to improve fiscal transparency (Wakerman-Lin et al., 2004).

The following sections reflect on the experiences of a few existing funds and provide guidance on designing and managing these funds for countries contemplating establishing such funds or reforming the management of existing funds.

Table 5.1 Non-renewable Resource Revenue Funds—Examples of large funds

Country	Name	Assets US billions (2013)	Year of inception	Source of funds	Transparency index
Norway	Government Pension Fund Global	$737 ($1338 in 2020)	1990	Oil	10
UAE-Abu Dhabi	Abu Dhabi Investment Authority	$627 ($709 in 2020)	1976	Oil	5
Saudi Arabia	SAMA Foreign Holdings	$533	1963	Oil	4
Kuwait	Kuwait Investment Authority	$342 ($738 in 2020)	1953	Oil	6
Russian Federation	National Welfare Fund	$175	2008	Oil	5
Qatar	Qatar Investment Authority	$115 ($450 in 2020)	2005	Oil	5
Algeria	Revenue Regulation Fund	$77	2000	Oil and gas	1
Libya	Libyan Investment Authority	$65	2006	Oil	1
Kazakhstan	Kazakhstan Nation Fund	$62	2000	Oil	8
Iran	National Development Fund of Iran	$52	2011	Oil and gas	5
USA-Alaska	Alaska Permanent Fund	$47	1976	Oil	10
Canada	Alberta Heritage Trust Fund	$16	1976	Oil	9
Chile	Social and Economic Stabilization Fund	$15	2007	Cooper	10
East Timor	Timor-Leste Petroleum Fund	$14	2005	Oil and gas	8

Source Sovereign Wealth Fund Institute website (http://www.swfinstitute.org/fund-rankings)

5.2 Selected Examples of Natural Resources Revenue Funds (NRRFs)

5.2.1 Industrial Country Examples

Three leading examples of NRRFs from industrial countries are Norway, Alaska, and Alberta trust funds as discussed below.

5.2.1.1 Norway: Government Pension Fund Global

This fund was established in 1990 to deposit all oil and gas revenues to serve as a financial reserve and as a long-term savings plan to shield the economy from instability of such revenues and for intergenerational equity. Market value of its assets as of June 15, 2022, stands at US$1,163 billion. The fund can invest only in foreign assets. Current investments include foreign stocks and bonds, physical assets such as commercial real estate, and infrastructure for renewable energy. Only the real return from the fund investments is available for budgetary support to the Government of Norway. This accounts for about 3% of fund assets and finances about 20% of government expenditures. Government budgetary surpluses are also invested in the fund enabling government access to more resources from the fund to finance deficits. A noteworthy feature of the fund is that the ownership of the fund is vested in Norwegian citizens at large on equal per capita share basis.

The Ministry of Finance has the overall responsibility for the fund's management and has issued guidelines for its management. It has also established an independent body, the Ethics Council to oversee fund investments. Norges Bank (the Central Bank of Norway) has been tasked with the management of the fund, and its Executive Board has delegated the operational management of the fund to Norges Bank Investment Management (see https://www.nbim.no/en/the-fund/about-the-fund/).

The Fund has been successful in achieving its objective of macroeconomic stability. On intergenerational equity, the Fund is skewed toward providing greater benefits to future generations.

5.2.1.2   USA State of Alaska Permanent Fund
In 1969, Alaska received US$900 million in Prudhoe Bay lease sales. This amounted to ten times the annual budget of the state. Initially, oil revenues were used to finance infrastructure and social programs. The second large windfall of oil revenues were expected with the construction of Trans Alaska pipeline in 1976. This motivated a Constitutional Amendment establishing the Permanent Fund to receive such revenues in 1976. Alaska Constitution Article IX, Section 15, Alaska Permanent Fund states that:

> At least twenty-five percent of all mineral lease rentals, royalties, royalty sale proceeds, federal mineral revenue sharing payments and bonuses received by the state shall be placed in a permanent fund, the principal of which shall be used only for those income-producing investments specifically designated by law as eligible for permanent fund investments. All income from the permanent fund shall be deposited in the general fund unless otherwise provided by law.

In 1980, Alaska Senate Bill 161 created the Alaska Permanent Fund Corporation (APFC), an independent agency, to manage and invest the assets of the Permanent Fund. The APFC is governed by six trustees with two of them government officials (the Commissioner of Revenue and a cabinet member) and the remaining four are private citizens. The APFC was expected to ensure that the Permanent Fund provides a means of conserving revenues from mineral resources to benefit all generations of Alaskans. Further, the Fund should maintain safety of principal while maximizing total returns. Senate Bill 122 created the Permanent Fund Dividend program. The fund assets as of FY20 totaled $65.3 billion.

The Fund invests in stocks, bonds, real estate, private equity, toll roads, and electric utilities. The Fund aims to achieve average annual real rate of return of 5% or higher.

The first use of the Fund's income is to pay dividends to every citizen in the state. All Alaska residents receive annual dividends from the Fund. The dividend calculation is based on the number of Alaskans eligible in a dividend year and half of the statutory net income averaged over the five most recent years. In 2021, each eligible resident received US$1114 as dividend from the Fund. The second use of the Fund's income is to provide inflation proofing based upon percentage change in the US Consumer Price Index multiplied by the principal balance at year-end.

After paying for dividends to citizens and inflation proofing, the rest of the net income is transferred to Earning Reserve. The decisions on the use of this reserve rests with the state legislature.

The Fund has been able to achieve its objectives of preserving resource revenues for intergenerational equity as well as growing its capital. However, the Fund rule provides relatively greater benefits to current generations.

5.2.1.3 Alberta Heritage Fund

The Heritage Fund was established by an Act of the Provincial Legislature in 1976 with the objective "… to provide prudent stewardship of the savings from Alberta's non-renewable resources by providing greatest financial returns on those savings for current and future generations of Albertans". It was initially managed as a savings vehicle for economic diversification and for future use. In 1997, the fund was restructured ending its economic and social development mandate and making it solely focused on savings for future generations. Initially, 30% of Alberta's non-renewable resource revenues were deposited into the Fund. As Alberta experienced difficult economic conditions in the early 1980s, this percentage was reduced to 15% in 1982 and deposits were stopped in 1987. The Alberta Government restarted depositing government budget surplus into the Fund in 2005. The financial assets of the fund as of March 31, 2021, were valued at US$17.8 billion. The Fund's investment income after deducting inflation proofing is transferred to the general revenue fund.

The Fund is overseen by the President of the Alberta Treasury Board and the Provincial Minister of Finance. The Department of Treasury Board and Finance develop its policies and investment allocation rules and subsequently monitor its performance and prepare annual and quarterly reports. The operational management of the Fund is entrusted to an autonomous body, the Alberta Investment Management Corporation. A multi-party standing committee of the provincial legislature reviews and approves Fund reports and its performance and reports its evaluation findings annual to the provincial legislature.

The Fund has averaged a 10-year average annual rate of return of 8.9%. As most of the Fund's investment income after retaining inflation proofing is returned to the General Revenue Fund, the Fund has accumulated assets at a slow pace. Since inception, the Fund has made cumulative transfers of US$46.2 billion to the General Revenue Fund.

5.2.2 Transition Economies Examples

Two interesting examples of NRRFs from transition economies are presented below.

5.2.2.1 State Oil Fund of Azerbaijan Republic

This fund was established in 1999 as an extra-budgetary institution with the stated objectives of "collective and effective management of foreign currency and other assets generated from the activities in oil and gas exploration and development as well as from the Oil Fund's own activities in the interests of citizens of Azerbaijan Republic and their future generations" (cited by Petersen and Budina, 2004 from Charter of the State Oil Fund of Azerbaijan Republic, Section 2). Total fund assets as of 2013 stood at US$33 billion.

The fund is accountable to the President of the Republic and the supervisory control over the fund is exercised by a supervisory board appointed by the President. The Board comprises 10 members with two legislative representatives and eight top government officials and academics. The Board sets the inflow and outflow rules and the investment strategy. The operational management is exercised by an executive director who makes regular reporting of its investments to the President. The management of a part of the portfolio can also be contracted out to an external manager. The fund invests primarily in highly secure investments such as government bonds and securities. Equity investment is not allowed. Only up to 60% of fund assets can be in long-term investments.

The fund assets accumulate in a special account of the National Bank (Central bank). The spending is carried through the Treasury system and subject to procurement laws. Such spending should also be consistent with the government's fiscal policy stance. The fund is audited by an external audit firm annually.

Petersen and Budina (2004) find that the fund's governance framework is sound and has worked well in achieving its savings and stabilization objectives.

5.2.2.2 National Fund of the Republic of Kazakhstan

This fund was established in 2001 by a presidential decree. It is intended to serve savings and stabilization objectives and to ensure transparent and prudent oil revenue management for the welfare of current and future generations. As of 2016, the fund had total assets valued at US$65 billion.

The fund is managed by the central bank—National Bank of the Republic of Kazakhstan. A management council appointed by the President of the Republic provides an oversight over its operations. The Council is presided over by the President, and comprises President of the Senate, Speaker of the House, Prime Minister, Governor of the central bank, Deputy Prime Minister, Minister of Finance, and the Chairmen of the Public Accounts Committee. The Council decides on the inflow and outflow rules and the investment strategy. 10% of oil revenues are diverted to the fund use. The fund is allowed to make investment in reliable and reputable liquid foreign assets. The management of foreign assets can be contracted out to an external fund manager. The Ministry of Finance controls the use of fund outflow. The fund makes regular periodic reports and is externally audited annually (Petersen and Budina, 2004).

The fund has served its objectives of fiscal transparency, savings, and stabilization well (Wakerman-Lin et al. 2004).

5.3 National Resource Revenue Funds: Are They Effective?

Many resource-rich countries have established stabilization funds to overcome volatility of resource revenues and to advance national development and intergenerational equity. However, several countries, for example, Oman, Papa New Guinea, Mexico, Venezuela, Chad, Ecuador, Nigeria, and the Province of Alberta, Canada, have over time tinkered the rules of these funds for political expediency (see Smith and Landon, 2012). Empirical evidence on the effectiveness of such funds in achieving stated goals, however, is scant. This section synthesizes available empirical evidence on this subject.

Wagner and Elder (2005) and Sobel and Holcombe (1996) find that such funds do contribute to macroeconomic stabilization. Smith and Landon (2012) find that a rule-based fund is welfare improving but the magnitude of potential gains in welfare depends upon their design. Using Alberta data, they find that the highest welfare is obtained by a fixed rule fund with a fixed deposit rate of 50% and a fixed withdrawal rate of 25%. This they find superior to the Norwegian fund rule of deposit and withdrawal rates of 100 and 4% respectively. They argue that the lower welfare of the Norwegian rule is that it leads to greater benefits for future generations while reducing the welfare for more recent generations. Further, a deposit/withdrawal rule of 50/25 would garner greater political support making it more sustainable.

Norway/s fund has enabled the government to adopt a counter-cyclical fiscal stance, dampen the appreciation of the real exchange rate in the face of rising oil revenues thereby protecting the competitiveness of the non-oil sector and achieving stabilization objectives. Chile's Copper Stabilization Fund has been successful in fiscal stabilization. Venezuela's Macroeconomic Stabilization Fund and Oman's Oil Funds have not been successful in achieving their stabilization objectives due to frequent discretionary changes in fund rules. Kuwait's Oil Funds has facilitated countercyclical fiscal policy and its income has supported financing government budgetary deficits (Fasano, 2000). Oil funds in Azerbaijan and Kazakhstan are under presidential control and lack any formal stabilization mechanism but have been successful in fiscal stabilization and growing fund assets to benefit future generations (Lucke, 2010; Davis et al., 2001).

Wakerman-Lin et al. (2004) find that the oil fund in Azerbaijan had a positive impact on fiscal discipline and transparency in revenue management. Prior to the establishment of the fund, such revenues were treated as off-budget revenues and deposited in undisclosed offshore accounts beyond the purview of the Parliament and Chamber of Accounts. They also observe that the Azerbaijan fund places greater emphasis on savings for future and the Kazakhstan fund places relatively greater concern for stabilization. Overall, both funds have contributed to improved fiscal discipline, and transparency and accountability in oil revenue management.

These funds have also advanced public and political awareness for maintaining long-term fiscal stability and facilitating diversified economic growth (Petersen and Budina, 2004, p. 125). These findings on transparency and accountability are further corroborated by Tsani et al. (2010).

In Iran, Trinidad and Tobago, Venezuela, and Nigeria, oil funds have not been successful in overcoming the adverse effects of abundance of oil revenues on the manufacturing sector (Natural Resources Governance Institute, 2015).

Astrov (2007) finds that the Russian Oil Stabilization Fund was successful in achieving both of its objectives: to reduce the vulnerability of budget to the oil price volatility and to sterilize the impact of oil-related foreign exchange inflows on money supply. It contributed to macroeconomic stability and has helped to decouple the GDP growth rate from oil price dynamics.

Davis et al. (2001) and Engle and Valdes (2000) find that expenditures in countries with NRRFs have shown lesser variability with fluctuations in oil prices than countries without NRRFs. They cautioned however that the same countries typically followed prudent fiscal policies even before the establishment of such funds. Devlin and Lewin (2004) based upon panel data for 71 countries over the period 1970–2000 find that NRRFs are associated with decreased government operating expenditures but higher investment expenditures.

Shabsigh and Ilahi (2007) use 30-year panel data for 15 countries with and without oil funds and find that oil funds are associated with reduced volatility of broad money and prices. However, there is a statically weak negative association between the presence of an oil fund and the volatility of the real exchange rate.

Bagattini (2011) studied fund performance by analyzing cross-section and time series data for 30 countries. He concluded that presence of funds contributed to better fiscal outcomes. He finds that autonomous governance devoid of political interference of these funds is the most important factor in determining their success. Also, fixed rules less susceptible to political pressures are important for success.

Heuty and Aristi (2010) study the experience with NRRFs in developing countries and find that they have generally failed to provide an enabling framework for economic diversification and sustainable development at home. This is because of their investing resource windfall abroad. They argue that these funds need to recognize the commodity price volatility as the most pressing challenge while the accumulation of wealth for future generations should be subordinated to the creation of a diversified economy in the present.

A review of US states' non-renewable revenue funds suggests that permanent funds such as the one in Alaska were successful in creating a long-term endowment for future generations whereas stabilization funds in Oklahoma and Louisiana were more successful in short-term fiscal stabilization only (Franklin and Moore, 2017).

An important limitation of the current models of NRRFs noted by Lebdioui (2020) is that these funds are focused on fiscal stabilization and financial diversification to the neglect of the transformation of the domestic productive structure. He argues that resource-dependent economies have a greater need for economic diversification and NRRFs could be usefully directed to achieve those ends.

Overall, the empirical evidence on economic stabilization, fiscal discipline, and good governance is mixed. However, there is positive evidence on wealth accumulation, intergenerational equity, and on province building. Various studies point out that the success of such funds critically depends upon stability and quality of fixed rules and fiscal and budgetary institutions (Smith and Landon, 2012). Successful funds have institutional and legal frameworks that resist deviating from fixed rules and keep the politics and current government political and budgetary pressures from raiding the funds (Briere, 2010). Thus, ultimately it is good governance that determines the success of these institutions.

5.4 Key Considerations in Designing Natural Resources Revenue Funds

There are several important considerations in designing and managing natural resource revenue funds: fund governance, transparency and civil society oversight, total pool, and inflow (savings) rules and outflow (spending) rules. The NRRF should be integrated with the budget and fiscal policy framework. All revenues should be deposited in the fund and a portion made available for budget support annually. Spending rules should ensure stability of income for the government. Asset allocation rules should be designed to create a diversified portfolio of financial assets mostly held abroad yielding maximum risk adjusted return (see Berkelaar and Johnson-Kalari, 2008; Das et al., 2009). Of critical importance are the fund governance framework and inflow and outflow rules. These are discussed below.

5.4.1 Fund Governance

An important consideration in designing fund governance is to depoliticize the management of the funds to ensure integrity and accountability. The institutional arrangements should safeguard against direct and indirect raiding of the funds. Direct raiding implies that either required deposits are not made into the fund or funds are withdrawn for purposes that were not mandated. Indirect raiding refers to the use of funds for backing unsustainable debt creation. Further, these institutional arrangements should not place constraints on fund investment that would result in inefficient management and reduced returns.

First, there must be a legal framework specified by central or provincial government legislation. This legal framework should specify parliamentary oversight of the fund through a Standing Committee on natural resource revenue funds and Public Accounts Committee. The Cabinet should be accountable for legislative compliance by the Fund.

Overall management responsibility should rest with the finance minister or the treasurer. The Ministry of Finance/Treasury should issue regulations establishing inflow/outflow rules and links with the budget. These regulations would specify asset allocation and investment policy and guidelines. For example, the Norwegian fund is allowed to invest 60% in stocks, 40% in bonds, and 1% in real estate—all foreign only. These should also specify threshold for risk tolerance. Norway specifies moderate risk tolerance with high returns. Reporting and transparency requirements should also be clearly stated and enforced. Guidance on ethical standards and rules is helpful. Norway requires respecting human rights, ensuring integrity and investment in non-military assets only. Finally, these regulations should also clarify whether fund would be managed internally or externally.

The Ministry of Finance in issuing these regulations would have to grapple with a choice on operational management responsibility. These choices include internal management by a division/unit of the Ministry of Finance/Treasury. Another option is to entrust operational management to the central bank or to a special/separate unit within the central bank, For example, in Norway the fund is managed by a separate unit of the Norges Bank (central bank).

Similarly, operational oversight function could be conducted in several ways. One option is to have an in-house committee as done in Norway. Another option is to have an independent board of directors comprising government, legislature, experts, and civil society members as pursued in Alaska (Bauer, 2014). A third option is to have a professional board of directors comprising senior business and investment leaders/experts as done in Alberta.

There are multiple assets management options. The assets can be managed internally within the government or the central bank. Internal management implies that asset allocation rules and investment guidelines are developed internally, and assets are also managed internally. In

Norway, 96% of the assets are managed internally by the central bank and the rest 4% through external fund managers. Another option is to manage internally by the investment agency as done by the Alberta Investment Management Corporation and Alaska Permanent Fund Corporation. A third option is to hire, on a contractual basis, an external fund manager assisted by a team of professionals with expertise in trading and asset management with a mandate to maximize long-term risk adjusted returns. In such a case, asset allocation rules and investment guidelines are developed by the government. Also, the asset managers are selected internally and only management responsibilities are outsourced to the external manager. The asset manager must follow the investment guidelines. The experience of sovereign wealth funds suggests that funds managed by external managers with proper government and civil society oversight have typically shown higher risk adjusted real rate of return compared to those managed in-house by government bureaucrats (see also OECD, 2018).

A final option is to simply outsource fund management to the private sector on a performance contract. In this case, asset allocation and investment guidelines are developed internally, and private mangers have full discretion of investment within the specified guidelines. Further, the outsourcing of fund management can be voided if annual performance criteria are not met. Outsourcing of fund asset management would be justified in cases when government lacks internal capacity to manage financial assets and outsourcing is expected to yield better management and ability to deal on a timely basis with critical issues. Outsourcing should only be for management of financial assets and for risk management, but ownership of risk must remain with the government.

Transparency and civil society oversight requirements can be met through several steps. First, legislating public's right to know, i.e., access to information legislation. Second, reporting and auditing requirements and finally, having an oversight board as discussed earlier.

5.4.2   Inflow Rules

There are variety of inflow rules practiced by various oil funds.

- *Commodity Price Threshold.* Chile, Venezuela, Russia, and Oman follow a commodity price threshold rule. Any revenues arising from

oil prices rising above the specified price are designated to be deposited in the oil fund.
- *Revenue Level Threshold.* Chile, Trinidad, and Tobago follow a revenue level threshold rule and incremental revenues above the threshold level are deposited in the fund.
- *All Resource Revenues.* Norway diverts all resource revenues to the Government Pension Fund Global.
- *A Fraction of Resource Revenues.* Alberta, Canada Heritage Trust Fund was initially set up to receive 30% of all resource revenues but this rule has not always been followed. Alaska Permanent Fund receives 25% of lease rentals, royalties, and federal revenue sharing mineral payments with an effective rate of 10% of all oil revenues.
- *Other rules*: Other rules include, a fixed amount, a macro rule linked to non-oil primary budgetary deficit and having annual discretionary decisions.

5.4.3 Outflow Rules

Multiple outflow rules have been followed by various countries to manage outflow of funds. These include:

- *Bird in hand rule.* This rule requires that transfers to government budget not to exceed the real rate of return multiplied by the market value of the fund minus amount needed to inflation proof endowment. Alaska uses this rule for dividend distributions to state residents rather than to government budget. Alberta uses the same rule but requires public investments in infrastructure or economic diversification.
- *Two birds in the bush or total wealth rule.* According to this rule, the fund transfers to the budget could be equal to the expected real rate of return on the fund assets plus estimated present value of oil in the ground. This rule is implemented by East Timor, Mauritania, and Sao Tome.
- *Hybrid of the above or a mixed rule.* Alberta after inflation proofing returns all net investment income of the fund annually to the General Revenue Fund of the Government.
- *Macro rule.* According to this rule, non-oil budget deficit can be covered by such outflows as in Norway.

5.5 Conclusions

Natural Resource Revenue Funds, if properly designed and managed, can serve important tools to advance economic stabilization, diversification, wealth accumulation, common citizenship, and regional and local development goals. For batter management of fund assets, external contractual fund management with an independent oversight board may be helpful in maximizing returns while keeping risk within tolerable limits provided overall management responsibility, policy determination, and accountability rests with the government. Accountable fund governance is critical to its success. For this purpose, transparency, integration with the budget, and strong parliamentary and civil society oversight is a must. Citizens' right to know and access to information and external audit is important for restraining direct and indirect raising of funds to advance political and bureaucratic interests.

References

Astrov, Vasily (2007). The Russian Oil Fund as a Tool of Stabilization and Sterilization. *Focus on European Economic Integration* (online) Q1, 07: 167–178.

Bagattini, Gustavo (2011). The Political Economy of Stabilization Funds: Measuring Their Success in Resource Dependent Countries. IDS Working Paper No. 356, Institute of Development Studies, University of Sussex, Brighton, UK.

Bauer, Andrew (2014). *Managing the Public Trust: How to Make Natural Resource Funds Work for Citizens*. Revenue Watch Institute, New York.

Berklaar, Arjan and Jennifer Johnson-Calari (2008). Commodity Savings Funds: Asset Allocation and Spending Rules. Presentation at the World Bank, March 10.

Briere, M. (2010). Managing Commodity Frisk: Can Sovereign Funds Help? Solvay Brussels School of Economics and Management. CEB Working Paper No. 10/056, November.

Das, Udaibir, Yinqium Christian Mulder and Amadou Sy (2009). Setting Up a Sovereign Wealth Fund: Some Policy and Operational Considerations. IMF Working Paper No. WP/09/179. IMF, Washington, DC.

Davis, Jeffrey et al. (2001). Oil Funds: Problems Posing as Solutions? *Finance and Development*, 38(4, December).

Devlin, J. and M. Lewin (2004). Issues in Oil Revenue Management. World Bank/ESMAP Workshop on Petroleum Revenue Management Proceedings, Washington, DC.

Engel, Eduardo and Rodrigo Valdes (2000). Optimal Fiscal Strategy for Oil Exporting Countries. Unpublished Manuscript cited in Mehrara et al. (2012).

Fasano, Ugo (2000). Review of Experience with Oil Stabilization and Savings Funds in Selected Countries. IMF Working Paper WP/00/112, June.

Frankel, J. (2010). The Natural Resource Curse: A Survey. NBER Working Paper No. 15836. Cambridge, MA.

Franklin, Aimee, and Samuel Moore (2017). State Level Choices for Non-Renewable Resource Revenue Funds. *Oklahoma Politics* (December), pp. 2–25.

Gonzales, Adrian (2016). The Land of Black Gold, Corruption, Poverty, and Sabotage: Overcoming the Niger Delta's Problems Through the Establishment of a Nigerian Non-Renewable Revenue Special Fund (NNRSF). *Cogent Social Sciences*, 2(1): 1126423. https://doi.org/10.1080/23311886.2015.1126423.

Heuty, Antoine and Juan Aristi (2010). Fool's Gold: Assessing the Performance of Alternative Fiscal Instruments During the Commodities Boom and Global Crisis. Revenue Watch Institute. www.revenuewatch.org.

Lebdioui, Amir (2020). Economic Diversification in Middle East and North Africa: A Development Approach to Managing Non-renewable Resource Revenues. Paper Presented at the 26th Conference of the Economic Research Forum.

Lucke, Matthias (2010). Stabilization and Savings Funds to Manage Natural Resource Revenues: Kazakhstan and Azerbaijan vs. Norway. Kiel Institute for the World Economy Working Paper No. 1652, October.

Mehrara, Mohsen, Abbas Rezazadeh Karsalari and Fateme Haghiri (2012). Oil Funds and the Instability of Macro-Economy in Oil-Rich Countries. *World Applied Sciences Journal*, 16(3): 331–336.

Natural Resources Governance Institute (2015). The Resource Curse: The Political and Economic Challenges of Natural Resources Wealth.

OECD Development Centre (2018). *Policy Dialogue on Natural Resource-Based Development*. OECD, Paris.

Petersen, Christian and Nina Budina (2004). Governance Framework of Oil Funds: The Case of Azerbaijan and Kazakhstan. Petroleum Management Workshop Proceedings, pp 114–126, March 2004. World Bank, Washington, DC.

Sachs, Jeffrey and A. Warner (1995). Natural Resource Abundance and Economic Growth. National Bureau of Economic Research Working Paper Series No. 5398. Cambridge, MA.

Saeed, Khalid Adnan (2021). Revisiting the Natural Resource Curse: A Cross-Country Growth Study. *Cogent Economics and Finance* (online), 9(1): 2000555.

Shabsigh, G., & Ilahi, N. (2007). Looking Beyond the Fiscal: Do Oil Funds Bring Macroeconomic Stability. IMF Working Paper WP/07/96.

Schick, Allen (2000). Budgeting for Fiscal Risk. In Hana Brixi and Allen Schick, eds. Government at Risk. Chapter 3:79–98. Washington, DC: World Bank.

Scott, Lewis (2015). Are Natural Resources More of a Curse Than a Blessing? https://www.air.info/2016/07/03/are-natural-resources-more-of-a-curse-than-a-blessing/.

Smith, C.E., and Landon, S. (2012). Government Revenue Stabilization Funds – Do They Make Us Better Off? University of Alberta, Department of Economics Working Paper, Edmonton, Albertra, Canada.

Sobel, R., and Holcombe, R. (1996). The impact of state rainy day funds in easing state fiscal crises during the 1990–1991 recession. *Public Budgeting and Finance*, 16(3), 28–48.

Tsani, Stella, Ingilab Ahmadov, and Kenan Aslanli (2010). Governance, Transparency and Accountability in Sovereign Wealth Funds: Remarks on the Assessment, Rankings and Benchmarks to Dater. Public Finance Monitoring Center Working Paper, March.

Van der Ploeg, Frederick (2010). Natural Resources: Curse or Blessing? CESIFO Working Paper No. 3125.

Venables, A. J. (2010). Resource rents; when to spend and how to save. International Tax and Public Finance, 17(4), 340–356.

Wakerman-Lin, John, Paul Mathieu and Bert van Selm (2004). Oil Funds and Revenue Management in Transition Economies: The Cases of Azerbaijan and Kazakhstan. Petroleum Management Workshop Proceedings, pp. 103–113, March 2004. World Bank, Washington, DC.

Wagner, G., and Elder, E. (2005). The rtole of budget stabilization funds in smoothing government expenditure over the business cycle. *Public Finance Review*, 33(4), 439–465.

PART III

Environmental Federalism

Chapter 6

Green Federalism: Principles and Practice in Mature Federations and the European Union

Anwar Shah

6.1 Introduction

Environmental issues have, in recent decades, taken center stage in public discourse and policy discussions worldwide. In federal and decentralized unitary countries, decision-making on environmental functions involve multiple orders (used interchangeably with levels) of government with exclusive and shared responsibilities. This chapter explores the conceptual underpinning for assignment of environmental functions to various

---

The author is grateful to Sandra Roberts for assistance with a survey of the literature on this subject.

A. Shah (✉)
Brookings Institution, Washington, DC, USA
e-mail: shah.anwar@gmail.com

© The Author(s), under exclusive license to Springer Nature Switzerland AG 2023
A. Shah (ed.), *Taxing Choices for Managing Natural Resources, the Environment, and Global Climate Change*,
https://doi.org/10.1007/978-3-031-22606-9_6

orders of government under decentralized governance. In doing so, it draws upon a wide strand of literature from public finance, public choice, political science, and neo-institutional economics literature. It then uses this conceptual lens to compare practice in selected few OECD countries namely Australia, Canada, Germany, the United States, and the European Union. It then draws lessons from these experiences for developing countries aiming to strengthen decentralized environmental governance.

6.2 Conceptual Underpinnings of Assignment of Environmental Functions

As a prelude to assigning government functions, a discussion of the rationale for government interventions is in order. A role for government in setting environmental policies is necessary because market failures prevent price signals from accurately reflecting the social cost of using scarcer natural resources (clean air, clean water, etc.). Environmental quality is a public good and private economic activities can generate externalities. Which order of government should intervene depends on the type of public good and the extent of externalities. The objective in all cases is to achieve an efficient level of economic activity. For example, pollution abatement should increase until the marginal cost equals the sum of the marginal benefits to all affected parties. Assigning responsibility to a particular level of government involves determining which jurisdiction is of the appropriate size to consider all the relevant costs and benefits (Dalmazzone, 2004). In addition, the regulatory regime itself should not impose excessive costs on an economy. The design should minimize monitoring and compliance costs and economic distortions. Specific policy instruments should be evaluated in this light as well. A good system will minimize short-run costs and provide long-run incentives for polluters to improve technologies to reduce the cost of compliance. Flexibility to adapt to specific conditions is preferable. In the case of pollution, for example, the objective is to internalize the negative externality. An efficient outcome also requires that the marginal cost of pollution abatement be equalized across polluters. Differentiated taxes or subsidies can be adjusted to local conditions. Another option is to create a market in which firms trade emission permits (the right to pollute) (see Dalmazzone, 2004, Esty, 1996, Oates, 1999, 2001). The following paragraphs synthesize diverse perspectives on the assignment of environmental functions under multi-order governance offered by various strands of literature related to this subject.

6.2.1 Fiscal Federalism Literature

Three important principles have been advanced by the fiscal federalism literature drawing upon ideas from economics, political science, and public choice literature. These are as follows:

> *Decentralization theorem* (Wallace Oates, 1972) statutes that an environmental function should be assigned to the smallest jurisdiction that internalized benefits and costs of environmental management and controls. Oates (2004) has further refined this framework using examples of various types of public goods as outlined in the next subsection.
> *Subsidiarity Principle*: This concept was advanced by the European Union for assignment of functions to union members and the EU Council. According to this principle, environmental authority should be assigned to the lowest order of government unless a convincing case can be made for higher order assignment. This reflects the belief that local preferences and cost conditions are best known to local officials.
> *Fiscal Equivalency Principle*. This principle was advanced by Olson (1969). According to this principle, environmental functions should be assigned by taking into consideration that political and service area jurisdiction overlap for democratic accountability.

6.2.1.1 Federalism Principles and a Stylized Assignment of Environmental Functions

Given the externalities between local, regional, and national environments, for most environment functions, some degree of shared rule or coordination would be required. Further, integrating above principles for some public goods/bads would entail complex jurisdictional design involving multiple jurisdictions horizontally as well as vertically. Deviating from these complexities, Boadway and Shah (2009) by using typology of public goods and associated externalities have suggested a stylized view of the assignment of environmental functions as follows:

- *Global Functions—Assignment to Supranational regimes: Public Goods/Bads with global externalities*: greenhouse gas emissions, ozone depleting substances, and R&D;
- *National Government Functions or Interstate Agreements: Public Goods/Bads with national externalities*: acid rain, nuclear waste, cross-boundary water and air pollution, auto emissions, standards, SOX, NOX, and hydrogen emissions, ammonia, pesticides, R&D, and technical assistance; and
- *Local Government Functions—Public Goods/Bads with local externalities*: drinking water, garbage disposal, local water and ground pollution, local air pollution, particulate emissions from diesel engines, trace metal emissions, monitoring, and enforcement.

In the following paragraphs, we outline a formal conceptual framework introduced by Oates (2004) and some of the complexities discussed in the literature.

6.2.1.2 Oates' Conceptual Framework for Allocation of Environmental Functions

Oates (2004) distinguishes three benchmark cases. The first is when environmental quality is a pure public good, varying across regions but dependent only on the aggregate level of polluting emissions. An example is greenhouse gases that deplete the ozone layer. In this case, responsibility for environmental regulation falls to the central government.

The second benchmark case is that of a local public good with no externalities, in which local environmental quality depends only on local pollution levels. An example is local drinking water. In this case, responsibility falls to the local government. Decentralized decision-making allows greater responsiveness to local conditions.

The third case is that of a local public good with externalities, in which local environmental quality depends on the level of pollution in all jurisdictions. This case is the most common and the most complex. The specifics of each situation determine whether government intervention is necessary and, if so, at what level. If externalities flow in one direction only (pollution moving downstream in a river), an optimal solution without government intervention is unlikely. The polluting party has little incentive to negotiate with the affected jurisdiction. In this case, responsibility for environmental regulation must rest with a jurisdiction that encompasses all affected parties. If the externalities are reciprocal, the parties may have an incentive to negotiate without the intervention of a higher level of government. For example, jurisdictions sharing the shoreline of a polluted body of water may voluntarily cooperate to reduce pollution. In practice, regional cooperation can be difficult to achieve.

Within this theoretical framework, a large literature developed to address specific issues.

6.2.1.3 Issues Arising from the Implementation of the Oates' Conceptual Framework

Defining all relevant parties An economically efficient solution requires calculating the benefits and costs of all affected parties. In the case of natural resources, this may be difficult to determine. One problem is the spatial dimension: where are all the people who have a stake in a natural resource? National parks are an example of a resource in which interested parties may constitute a much larger and physically separated group than those who are directly affected by changes within the park's boundaries (Esty, 1996, p. 9). A second problem is determining the impacts of policy change through time. Our understanding of the links within and across ecosystems continues to evolve; policy changes may have effects which only show up with significant time lags. Dalmazzone (2004, p. 13) notes that the efficiency conditions for resource use are derived from the literature on non-renewable resources (such as oil), in which case changes in the stock had little impact on broader ecosystems.

Need for technical expertise The technical complexity of environmental issues requires specialized knowledge. In the design of any regulatory regime, the administrative costs of acquiring and disseminating information need to be considered. The need for expertise has been used as an argument against decentralizing environmental policy: there are economies of scale in research and development; adequate technical skill is lacking in smaller jurisdictions (Esty, 1996). A counter argument is to separate functions. The central government provides the technical expertise, undertaking research and disseminating information in the form of a set of choices on standards and policy instruments. For local public goods, the final choice rests with the appropriate local government (Oates, 2004).

Intertemporal Discount Rates The effects of environmental policies can occur years after the initial decision. The political science literature frames the issue in terms of the time horizon of legislators. In economic terms, the relevant issue is the choice of the discount rate with which the net present value of future outcomes is calculated. Dalmazzone argues that any differences in the intertemporal discount rates used by governments should be considered in assigning functions Dalmazzone (2004, p. 14). Oates notes a general argument that local residents, particularly in a mobile society, are likely to undervalue future benefits; this is considered a rationale for centralization. Oates argues in response that current property values will reflect the expected future costs or benefits of current environmental policy (Oates, 2004). This statement is based on a two-period model with a housing market in which individuals relocate at the end of the first period (Oates and Schwab, 1996).

Race to the Bottom In the case of local public goods without externalities (Oates' 2nd case), the argument is made that decentralization would produce harmful interjurisdictional competition. Localities competing for mobile capital would lower costs to firms by setting lax environmental standards, resulting in lenient and suboptimal pollution controls. There is no clear consensus on the theoretical support for this argument.

Theory: Oates and Schwab (1988) constructed a model in which interjurisdictional competition is welfare-enhancing. However, the model is restrictive: local governments are price-takers on the capital market; there is no strategic interaction between governments; and each local jurisdiction has a full range of policy instruments available (expenditures, taxes, environmental policy tools). Relaxing these restrictions can yield suboptimal results. Models restricting local governments to a tax on mobile capital yield results in which capital is misallocated and local public services, including environmental protection, are underprovided. Allowing strategic interaction between local governments creates a potential role for a central government. In game-theoretic models, where local governments use lax environmental standards as a competitive tool, the uncoordinated equilibria can be suboptimal. Central government regulations can play a coordinating role to improve welfare. This typically involves a trade-off between gains from coordination and losses from uniform standards (which ignore local conditions). Results are not clear-cut, however. List and Mason (2001) use a game-theoretic model with asymmetric players to generate numerical simulations of two scenarios: decentralized standard setting and a single, uniform national standard. The optimal outcome depends on initial conditions: decentralized standards were better if initial pollution levels were low, and jurisdictions had significant differences; otherwise, the uniform standard was preferable (Oates, 2004; Dalmazzone, 2004). Numerous studies are cited by authors. For example, analyses have shown that either over- or under-regulation can result from competition among states under imperfect market conditions. The outcome is a function of costs, either firm specific, plant specific, or transportation related (Revesz and Stavins, 2004).

Empirical Evidence: With theoretical debate inconclusive, attention turns to empirical studies (Oates, 2004). One strand of research investigates the effect of environmental regulations on industry location decisions. Initial studies concluded that environmental regulations were not a significant factor in the location decisions of polluting firms. McConnell and Schwab (1990) found no apparent effect of regional differences in regulations on location decisions for industrial branch plants. A survey by Jaffe et al. (1995) reached the broad conclusion that evidence did not exist to support the view that environmental regulations adversely affected a jurisdiction's competitiveness. More recent studies, however, making use of richer and more recent datasets and a wide range of econometric techniques have found evidence that environmental regulations affect location decisions. Henderson (1996) concluded that, if relocating, polluting firms are more likely to go to areas with less stringent control measures. List and Kunce (2000) found that foreign multinational corporations locating in the United States were sensitive to environmental regulations in location decisions for new plants. List and Kunce (2000) found that stringent environmental measures reduced job growth in polluting industries. Therefore, one can conclude that the existence of competition doesn't necessarily imply that outcome will be lax environmental standards. A second strand of empirical research looks at the history of environmental policy in the United States. Millimet (2003) evaluated the impact of the Reagan administration's decentralization in the 1980s. He studied environmental quality after the Reagan policy shift by comparing actual quality levels with levels projected from data before the policy shift. He found that "the data are consistent with decentralization leading to a race to the top in abatement expenditures" (p. 714). Fredriksson (2004) broadened the analysis of strategic interactions by considering competition across several policy choices simultaneously. Using US data from 1977–1994, they test a model in which states determine three policies: state-level taxation, infrastructure spending, and pollution control standards. They find evidence that policy decisions can be interrelated, both within and across jurisdictions. "For example, states respond to higher levels of governmental expenditure levels in neighboring states by lowering their own pollution standards" (p. 390).

Conclusion regarding the race to the bottom: Theoretical and empirical evidence is mixed. In Oates' opinion: (1) the magnitude of distortions from interjurisdictional competition matters. This is not well measured yet. (2) Even if decentralization leads to distorted outcomes, is there a better alternative? The choice could be framed as suboptimal local decision or inefficient national standards. Empirical studies to determine which situation yields higher social welfare are needed (Oates, 2004). Similar conclusion is reached by Dalmazzone (2004): "..it is the case by case magnitude of the distortions that would allow us to say more on how best allocating powers over environmental regulation". The theoretical literature is inconclusive. Models have shown both optimal and suboptimal results, depending on the assumptions made about local governments' strategic behavior, range of policy options, and influence in the capital markets (Oates, 2001), Fredriksson (2004), Levinson (1997), Esty (1996), and Oates and Portney (2001). Empirical work likewise yielded mixed results (see summary in Oates (2001), Oates and Portney (2001). Indeed, in the US data one can find a "race to the top" when environmental regulation is decentralized (Millimet, 2003). The overall conclusion is that interjurisdictional competition is not necessarily a problem; the case-by-case magnitude of any distortions would need to be compared to alternative policy options.

"States as laboratories" A separate argument for decentralization is the potential for experimentation by many subnational governments ["laboratory federalism"]. Policy measures can be tested in practice in smaller settings, allowing learning by doing. New solutions to problems may emerge from the larger number of jurisdictions trying a variety of approaches. The ability of smaller jurisdictions to experiment with alternative solutions is considered a separate benefit of decentralized decision-making; "states as laboratories" increases the chance of finding a cost-effective solution to a pollution problem (Oates, 2004).

Technical difficulties of Measuring and Monitoring A separate issue affecting the assignment of regulatory responsibility is the technical difficulty of measuring and evaluating the effects of pollution. The complexity of biological systems, and the uncertainty surrounding their long-run adaptation to pollutants, requires a high level of technical expertise. There may be economies of scale in having this expertise maintained at the central level. However, it does not necessarily follow that standard setting should be a purely federal responsibility. Information and expert opinions can be shared with lower-level governments, which can modify this advice considering local conditions (Oates, 2001), Oates and Portney (2001) and Esty (1996). The ability of smaller jurisdictions to experiment with alternative solutions is considered a separate benefit of decentralized decision-making; "states as laboratories" increases the chance of finding a cost-effective solution to a pollution problem (Oates, 2004).

While this federalism literature provides clear theoretical prescriptions, alternative analyses have been developed to consider the complexity of policy-making environments. Authors have relaxed the assumptions of a monolithic, social welfare maximizing government; perfect information; and the absence of transaction costs. Changing these assumptions shifts the focus to the process through which regulatory decisions are made and implemented. Instead of a single decision-maker, the process is modeled with several actors with competing interests. The focus is on the incentives faced by these agents.

The basic idea that institutional settings matter has been emphasized for practitioners. Litvack et al. (1998) underscore the importance of tailoring decentralization proposals to the specific institutions of a country, which they define broadly as the "rules of the game" (the incentives and constraints faced by actors) and the enforcement mechanisms. Papers in the political economy tradition have examined incentives within a decision-making body such as a legislature. Other authors, drawing on the theory of the firm in cases of imperfect information, have studied regulatory regimes from the perspective of principal-agent relationships. These ideas are elaborated in the following sections.

6.2.2 Public Choice and Political Economy Perspectives

Public choice literature is concerned with influence of interest groups in policy choices and outcomes as it argues that such choices may not be guided by public interest but by interest group politics. Achieving an optimal outcome requires that the political process accurately translate people preferences into policy outcomes (Esty, 1996, p. 9). The ability of special interest groups to organize and lobby, exerting undue political influence, through regulatory capture is an important feature of most political systems around the world. However, authors differ in terms of whether this problem is greater locally (Dalmazzone, 2004; Esty, 1996) or at the central government level (Revesz and Stavins 2004). Esty (1996) has argued that there would be lesser influence of the interest groups at the federal level and therefore lesser under-regulation. Esty argues that environmental groups at the federal level would be as well organized as industrial interest groups and able to counter their influence. Stewart says that due to high transaction costs, it would be difficult for environmental interest groups to get organized in multiple local jurisdictions to provide a counterweight to local industry groups. Sarnoff (1998) supports this position and argues that diffuse environmental interest groups due to lower transactions costs and greater economies of scale would be more effective at federal than at the state level.

6.2.3 Organizational Theory and Neo-Institutional Economics Perspectives

Organizational theory is concerned with the study of the structure, functioning, and performance of organizations and the behavior of individuals and groups within it.

Neo-institutional economics (NIE) is concerned with asserting government accountability to citizens as governors. According to the NIE framework, various orders of governments (as agents) are created to serve the interests of the citizens as principals. The jurisdictional design should ensure that these agents serve the public interest while minimizing transaction costs for the principals.

The existing institutional framework does not permit such optimization, because the principals have *bounded rationality*; that is, they make the best choices based on the information at hand but are ill informed about government operations. Enlarging the sphere of their knowledge entails high transaction costs, which citizens are not willing to incur. Those costs include participation and monitoring costs, legislative costs, executive decision-making costs, agency costs, or costs incurred to induce compliance by agents with the compact, and uncertainty costs associated with unstable political regimes (see Horn, 1997; Shah and Shah, 2006). Agents (various orders of governments) are better informed about government operations than principals are, but they have an incentive to withhold information and to indulge in opportunistic behaviors or "self-interest seeking with guile" (Williamson, 1985, 7). Thus, the principals have only *incomplete contracts* with their agents. Such an environment fosters *commitment problems* because the agents may not follow the compact.

The situation is further complicated by three factors—weak or extant *countervailing institutions, path dependency, and the interdependency* of various actions. Countervailing institutions such as the judiciary, police, parliament, and citizen activist groups are usually weak and unable to restrain rent-seeking by politicians and bureaucrats. Historical and cultural factors and mental models by which people see little benefits and the high costs of activism prevent corrective action. Further empowering local councils to act on behalf of citizens often leads to *loss of agency* between voters and councils, because council members may interfere in executive decision-making or may get co-opted in such operations while shirking their legislative responsibilities. The NIE framework stresses the need to use various elements of transaction costs in designing jurisdictions for various services and in evaluating choices between competing governance mechanisms. There is an emerging literature that draws upon some of the ideas from the organization theory or the NIE framework. In the following paragraphs, we summarize a couple of these studies which deal with regulatory functions.

Estache and Martimort (1998) view regulatory environment as a series of principal-agent relationships with asymmetric information in each link. The inability to write a complete contract covering the entire regulatory relationship creates the potential for inefficiencies in each stage. The authors argue that regulatory design should recognize the full complexity of this situation. Analyzing transaction costs at each link with the goal of minimizing inefficiencies leads to a set of recommendations on the optimal design of a regulatory structure.

The authors distinguish between the structure and the process of regulation. The structure includes the assignment of responsibilities across levels of government, the objectives given to each actor, and the method by which political principals are chosen. The process of regulation refers to the timing of government intervention, the extent of control exercised by a regulatory body (length of time as well as scope), and the communication channels within the governmental hierarchy.

The basic model has a four-layer hierarchy: voters, political principals, regulators, and regulated agents. It is a vertical series of principal-agent relationships. A further complicating factor is the existence of multiple principals. Agents are not simply responding to a single, unified voice. The legislative body consists of many political principals; the regulated entity has an internal board of control in addition to the external regulators; and the regulating bureaucracy can consist of many competing entities.

There are numerous informational asymmetries. The voters lack complete information on the effort and efficiency of politicians. The political principals are less informed about the industry than the regulators (in terms of either technical expertise or institutional memory). Regulators themselves don't know the technical details of the regulated firm. In addition, there are limited means for principals to influence agents. The only constraining factors on the actions of political principals are career concerns and the loss of political power. Regulators have a limited number of instruments to apply to the regulated entity.

Transaction costs result from a variety of failures: lack of commitment by the government; the multiple-principal character of government; and discretion by various decision-makers, which creates opportunities for collusion and corruption.

A basic commitment issue is that governments' tenures are limited by election cycles. The possibility of being voted out of office makes it difficult for elected political principals to commit credibly to lengthy regulatory regimes. Creating an independent agency with a longer lifespan is one option.

In terms of the relationship between regulator and firm, the lack of commitment by regulators (the possibility of renegotiating terms after firm-specific information has been revealed) leads to inefficient behavior in the initial period of regulation. The ability to commit is improved if clear rules and procedures exist at the renegotiation stage. One method to achieve this is to separate the functions of regulators, creating agencies with different objective functions. This separation is a complicated issue discussed in more detail below. To the extent that separation improves commitment it will lead to a more efficient dynamic solution, but it will involve static inefficiencies (e.g., competition between multiple regulators). The theoretical conclusion may be an impractical result of an initially unified regulatory structure with responsibilities separated at the renegotiation stage. The authors suggest a possible implementation strategy of sequential regulation, where the lead agency in the renegotiation is federal with a complementary role played by the states. The overall benefit of regulatory separation is unclear but the issue bears consideration.

Separating regulatory functions adds to the multiple-principal nature of government. It can increase the allocative inefficiency of a regulatory scheme. Multiple regulators pursue individual objective without considering the actions of others. If the regulated activities are complements, too much informational rent is taken from the firm. When the regulated activities are substitutes, the regulated firm can play the principals against each other; the total amount of informational rent retained by the firm is too large. Retaining rent is easiest for efficient firms. Inefficient firms fare worse under separation, which is the source of societal benefit. As the number of regulating entities increases, the inefficiency increases. Allocative distortions are also greater when the regulators act sequentially rather than simultaneously.

The amount of discretion exercised by decision-makers affects the overall efficiency of the regulatory scheme. Arguments to limit discretion are based on this point. In terms of political principals, inefficiency can arise from a limited perspective on social welfare. Political principals chosen through a voting procedure will try to ensure re-election by implementing the preferences of the median voter in their constituency, rather than looking at overall social welfare.

The implementing agency (regulator/bureaucrats) has real power over the regulated firm, which creates opportunities for collusion. The transaction costs of these collusive agreements (side contracts) are endogenous: a better-informed regulator finds it easier to enforce a side contract. Limiting these opportunities can involve reducing the informational rent earned by the firm (the source of bribes), reducing the information available to regulators, or reducing the discretion of the regulator. Competition between regulators is one way to limit their information and reduce the overall cost of capture. A set of asymmetrically informed regulators, each controlling one aspect of a firm, will lack adequate information to extract full rent. Sequential regulatory interventions can also limit the information available to regulators, although this applies most forcibly to the first (least informed) regulator. In all cases, the potential for collusion is an argument in favor of limiting discretion. This can apply to the choice of regulatory tools. Simple contracts leave less room for interpretation; instruments such as quotas are not sensitive to revealed information.

Another approach to limit collusion is to ensure that the regulator's stake in the industry is not too large. A regulator who expects future employment within the industry may have a greater potential for collusion. This can be an argument for using politicians instead of regulators, as their career interests lie elsewhere (although elected officials may not be more independent in the absence of well-functioning voting procedures). Alternatively, the degree of regulation can be reduced. An arms' length approach can be adopted, limiting day-to-day regulation, or the regime can be shifted toward ex post regulation (as is done in anti-trust cases).

Concerning the debate over decentralizing regulatory responsibilities, the authors note three arguments that favor decentralization to complete regulatory contracts. The first argues that the main benefit of decentralization comes from the ability of the local governments to collude with specific interest groups at the local levels. This collusion is indeed socially optimal because it allows the overall contractual arrangement to use the shared information of the local behavior to improve on the centralized arrangement. Capture is not a curse for society, but on the contrary, it allows regulatory contracts to be completed. A second solution is to recognize that implicitly, communication between the local and the centralized governments is assumed to be limited because, for instance, the so-called revelation mechanisms are not available. Contracts are then incomplete. The mere existence of a communication constraint creates the scope for collusion at the local level between the regulated firm and its regulator. The structural asymmetry that decentralization introduces between these two layers of the hierarchy helps the central level use the local regulators to complete their regulatory contract. A third, less normative, perspective about the optimal organization argues that delegation may also help in the case of non-benevolent political principals elected through voting procedures. The basic idea is that the optimal organization trades off the incentive costs of decentralization (modeled as a moral hazard problem between the local regulators and the centralized one) and the benefits of decentralization, which is a better representation of the preferences of the local median voter by the local elected principals (p. 20). The authors conclude by noting the general agreement among practitioners and theorists that regulatory systems should aim for efficiency; recognize the trade-off between commitment and flexibility; and aim for independence, autonomy, and accountability. They point out that the way these criteria are met is determined by how the transaction costs are minimized. Their specific recommendations include:

- if there are commitment problems, short-term institutional contracts between the various players are more likely to ensure independence and autonomy.
- [autonomy implies that] agencies need to have access to their own funding sources; the regulator can recruit its own staff.
- To be effective in his role, the regulator must be able to impose penalties according to clearly defined rules. This is consistent with the emphasis on simple transparent rules that emerged from the literature reviewed in the paper. The other theme addressed here is that there is an ideal sequence in the decision-making process that depends on the distribution of information among the actors.

Aubert and Laffont (2000) evaluate regulatory frameworks with multiple regulators. They discuss decentralization, in which regulators have limited geographical jurisdiction, and the issue of multi- vs. single-industry regulators. Their evaluation of organizational theory starts with the assumption of a benevolent government. They first examine the problems of bounded rationality, which include evaluating the quality of decisions and decision-makers, as well as the difficulty of transmitting information. Looking first at decisions, bounded rationality is defined as having a positive probability of making two separate types of error: accepting a bad decision (project or manager or rule) and rejecting a good one. Centralization is modeled as a hierarchical decision process, in which each of the two decision-makers must accept a project. Decentralization is modeled as a polyarchic decision process, in which case a project rejected by one decision-maker can still be reviewed and accepted by the other. Expected social welfare is evaluated under each regime, with a good decision having a positive social value and bad decisions a negative one. The authors conclude that centralization is better in cases where bad projects are costly (high negative value) and common (high probability of accepting a bad project). Conversely, decentralization is better when good projects are common and have a high value. In terms of the varying quality of decision-makers, the authors note that decentralization is preferred if decision-makers are drawn randomly from the population: a larger number of decision-makers diversify the overall performance of the system. However, it is rarely the case that the selection is random. In the presence of a careful selection process, centralization (a single, well-chosen decision-maker) is better. Finally, the authors note that transmitting information is difficult, which argues for limiting information flows by decentralizing decisions.

The second issue evaluated in the context of benevolent government is private information by regulated agents in the economy. The focus is on the incentives of the regulators to search for and use this private information, and the question addressed is whether it is better to have a unified regulator or to separate the functions across agencies. If multiple regulators do not collude with each other, separation can lead to better searches for information and can make collusion with the regulated agent (corruption) more difficult. To limit the possibility of collusion between regulators, the government can create informational asymmetries through its control of the information available to each regulator.

In the third analysis, the authors maintain the assumption of benevolence and focus on issues that prevent complete contracts. The first is the inability of the center to obtain complete information on local areas. This argues in favor of decentralization of decision-making when local information is good (water resources); however, in cases where broader expertise is needed (health care) centralization may still be preferred. A second issue is the inability of the government to commit not to renegotiate after a firm has revealed its type. Delegating authority to separate regulators can help address this problem. Collusion is viewed as a problem in controlling information flows within the regulatory framework. The literature has shown that decentralization is preferred (e.g., the optimal contract between the center and a colluding supervisor and agent is equivalent to granting the supervisor control of the agent).

Finally, the authors consider the case of a non-benevolent government. Their review of the literature shows that there is no clear theoretical answer to the probable existence of non-benevolence at all levels of government; centralization or decentralization can be preferred depending on assumptions.

The authors review the experiences of both developed and developing countries. Among their conclusions are the following points.

For small developing countries:

> From the results of organization theory ..., a number of reasons can be given why small, less developed countries should rather have multisectoral agencies and centralized regulation. Lack of expertise, the high cost of setting up regulatory agency, as well as the difficulty of avoiding collusion

between separated regulators are factors in favor of not separating regulators but rather have most industries controlled by the same body. The same reasons and in addition the high cost of public funds in less develop countries make centralization likely to be more efficient and less costly than a decentralized system. (p. 45)

For larger developing countries

Organization theory indicates that the advantages of separating regulators (use of yardstick competition, negative externalities between corrupted supervisors, better focus on specific issues) can outweigh the costs when the costs of setting up separated agencies becomes smaller. (p. 54)

Decentralization also becomes more feasible when the country is more developed, according to organization theory. Its main benefits are a better accountability of regulators, more diversification of regulation and experimentation, decisions more adapted to local conditions when communication costs are high (which is likely to be the case). The size of the country is of course a major variable in determining the benefits of decentralization. The main costs of decentralizing responsibilities (lack of coordination, local capture of governments, loss of power of the central government) may be lower than those benefits at an intermediary stage of development. Notice that when countries get to a later stage of development, centralization may again be preferred since communication costs decrease and coordination of actions may become a more crucial issue in more complex regulatory policies. (p. 55)

Addressing decentralization specifically,

The benefits of decentralization being larger when information on local conditions and preferences matters a lot, when externalities across regions or states are relatively unimportant, and when the industry is not evolving rapidly according to technological changes or scientific discoveries, we can expect to see more decentralized systems of regulation for industries such as local transportation, roads, ports management and water management. These sectors can be considered as 'stable', in the sense that they are not the object of frequent technological change. They are 'local' in the sense that local knowledge appears as more important than national expertise and that regulation in a given region has little impact on the industries in another region. This last argument does not always apply in the case of water, as we will discuss later.

The theoretical prediction is confirmed by the fact that even centralized countries have chosen relatively decentralized structures in those sectors. (p. 63)

The caveat on local water management is the recognition that in countries where water is scarce or important for critical production processes, control remains centralized (e.g., Israel).

The authors' concluding section on the centralization/decentralization debate includes the following points. An argument given in favor of decentralization is that it allows differentiated responses to local preferences. This implies that a centralized regulator cannot differentiate in its choice of instruments. Two arguments could support this view. One is bounded rationality: it is costly and difficult for the center to obtain accurate information, so uniform rules work best. The second argument is that bureaucratized, uniform rules are necessary for the center to limit potential capture at the local level. Neither of these arguments is unambiguously correct. Local information may not always be best for regulatory decisions, especially if expertise is limited to the central level. Corruption may be as much a problem at the central level as in the localities. In sum: detailed knowledge of the type of information required for decisions, and of the political realities at the central and local levels, is needed to determine whether a centralized or decentralized framework is best.

The need for detailed knowledge of each country is reiterated in the conclusions on implementation. "[O]ne must take into account the initial allocation of responsibilities that political bodies have acquired. Even if one may have in sight a reallocation of powers, the priority may often be to improve the regulations themselves, to favor horizontal or vertical cooperation of existing authorities to prepare the ground for politically acceptable reforms of institutions" (p. 74).

6.3 A Comparative Perspective on the Assignment of Functions in Practice

In the following paragraphs, we provide a brief overview of the assignment of environmental functions in the federal systems of Australia, Canada, Germany, the United States, and the European Union (For a comprehensive review on constitutional, intergovernmental relations and judicial aspects see Holland et al., 1996).

6.3.1 Australia

Table 6.1 provides a summary of the allocation of environmental functions in Australia. Environmental protection is primarily a state function but the federal (The Commonwealth Government) can intervene in most environmental functions if it deems necessary to do so. In the event of inconsistency between the Commonwealth law and a state law, federal legislative paramountcy in the Constitution ensure that the conflict is resolved in favor of the Commonwealth. Most responsibilities under the Constitution are shared/concurrent responsibilities, but the federal government typically leaves shared rule in more populous states to the state governments and is more likely to get involved in assisting less populous states. The federal government typically does not impose unfunded mandates on the states in environmental protection. Instead, it promotes cooperative federalism through intergovernmental agreements. Such agreements led to the creation of the National Environmental Protection Authority and an inter-ministerial council on the environment.

Overall, environmental federalism has worked well in Australia to advance environmental protection objectives. Strong institutions of executive federalism and the spirit of cooperative federalism ensure timely action to respond to emerging environmental concerns.

6.3.2 Canada

The preponderance of environmental regulation in Canada is at the provincial level (see Table 6.2). The Canadian Constitution has assigned provinces the primary role in environmental regulation. The federal government has a consultative/coordination role only and does/cannot impose any federal mandates over provinces. Federal-provincial and inter-provincial consultation/negotiation play an important role in determining jurisdictional responsibilities. The ability of the federal government to enact environmental regulation imposing uniform national standards is restrained by the Constitution and provincial opposition. The provinces are willing and powerful to resist any attempt toward centralization of environmental policy as a result of transboundary nature of new environmental concerns. Provinces, especially, Quebec, Alberta, and British Columbia, fear of federal influence under the garb of environmental

Table 6.1 Australia: assignment of environmental functions

Environmental functions	Federal	State/provincial/territorial	Local
Water pollution control	The Commonwealth can enact legislation controlling pollution or waste by manufacturing companies. The federal laws have been enacted to control sea dumping, establishing marine parks, and preventing pollution from ships	The use and management of the waters of the River Murray is administered by a Ministerial Council which serves as forum for consultation and coordination among the commonwealth, state, and territorial governments on environmental matters	
Vehicle emissions		The Commonwealth has a shared role in regulating automobile emissions	
Land	The Commonwealth is involved in land conservation and projects to overcome soil degradation by making conditional grants to states (Soil Conservation Act 1985)	The Australian state own and manage public lands	
Mining	Federal law regulates mining activities		
Fish and wildlife	The Australian Constitution assigns legislative power over fisheries to the federal government. The Commonwealth passed the National Parks and Wildlife Act in 1975	The states assume prominent role in conservation of wildlife	
Endangered species	The Commonwealth enacted the Endangered Species Protection Act in 1992. Federal agencies implement this law		
Hazardous waste disposal	The federal government regulates industrial chemical and nuclear waste		
Air Pollution control	Both Federal and states can legislate	Shared with the Federal government	
Global climate change	Federal government empowered to enter into international agreements	States implement federally negotiated agreements	

Source Government of Australia website and Shah (2017)

protection. However, during the last decade, the scope of federal influence over environmental matters has been extended without undermining provincial legislative powers over the environment.

Canada has well-developed institutional apparatus to address any intergovernmental conflicts over environmental protection. The institutions of executive federalism facilitate intergovernmental consultation and cooperation among environment ministers and officials. The Canadian Council of Ministers of Environment (CCME) renders an important forum within which the two levels of government discuss environments matters. Complementing CCME is the work of the Federal-Provincial Advisory Committee (FPAC) responsible for advising the federal Minister of the Environment on how federal and provincial regulations can be harmonized so as to avoid duplication and conflict.

Overall Canada's decentralized federal system has worked well to secure environmental protection. Nevertheless, such a complex system with independent (self-rule) and shared rule, and asymmetrical federalism can encounter delays or inaction in addressing emerging crisis when majority or unanimous consent of all governments is needed to affect a change. Further, such a system is prone to piecemeal solutions when a comprehensive policy response may be required.

6.3.3 Germany

Germany has an integrated/collaborative/hierarchical multi-order governance system with the Federal Government (*Bund*) at the apex and states/*Laender (Land)* as the lower order with municipalities (cities and districts) as the constituent parts of the *Laender* governments. The division of legislative powers consists of three categories: exclusive powers of the Bund, the shared powers of Bund and Laender, and exclusive powers of the Laender. Prior to 2006, Bund also enjoyed framework powers for Laender legislation (Alberton, 2012). For most functions except education, health, and security, the federal bicameral parliament has the exclusive legislative powers and Laender assume the responsibility of implementing federal legislation. To ensure Laender governments' input in federal legislation, the upper house of the parliament Bundesrat (Senate) comprises representatives of the 16 lander governments. The Constitution requires that every bill passed by the Bundestag (House) must be submitted to the Bundesrat for consent or objection. There are also strong institutions of executive federalism

Table 6.2 Canada: assignment of environmental functions

Function	Federal	Provincial	Local
Drinking water quality	Legislation on National drinking water standards	The provinces are allowed to enforce national standards through bilateral federal-provincial accords	
Water pollution control	Great Lakes water, coastal waters beyond provincial boundaries, and interprovincial rivers are subject to the federal legislation The Federal Government adopted the Pulp and Paper Effluent Regulations in 1971 and delegated enforcement of the rule to provinces via bilateral agreements but the law was amended in 1990 to give federal officials more direct access to enforcement procedures by allowing Environment Canada to issue separate pollution control permits	Each province has enacted its own water pollution control laws	
Sewerage			Discharge of sewerage
Vehicle emissions	Auto emission standards		
Air pollution control	Federal Government enacted the Clean Air Act in 1971 and delegated enforcement to provinces via bilateral agreements Acid rain regulations	Each province has enacted its own air pollution control laws	
Land		Canadian Constitution grants the provinces the legislative powers over public lands. Land use is subject to provincial legislation	

Land- soil conservation	A federal-provincial program of Environmentally Sustainable Initiatives is supervised by a joint federal-provincial management committee. The responsibility for delivery varies, sole delivery by the provincial government in Quebec, sole federal delivery in New Brunswick and joint delivery in other provinces
Mining	Nuclear power
Forestry	All natural resources including mining
Fisheries	Exclusive provincial jurisdiction
	The Federal Government has the authority to protect the environment of fish, both at seacoast and inland
	Despite the explicit grant of power over fisheries to the federal government by the Supreme Court judgment, fisheries remain an area of concurrent jurisdiction. To minimize conflicts, the federal government has delegated administration of federal regulations to provincial officials charged with administering provincial legislation
Wildlife Protection	Wildlife habitat protection is a joint federal-provincial program. The responsibility for delivery varies, sole delivery by the provincial government in Quebec, sole federal delivery in New Brunswick and joint delivery in other provinces
Endangered species	Federal responsibility

(continued)

Table 6.2 (continued)

Function	Federal	Provincial	Local
Hazardous waste disposal	The federal government has the authority to regulate and discharge of wastes from a nuclear power plant. The federal law (CEPA 1988) provides a comprehensive regulatory scheme for tox substance, a domain until then predominantly controlled by the provinces. To quell provincial disagreement, the federal government has subsequently agreed to provincial demands for equivalency agreements	Provinces whose laws provide for standards equivalent to those of the federal government CEPA Act are allowed to have their own provincial laws applied	
Pollution prevention	The federal government has the authority to conduct environmental impact assessment for development projects that are financed by the federal government, on federal lands, or are subject to federal legal authority (CEAA 1992). Provinces lobbied to have an equivalency provision incorporated into CEAA similar to that which existed in CEPA. In place of equivalency, CEAA allows for joint federal-provincial review panels	Each province has enacted its own set of pollution control and environmental assessment laws. Environmental impact is a concurrent responsibility with respect to most projects	
Public health	Global pandemics. National Parliament shares its jurisdiction with the provinces over the manufacture of drugs to safeguard public health	Canadian Constitution grants provinces the legislative power over public health. Ontario enacted its first Public Health Act in 1884	

Global climate change	Federal Government regarding combatting greenhouse gas emissions, preventing depletion of the atmosphere's ozone layer. However, provinces exercise strong influence/restraint on its treaty making powers	The provinces constrain national parliament's ability to meet treaty obligation though provincial legislation. All foreign treaties must be endorsed by provincial legislatures

Source Government of Canada website

for continuous dialog among federal and Laender officials. Therefore, all regulatory processes are subjected to deliberations among the two orders. Environmental federalism in Germany is shaped by these institutions of executive and legislative federalism. As the role of EU directives in environmental protection increased, Laender (states) sought to maintain their role. Laender representatives (in their capacity as members of the Bundesrat) participate with Federal Government representatives to working groups of the EU Council of Ministers. Nevertheless, EU directives have constrained Laender legislative and implementation flexibility as EU directives are more detailed with procedural requirements and deadlines, which get transposed into federal legislation. The Federal Government also becomes liable to infringement actions by the EU if compliance lags; as a result, deadlines are included in regulations (Keleman, 2000, 2004).

Federal Government has over the years assumed an over-arching role in environmental legislation with few exceptions such as nature, landscape, and water resource conservation. In the latter areas, the Federal Government provides the framework legislation whereas Laender governments enact detailed legislation for own jurisdictions and provide regulations (von Ritter and Capcelea, 2002). Laender retain their primary role in implementing environmental legislation but delegate some functions to cities and districts such as urban planning, traffic, and waste management. An important exception is that nuclear waste management function is retained exclusively by the federal government. In fields of concurrent legislative powers (animal and plant protection, waste disposal, air pollution control., noise abatement), the Federal Government typically specifies both framework laws and detailed regulations. For example, Federal Air Pollution Control Act defines principles, technical guidelines, technical standards for individual facilities, and ambient air quality standards.

To date important federal legislation has been in the areas of water management, emissions control, nature conservation, wastewater charges, energy management, chemicals and pesticides, atomic energy and radiological protection, and waste avoidance and management. Important federal institutions for environmental protection include Ministry of the Environment, Federal Environmental Agency, The Federal Research Center for Nature conservation and Landscape Ecology, and the Federal Office for Nuclear Safety. The Conference of Environment Ministers serves to coordinate policies among federal and Laender levels as well as among the Laender (von Ritter and Capcelea, 2002).

6.3.4 USA

The US Constitution is silent on the assignment of environmental functions and all residual functions are state responsibility. Environmental protection being a residual function, therefore, belongs to states. Till 1980s environmental regulations primarily emanated from state and local laws and the federal role in this area was quite passive (Percival, 1995). However, the federal government has, over time since 1980s, assumed a commanding role in environmental protection (see Table 6.3). The federal government carved out this role by asserting the interstate commerce clause of the Constitution. It argued that due to transborder migration of migratory birds e.g., Canada geese, transboundary pollution such as acid rain, water, and air pollution, it should have a predominant role in regulating migration and pollution activities. Judicial activism of courts and environmental activists played a catalytic role in encouraging the strengthening of the federal role in environmental protection in the United States.

To date federal laws and regulations constitute the core of environmental protection regulations including local safe drinking water standards. The Federal Government mandates state and local compliance with federal rules with or without any financial support. It typically imposes minimum national standards. The US Congress has followed three types of approaches to environmental regulation: cooperative, conjoint, and national. Most Federal Government statutes fall into conjoint category, which means that Congress or the relevant federal executive agency, e.g., a line ministry or Environmental Protection Agency's established standards must be implemented by state and local governments through an approved plan.

Historical background. Prior to the 1960s, state and local governments had primary responsibility for environmental policy. A strong central government role was created in 1968 with the establishment of the Environmental Protection Agency. Fear of a race to the bottom was one of the driving forces behind this change (Millimet, 2003, Fredriksson et al., 2004 and Engel, 1997). Major pieces of legislation include: Clean Air Act Amendments (1970); Clean Water Act Amendments (1972); and Resource Conservation and Recovery Act (1976). Initial guidelines were

Table 6.3 USA: assignment of environmental functions

Function	Federal	State	Local
Drinking water quality	The Federal Government established procedures and rules for setting drinking national water standards as well as monitoring requirements for public drinking water systems. It also requires states to assist schools in testing and removing lead in drinking coolers	Follow federal guidelines	
Water pollution control	Congress has the authority to regulate pollutant discharges into any waters of the U.S.	Compliance with federal laws	Compliance with federal laws
	The federal government established grants and rules for states to identify and control nonpoint pollutants, required states and localities to control direct industry discharges into public waters; authorizes states to issue discharge permits and monitor effluents	Administer federal laws	Compliance with federal mandates

Sewerage	The federal law directed EPA to develop regulations for toxic waste in sewerage sludge; set rules to test and permit municipal storm sewerage discharges; prohibits dumpping of municipal sewage sludge in ocean waters; recognized funding for municipal waste treatment plants		Compliance with federal mandates
Vehicle emissions	The Federal law imposes rules governing toxic emissions	States' own emission standards (typically more stringent than federal standards) and compliance with federal mandates	Compliance with federal and state laws
Air pollution control	The federal law imposes rules governing urban smog, municipal incinerators; requires states implementation plans and permits for all major sources of air pollution; requires nonattainment cities to take corrective measures of varying stringency	Administer the federal law	Administer the federal law
Land	The federal government owns two-thirds of the land west of Mississippi River and exercises plenary power over the resources of public lands	The ability of state governments to enact land-use regulations to protect the environment is limited by past Supreme Court judgments	
Mining	Congress has the authority over surface mining and setting national standards	The State of Pennsylvania justifies its statute requiring mining companies to leave in place some subsurface coal	

(continued)

Table 6.3 (continued)

Function	Federal	State	Local
Fish and wildlife	Regulations over wildlife located on federal public lands	Primarily a state function	
Endangered species	The federal law requires state and local compliance with federal regulations governing the conservation, protection, restoration and propagation of species of fish, wildlife and plants deemed in danger of extinction and requires state and local monitoring of potential destruction of natural habitats by public or private projects	Compliance with the federal law	Compliance with the federal law
Solid waste	Solid waste disposal is a federal regulatory responsibility	States must comply and cannot exclude out-of-state waste	
Hazardous waste disposal	Federal legislation. States must deal with low level radioactive waste within their borders following federal guidelines. The federal government provides limited federal funds for clean up of toxic waste sites but local governments financial liable for cleaning sites contaminated by hazardous industrial waste even if local governments deposited only normal waste at the site	Administering federal programs	Compliance with federal mandates

Pollution prevention	Local governments are required by the federal government to regulate a wide range of solid and hazardous waste activity, to regulate underground storage tanks in their jurisdiction, and to be liable for publicly owned and operated storage tanks The federal law requires annual EPA inspection of hazardous waste sites operated by state and local governments	Compliance with the federal law
Public health		Compliance with the federal law
Packaging waste		State responsibility. For example, Minnesota has banned retail sale of milk in plastic non-returnable containers
	The federal government has regulations governing pesticides	Primarily state responsibility
		States laws and regulations
Global climate change	Federal government enters into international treaties	States' own regulations. Implementation of federal laws
		Local initiatives and state mandates
		Local initiatives and implementation of states and federal mandates

Source USA Federal Government website

set with little reference to economic analysis. For example, the Clean Air Act required standards such that "no one anywhere in the United States would suffer any adverse health effects from air pollution" (Oates, 1999). Courts interpret the legislation to mean that costs cannot be considered when setting standards. The Clean Water Act sought broad goals of "complete elimination of 'all discharges into the navigable waters'" (ibid.).

The above approach to regulation changed in 1980s and no executive order under the Reagan administration required benefit/cost analysis for all new major regulatory measures (Oates, RFF Reader, 1999). Two major environmental statutes now require that standards be set through a comparison of costs and benefits (Toxic Substances Control Act; Federal Insecticide, Fungicide, and Rodenticide Act) (Oates, 2001). (2) Decentralization. The Reagan administration shifted responsibility to states, "whenever feasible" (Oates, 2001). Federal role reduced through employment and budget cuts (EPA lost 22% jobs and 45% R&D budget in 1981–1983). Enforcement responsibilities were delegated to states. Federal research, information dissemination, monitoring, and enforcement fell sharply and "…much of the institutional structure that had been established in the previous decade was undone over such a short time" (Millimet, 2003, 714–715). There was significant progress in restoring these institutions under Presidents Clinton (1993–2001) and Obama (2009–2016), but most of this progress was undone by President Trump (2016–2020).

Current practice is a mix of policies. The EPA sets nationwide standards for air quality (Clean Air Act Amendment) and safe drinking water; the states set standards for water quality within their boundaries (Clean Water Act Amendments) (Oates, 2001). Benefit/cost studies are prohibited by law for some programs and required for others. The piecemeal changes in legislation have created situations in which regulators cannot consider costs in their decisions but must analyze the cost and benefits of major new rules. (Oates, 1999).

Examples of policies Learning from state experiences: Emissions trading systems introduced at the state level in the 1970s and 1980s proved the system was feasible, highlighted benefits and problems. The 1990 Clean Air Act Amendments created a national system of tradable sulfur allowances (Oates, 2001).

Inappropriate national standard Safe Drinking Water Act (1974) mandated federally enforceable standards for drinking water. EPA set maximum contaminant standards at level such that no known or expected adverse effects occurred, leaving an adequate margin for safety. Amendments to the legislation allowed for some consideration of costs. However, the uniform national standard does not consider the broad range of treatment costs across jurisdictions. Economic analysis has shown that local drinking water is almost purely local public good; heterogeneous cost structures across jurisdictions; and national uniform standard imposes higher welfare cost than efficient outcome under decentralized cost-benefit decision-making (Oates, 2001).

Regional cooperation Experience is mixed on such cooperation. Examples of clear failure include Delaware River Basin Commission (1961) to manage water quality on the river with four states (PA, NY, NJ, and DE) and federal government. Participants never moved beyond parochial concerns. Better examples include: (1) consortium managing the Chesapeake Bay, voluntary agreement in 1987 (MD, VA, PA, DC, and federal government) led to improved water quality, and (2) ozone pollution in NE. In 1990, US Congress created ozone transport region and a commission but the group was enlarged when it became clear that physical boundaries of the region did not cover the problem leading to the establishment of the Ozone Transport Assessment Group. The member states reached agreement on control of stationary source emissions: each state has aggregate allowance for total level of emissions, which can be reallocated through trading among members. The system has potential for savings over command-and-control options as trading has been substantial (Oates, 2001).

6.3.5 *The European Union*

Environmental protection and sustainable development are recognized in the *Charter of Fundamental Rights of the European Union* (EU) and EU strives to harmonize environmental policies of member states. To this end, it has issued directives on environmental impact assessment of projects and programs; packaging waste; greenhouse gas emissions; auto emissions; industrial emissions; the EU emissions trading system (EU ETS); fuel efficiency standards; regulations concerning the registration, evaluation, authorization, and restriction of chemicals (REACH); and the protection of species of wild fauna and flora through the regulation of their trade. Moreover, Member States (MS) are obliged to implement those directives through national (federal) statutory laws, and comply with these regulations (Cartaxo, 2018, Vogel et al., 2010, Keleman 2000, 2004).

EU directives are typically very detailed and specify deadlines regarding implementation and punitive measures for non-compliance. These directives are framed with member states participation to ensure that the required standards are feasible to implement within the defined deadlines and acceptable to member states. For auto emissions, EU does not allow member states to impose more stringent standards whereas in some other areas, such as greenhouse gas emissions, industrial and wastewater pollution, packaging waste, etc., EU directives merely establish a floor and member states may impose higher standards if they so choose (Oates, 1998b; Knill and Lenschow, 1998, Alberton; 2012).

6.3.6 Comparative Perspectives

In green federalism, in Australia and Canada states/provinces have retained their predominant role both in legislation and in implementation whereas in the United States and Germany, there has been a trend toward legislative centralization of environmental policies. In Germany, this trend toward legislative centralization has been accelerated by EU directives whereas Laender still retain their primary role in implementation of these policies. In the United States, while there is a general trend toward centralization, states retain their rights to impose more stringent standards in some areas of environmental protection such as auto emissions and variable standards in other areas such as packaging waste, air, and industrial pollution. The United States has lagged significantly behind other federal countries and the EU, in initiatives relating to global climate change, as it ratified the Kyoto protocol only in 2021.

6.4 Lessons

The following broad lessons emerge from a review of practices in industrial countries.

- Cooperative or competitive federalism are better approaches to environmental management under multi-order governance.

Federal/national command and control approached prove costly and ineffective.
- Federalism principles are useful guides to practice. Applying "one-size-fits-all" approaches to local environmental quality leads to costly administration, jurisdictional conflict, and compliance failures. A leading example is the USA Federal Arsenic Rule for Drinking Water, 2001 providing uniform standards—turned out to be costly and difficult to comply with by small local jurisdictions.
- Federal spending power could be used to have national minimum (possibly variable) standards, but unfunded mandates must be avoided.
- Decentralized governance for "local" environmental functions does not lead to a "race to the bottom". It may in fact enhance economic efficiency and environmental quality through jurisdictional competition and innovation (e.g., emissions trading at state level in the United States in 1970s)
- Democratic participation ensures safeguards for environmental protection. Civil society groups help in ensuring compliance.
- Local governments typically link environmental quality with economic development and therefore there is some evidence of "race to the top" in ensuring environmental protection.
- For some cross-border externalities, subnational agreements may prove less costly alternative to centralization.

In conclusion, environmental federalism represents a dynamic influence of complex interactions of societal consensus and government commitment to environmental protection, constitutional-legal framework, political system, party competition, institutions of intergovernmental relations, judicial system and traditions, public interest groups, private lobbyists, bill of rights including protection of property, and international agreements and influences. Any efforts at a reform of this system must pay attention to all these elements.

References

Alberton, Mariachiara (2012). "Environmental Protection in the EY Member AStates: Changing Institutional Scenarios and Trends". Revue L'Europe en formation, no. 363, Printemps 2012: 289–300.

Aubert, Cecile and Jean-Jacques Laffont (2000). "Multiregulation in Developing Countries". Universite de Toulouse, France, Unpublished paper.

Boadway, Robin and Anwar Shah (2009). *Fiscal Federalism: Principles and Practice of Multiorder Governance*. New York: Cambridge University Press.

Cartaxo, Tiago de Melo (2018). "Environmental Subsidiarity in the EU: Or Halfway to Green Federalism". *Perspectives on Federalism* 10, 3: 303–324.

Dalmazzone, Silvana (2004). "Decentralization and the Environment". In Ehtisham Ahmad and Giorgio Brosio (eds), *Handbook on Fiscal Federalism*. New York: Edward Elgar Publishing.

Engel, Kirsten H. (1997). "State Environmental Standard Setting: Is There a 'Race' and Is It 'to the Bottom'?" *Hastings Law Journal* 48: 271–398.

Estache, Antonio and David Martimort (1998). "Transaction Costs, Politics, Regulatory Institutions and Regulatory Outcomes". EDI Regulatory Reform Discussion Paper, World Bank, Washington, DC.

Esty, Daniel C. (1996). "Revitalizing Environmental Federalism". *Michigan Law Review* 95: 570–590.

Fredriksson, Per G., John A. List, and Daniel Millimet (2004). "Chasing the Smokestack: Strategic Policymaking with Multiple Instruments". *Regional Science and Urban Economics* 34: 387–410.

Henderson, J. V. (1996). "Effects of Air Quality Regulation". *American Economic Review* 86, 4: 789–813.

Holland, Kenneth, F. Morton and Brian Galligan, eds. (1996). *Federalism and the Environment. Environmental Policymaking in Australia, Canada and the United States*. Westport, CT and London: Greenwood Press.

Horn, Murry (1997). *The Political Economy of Public Administration*. Cambridge, UK: Cambridge University Press.

Jaffe, A. B., S. R. Peterson, P. R. Portney, and R. N. Stavins (1995). "Environmental Regulations and the Competitiveness of US Industry". Illumina Technology Recods Report 03/1995. Brandeis University.

Keleman, Daniel (2000). "Regulatory Federalism: EU Environmental Regulations in Comparative Perspective". *Journal of Public Policy* 20, 3:133–167.

Keleman, Daniel (2004). *The Rules of Federalism*. Institutions and Regulatory Politics in the EU and Beyond. Cambridge: Harvard University Press.

Knill, Christoph and Andrea Lenschow (1998). Change as "'Appropriate Adaptation': Administrative Adjustment to European Environmental Policy in Britain and Germany". European Integration Online Papers. Vol. 2, No. 1. http://aetop.or.at/etop/texte/1998-001.htm

Levinson, Arik (1997). "A Note on Environmental Federalism: Interpreting Some Contradictory Results". *Journal of Environmental Economics and Management* 33: 359–366.

List, John and Mitch Kunce (2000). "Environmental Protection and Economic Growth: What Do the Residuals Tell Us". *Land Economics* 78, 2: 267–282.

List, John and Charles Mason (2001). "Optimal Institutional Arrangements for Transboundary Pollutants in a Second-Best World: Evidence from a Differential Game with Asymmetric Players". *Journal of Environmental Economics and Management* 42, 3: 277–296.

Litvack, Jennie, Junaid Ahmad and Richard Bird (1998). "Rethinking Decentralization in Developing Countries". World Bank Sector Studies Series Paper. Washington, DC.

McConnell, Virginia and Robert Schwab (1990). "'The Impact of Environmental Regulation' on Indiustry Location Decisions: The Motor Vehicle Industry". *Land Economics* 66: 67–81.

Millimet, Daniel (2003). "Assessing the Empirical Impact of Environmental Federalism". *Journal of Regional Science* 43, 4: 711–733.

Oates, Wallace (1972). *Fiscal Federalism*. New York: Harcourt Brace Jovanovich.

Oates, Wallace (1998a). "Environmental Federalism in the United States: Principles, Problems, and Prospects". Mimeograph, Department of Economics, University of Maryland.

Oates, Wallace (1998b). "Environmental Federalism and European Union: Some Reflections". University of Maryland Department of Economics Working Paper Series No. 98-04, July.

Oates, Wallace (1999). "An Economic Perspective on Environmental and Resource Management: An Introduction". In W. Oates (ed), *The RRF Reader in Environmental and Resource Management*. Washington, DC: Resources for the Future.

Oates, Wallace (2001). "A Reconsideration of Environmental Federalism". RFF Discussion Paper 01-54. Washington, DC: Resources for the Future.

Oates, Wallace (2004). "Toward a Second-Generation Theory of Fiscal Federalism". Preliminary draft.

Oates, Wallace and Paul R. Portney (2001). "The Political Economy of Environmental Policy". RFF Discussion Paper 01-55. Washington, DC: Resources for the Future.

Oates, Wallace, and Robert Schwab (1988). "Economic Competition Among Jurisdictions: Efficiency Enhancing or Distortion Inducing?" *Journal of Public Economics* 35: 333–354.

Oates, Wallace, and Robert Schwab (1996). "The Theory of Regulatory Federalism: The Case of Environmental Management". In W. Oates (ed), *The Economics of Environmental Regulation*. Aldershott, UK: Edwartd Elgar, 1996, pp. 319–331.

Olson, M. (1969). "The Principle of Fiscal Equivalence: The Division of Responsibilities Among Different Levels of Government". *The American Economic Review* 59, 2: 479–487.

Percival, Robert (1995). "Environmental Federalism: Historical Roots and Contemporary Models". *Maryland Law Review* 54: 1157–1178.

Revesz, Richard and Robert N. Stavins (2004). "Environmental Law and Public Policy". RFF Discussion Paper 04–30. Washington, DC: Resources for the Future.

Sarnoff, Joshua (1998). "The Continuing Imperative (But Only From a National Perspective) for Federal Environmental Protection". *Duke Environmental Law and Policy Forum* 7, 2: 225–318.

Shah, Anwar (2017). *Horizontal Fiscal Equalization in Australia: Peering Inside the Black Box*. Australian Productivity Commission, Submission DR# 103, November 2017.

Shah, Anwar, and Sana Shah (2006). "The New Vision of Local Governance and the Evolving Role of Local Governments". In A. Shah (ed), *Local Governance in Developing Countries*. Washington, DC: World Bank, pp. 1–44.

Vogel, David, Michael Toffel and Nazli Aragon (2010). "Environmental Federalism in the European Union and the United States". *SSRN Electronic Journal*, March 2010, pp. 310–356.

Von Ritter, Konrad and Arcadie Capcelea (2002). "Environmental Management in Federal Systems". Unpublished Paper.

Williamson, Oliver (1985). *The Economic Institutions of Capitalism*. New York: Free Press.

CHAPTER 7

Environmental Federalism in Brazil

Jorge Jatobá

7.1 INTRODUCTION

The literature on environmental federalism is almost inexistent for the Brazilian case. The theme has not brought the attention of researchers despite the fact that Brazil is one of the largest federations in the world. It must be added that it is a federation with singularities that are not found elsewhere. By contrast, the literature on environmental federalism in the United States (US) and in the European Union (EU) is abundant (Vogel et al., 2003; Oates, 1998). This is so because environmental policymaking in those countries takes place in the context of several levels of government with similar implications to the ones that apply to fiscal

---

This chapter is a revised version of a paper commissioned in 2005 by the World Bank Institute. The author is grateful to Dr. Alexandrina Sobreira de Moura for the discussions, information, and criticism.

J. Jatobá (✉)
Consulting and Economic Planning (CEPLAN), Recife, Brazil
e-mail: jorgejatoba@gmail.com

federalism. The latter has laid out the foundation for sounding principles regarding the effectiveness and the welfare consequences of decentralized public expenditures (Oates, 1997).

In the US and in the EU, environmental policy has brought tensions and contradictions that escape from the logic of federalism. Oates (2001) states that it is not clear why in the US air quality standards are set at a national level whereas the water quality standards are set at the States level. Similarly, the European Union has experienced a tension between the principle of subsidiarity that calls for decentralized policy-making and some trends toward centralization founded in the argument that some Europe-wide standards are needed for an effective control of environmental threats.

It is rather common that the formulation of standards be established at the national level whereas States and municipalities implement and enforce them. Thus, environmental policy is, in a sense, a joint product. However, should the lower levels of government take as given the stringency of the national standards? Is there any degree of freedom or situations that would justify that the standards be set at a decentralized level? Oates (2001) tries to answer to these questions by posing three benchmark cases.

The first case is when environmental quality is a pure public good for the country as a whole. This means that a unit of polluting emissions has the same effect on national environmental quality irrespective of where it occurs. In such situation, Oates argues that the central setting of standards is recommended. Examples of such situations would be global climate change and the degradation of the ozone layer.

The second case is one for which environmental quality is a purely local public good. This means that a unit of polluting emissions affects only that jurisdiction. This situation calls for a decentralized setting of environmental standards, being a situation where the principle of subsidiarity is fully applicable. It is also a situation where prevails the interest of the municipality. Drinking water is a good example of a pure local public good.

The third case involves local spillover effects, the most common situation. A unit of polluting emissions affects the locality and other jurisdiction, close or far away, bringing the phenomenon of negative externalities to the forefront of the analysis and of policymaking. Further, it complicates environmental federalism. Either decentralized or centralized determination of environmental quality will bring inefficiency and welfare losses. There are two classes of spillovers. The most complicated one is when there is a unidirectional flow of pollution. Thus, polluting activities in one jurisdiction affect the other but not the converse. In the second case, the flow of polluting emission goes in both directions. The best alternative is horizontal cooperation through agreement among the involved jurisdictions. In the first class, one party loses (the one which controls the emissions) while the other gains from the reduced pollution. Cooperation, under this circumstance, may entail some compensation for the municipality responsible for the emission of pollutants because the local government has no incentives to cut back on its activities. The second class is easier to resolve since both municipalities will have to internalize costs. Further, there are mutual gains from cooperation and no need for compensation. Whether centralized, decentralized, or cooperative decision-making is adopted it will depend on the circumstances of each case.

These theoretical considerations are useful for understanding the foundations for the determination of which tier of environmental should be in charge of setting the standards according to the public good nature of the environmental asset. However, they are quite limited for analyzing other federative dimensions of environmental policies.

Decentralization has many adepts and some foes. One of the fears of decentralization is the possibility of a race to the bottom since local governments would be exposed to political lobbies to lax environmental standards in order to bring about private investment. That is, a predatory competition among local governments could bring down environmental quality and impose heavy welfare losses to local communities. However, the subject is controversial. Oates (2001) examining the US case concludes that there is no overwhelming historical evidence to support the view of a race to bottom environmental federalism. The two other arguments against decentralization is the lack of expertise of local governments and local political clout. Decentralization, however, is the basic norm according to the principle that the responsibility for providing a particular service should be assigned to the smallest jurisdiction whose boundaries comprise the various benefits and costs associated with the provision of the service. It can be theoretically demonstrated (Oates and Portney, 2001) that there are welfare gains by adopting "the principle of subsidiarity." This norm is explicitly integrated into the Maastricht Treaty and permeates some Directives emanated from the European Union.

Thus, an ideal setting would be one in which the central government defines standards and oversees measures to deal with clear-cut national pollution problems and only intervenes when there are negative externalities across States and municipalities. Further, the central government should provide knowledge derived from sound research and disseminate information on environmental issues and policies. The lower levels of government would set their own standards and define their own policies and programs to deal with environmental problems located in their jurisdictional boundaries.

Another way to assign responsibilities is to define which level of interest prevails (Martins de Castro and Fernandes, 2005). Thus, in this sense, the concept of local interest is, indeed, that of the prevalence of local interest over those of the State and Federal Governments. When local interest prevails, licensing, and decentralization and implementation of environmental standards would be the responsibility of the local government. However, how to assess which interest prevails? Environmental studies would evaluate the direct and indirect impact of an undertaking and, therefore, would determine the prevailing direct interest of the federative entity. Responsibilities would be assigned accordingly. A local interest prevails when the direct and indirect impact of an undertaking is restricted to the territorial limits of the municipality. This would also be the least cost solution. Under this approach, the setting of assignments would depend on studies and it would be applicable on a case-by-case basis.

These principles are to be kept in mind when analyzing the Brazilian Environmental Federalism.

7.2 Constitutional Assignment of Environmental Functions

The National Environment System (Sistema Nacional de Meio Ambiente-Sisnama) was set up in the early eighties by means of the Law No. 6938 of August 31, 1981 (National Environment Policy Act), that instituted its structure and defined the main instruments of environmental policy.[1] Institutions responsible for the conception of environmental policy as well as the creation of environmental control agencies were

---

[1] This Law was regulated by Decree No. 99 274 of June 6, 1990. Since then, the Law has been amended by subsequent legislation. However, the core of the National Environment System remains untouched in the original Law.

established shortly afterward the federal legislation in the majority of the Brazilian States (Órgãos Estaduais de Meio Ambiente-OEMAS). Later on, Brazilian Municipalities that have a constitutional-designed federative status, start creating their own institutions, at first in the state capital cities and afterward in mid-sized cities.[2] In fact, many but not the majority of the Brazilian municipalities have established some sort of an administrative apparatus to deal with the challenges posed by the environment.

The Brazilian System of Environmental Management is largely decentralized across tiers of government. The Federal Government is responsible for the setting of general norms and guidelines as well as for its enforcement through its operational organs whereas States and municipalities formulate and implement supplemental legislation and its application.

The new Constitution of 1988 defined broadly the competence of the tiers of government regarding environmental issues and dedicated a specific chapter to the environment.

Thus, the Brazilian Federal Constitution of 1988 states in Article 23 (VI):

> The common competence of the Union, States, Federal District and Municipalities to protect the environment and to fight pollution in any of its forms. (Constituição da República Federativa do Brasil, 2001)[3]

Further, Chapter VI of the Federal Constitution is entirely devoted to the environment. It states in the Caput of Article (225) that:

> All (citizens) have the right to an environment ecologically balanced, common good for the use of the people and essential to a healthy quality of life, being the duty of the State and of the community the responsibility for its defense and preservation for the present and future generations.[4] (Constituição da República Federativa do Brasil, 2001)

Article 225, with one single exception, attributes to a general entity named the Government ("Poder Público") the enforcement of these

---

[2] Brazil is a singular case among federations. Municipalities are autonomous federative entities that have both an Executive and a Legislative branch.

[3] Author's translation.

[4] Idem.

rights. There are no specific assignments to the three tiers of Government except for the case of nuclear reactors the location of which depends on Federal Law.

In the Brazilian Federative System, States and municipalities are endowed with legislative and executive powers. States have their own Constitutions and municipalities their own organic laws. Both statutes have to conform to the principles of the Federal Constitution. Municipalities are, therefore, full federative entities with great autonomy provided they don't confront the fundamental principles stated in the Nation's Constitutional Charter.

Despite recent advances, there are still unclear areas in relation to the assignments of environmental responsibilities across the tiers of government. Article 23 (VI), as it will be seen later was regulated, providing clearer statements to which level of government is responsible for what. However, there is still some ambiguity what makes it a source of confusion and of federative conflicts, especially between the Union and the States. The lack of clarity with respect to the Constitutional assignments of environmental functions across tiers of government is also present in the structure and functions of Sisnama.

Sisnama encompasses six entities that are responsible for the formulation, implementation control, and evaluation of policies designed to protect and to improve the quality of the environment. The current Sisnama's structure comprises:

1. Highest Organ: The Government Council that advises the President of Brazil on the formulation of national policies and Government guidelines for the environment and environmental resources;
2. Consultative and Deliberative organ: the National Environment Council (Conama) is the leading institution of the Brazilian National Environment System. Its main objective is to advise the Government Council and the President of the Republic in the formulation of the National Environmental Policy Directives. It is composed of representatives from all ministries, many federal Secretariats and Agencies, States, municipalities, civil society, and the private sector. The Conama mandate comprises the setting up of norms and criteria for the environmental licensing system, the establishment of environmental norms and standards, the definition of guidelines and standards concerning protected areas and the elaboration of criteria for determining critical polluted areas;

3. Central organ: the Ministry of the Environment (MMA) created in 1992 is responsible for policymaking, coordination and implementation of the Directives issued by the National Environment Council (Conama). The Ministry of the Environment is responsible for the preservation, conservation, and sustainable use of ecosystems, forests, and biodiversity. It is also in charge of developing strategies and of designing tools aimed at improving environmental quality and the sustainable use of natural resources and of promoting and implementing policies to integrate the economy and the environment, especially in the Amazon[5];
4. Executive/Operational Organ: the Brazilian Institute for the Environment and Renewable Natural Resources (IBAMA) is a key federal agency subordinated to the MMA, being responsible for the implementation of the environmental policy conceived by the MMA and regulated by Conama. It operates in the fields of environmental control, use of renewable resources, including wood production, and trade licensing and in the provision of environmental information. Licensing is a fundamental tool operated by IBAMA and a source of relevant conflicts with the private sector, States, and municipalities.
5. State Environmental Systems: States are responsible for implementing federal legislation and for executing the Conama Directives. However, they also conceive and implement their own policies that cover issues ranging from licensing, protection of natural resources to control of activities that are potentially harmful to the state environment. The majority of States have a Secretariat for the Environment and also an executive organ similar to IBAMA in addition to a council that replicates at the State tier the function of Conama;
6. Municipal Systems: The Brazilian Constitution also assigns environmental responsibilities to municipalities. Nevertheless, with the exception of a few big cities, the majority of local governments have not developed institutional capacity to formulate and implement environmental policies. Municipalities are in charge of zoning, solid waste management, and the control of noise pollution. They

---

[5] The MMA now comprises five Secretariats: Environmental Quality, Biodiversity, Amazon and Environmental Services, International Relations and Climate, Protected Areas and Ecotourism. The Forestry Secretariat is now under the responsibility of the Ministry of Agriculture.

want a more prominent role in licensing as part of a request to have an increasing participation in the National Environment System.

In 2007, it was created the Chico Mendes Institute for the Conservation of Biodiversity.[6] It took over some of the previous Ibama's functions such as the management of federal protected areas, research and development, and the protection of biodiversity.

> **Box: 7.1 What is CONAMA?**
> The National Environment Council (Conama) is the key institution of the National Environment System. Its objective is to advise the President of Brazil on the formulation of the National Environmental Policy. Representatives from all the Ministries and States, eight municipalities, Federal Agencies and Secretariats, private sector and civil society compose the Conama. It has a mandate for; defining environmental norms and standards for the entire country; issuing directives and standards applicable to protected areas; establishing norms and criteria addressed to the environmental licensing system, and; defining criteria for critical polluted areas. It is also the institution of last resort for those that appeal from sanctions and fines applied by the Federal Environmental Agency-IBAMA. Conama can create technical committees for studying, assessing and suggesting standards, norms and complementary regulation. In 2002, the number of Conama's members increased from 73 to 109, making more difficult to build consensus. In 2019, the Federal Government reduced the number of members to 23. The Council issues environmental standards and norms that are very detailed instead of devoting efforts to formulate strategies and policies to deal with the country's huge environmental challenges.

Water is a key environmental asset. Thus, a few lines will be devoted to describe how it is managed and how the legislation attributes functions to the different levels of government. The National Water Resources Management System (SNGRH) was created in 1997 (Law 9433). Its objectives are to coordinate and integrate the water management system and to implement the National Water Resources Policy. The SNGRH comprises the National Water Resources Council (CNRH), the National Water Agency (ANA), State Water Resources Councils, River Basin

---

[6] The Chico Mendes Institute for the Conservation of Biodiversity (ICMBio) was created by Law 11.516 (Brasil, 2007).

Committees and federal, state and municipal institutions involved with water management issues. In some States, there is a specific Secretariat for Water Management. In others, water and environmental management share the same Secretariat.

The National Water Resources Council (CNRH) is composed of representatives from the Federal and State Governments, civil society, and water users. It is responsible for complementary Directives designed to implement the National Policy on Water Resources.

The Water Resources Secretariat (SRH) provides the technical support for the CNRH as its executive arm. It is also responsible for the formulation of the National Policy on Water Resources and for the integration of water resources and environmental management.

The National Water Agency (ANA), an executive agency of the Ministry of the Environment, regulates the utilization of water resources that is subject to conflicting uses (economic, human and recreational). Its mandate is to carry out the management of the national water resources according to the guidelines and Directives emanated from the National Policy on Water Resources.

The State's Water Resources Secretariat or similar institution is a component of the national water management system. In fact, the water resource management system is highly decentralized. These institutions are responsible for the management, formulation, and implementation of water resources policies at the State's level as well as for the enforcement of the legislation in accordance with the federal norms.

Actually, state-level institutions similar to the federal ones exist both for the environmental and water resources management systems. They are supposed to complement each other although many conflicts of competence emerge in every day practice.

Thus, together with Sisnama, the SNGRH completes the institutional architecture of the Brazilian State designed to provide the Executive branch of the three levels of Government with tools to preserve the country's natural capital, to improve environmental quality, to promote sustainable development, and to mainstream environmental policy (Carvalho, 2003).

However, despite its systemic conception, the National Environment System (Sisnama) does not have the necessary capillarity to place the environment as a transversal theme in the formulation and implementation of sectoral policies. Thus, mainstreaming environmental policy is one of the challenges posed to Sisnama.

7.3 Specification of Assignments

As stated in the previous section, the assignment of functions across tiers of government is not clearly defined. In fact, there are many instances in which the same competence is shared among different levels of government.

The Federal Government has exclusive competence in areas such energy, inclusive of nuclear origin, minerals and mining, indigenous people, and all aspects related to nuclear waste and materials. Notwithstanding this, States may still conceive and implement supplementary legislation in these fields. In other areas such as nature conservation, forests, pollution control and environmental liability, the competences are concurrent, i.e., the responsibility of the Federal Government does not exempt supplementary responsibilities by States.[7] However, the general rule is that lower levels of government can only impose rules and standards that are more stringent than the ones set by higher levels.

Regarding target setting, standards for State ecosystems should be at least as stringent as national standards. In the same way, standards for municipal ecosystems must be at least as stringent as State standards. The Federal Government defines national air quality standards and water quality classification. States and municipalities can do the same provided they are more stringent. The Federal Government can define National Ecological Economic Zoning but so can the state and the municipality for their respective jurisdictions.

With respect to guidelines and Directives, the Federal Government defines criteria for licensing, including environmental impact assessments (EIA); criteria and methods for sampling and analyzing the air, water, and soils; classification and procedures for handling hazardous materials; legislation for nuclear energy and waters; parameters for clear-cutting vegetation in private properties and for permanent protection areas (APP) and classification of fauna and flora species that can be managed or fully protected. Monitoring and enforcement by the Federal Government applies to federal ecosystems, international agreements, interstate ecosystems, and policies on forests, biodiversity, and national protected areas. The federal government can also settle interstates disputes.

---

[7] In Brazil's case, some states like São Paulo and Rio de Janeiro have taken a lead in the management of the brown environmental agenda, especially urban pollution.

The States establish complementary guidelines and procedures for licensing regarding the application of environmental management instruments. It monitors air, water, and soils in state ecosystems as well as it inspects and apply fines and sanctions to all polluting activities not under federal control. In practice the States, with a few exceptions, deals with pollution management.

Municipalities are responsible for the setting of noise standards and for providing Directives regarding solid waste practices and criteria and procedures for monitoring and enforcing noise pollution. It grants permits for location of polluting activities; monitors and applies sanctions/fines on sources of noise pollution, and controls the collection and disposal of domestic solid waste. Thus, the two basic, but not the only ones, environmental functions of municipalities are the control of noise pollution and solid waste.

Table 7.1 displays the Division of Responsibilities for Environmental Protection among Federal, State and Local Government. It is clear that there are some situations in which the responsibilities are shared by all federative entities. Water pollution control is one case but so is biodiversity and public health. There are many instances in which one sector or pollutant depends on the action or is under control of more than one tier of government.

Summing it up, the system of assignments is still confusing because of overlapping and ill-defined functions. Under these circumstances, conflicts of competence are established because either the limits are precariously set or one entity takes over the competence of the other.

Therefore, it seems likely that federative conflicts be established unless there is full cooperation among the entities or yet a new legislation is able to draw unequivocally clear the borders of competence across the three levels of government. Further, it turns out to be more difficult to coordinate vertically environmental policies under these circumstances. Vertical coordination requires integration and coherence across tiers of government what is hard to achieve when the assignment of functions is ambiguous.

Table 7.1 Division of responsibilities for environmental protection among federal, state and local government

Sector/polluter	Policymaking	Implementation	Financing	Monitoring & Evaluation
Drinking Water	Federal (for quality)	State/Municipalities	All	All
Water Pollution Control	All	All	All	All
Sewerage	All	States/Municipalities	All	States/Municipalities
Vehicle Emissions	Federal/States	States	Federal/States	Federal/States
Air Pollution Control	Federal	All	All	All
Soils (use and occupation)	Municipalities	Municipalities	Municipalities (sic)	Municipalities (sic)
Mining	Federal	States/Municipalities	All	States/Municipalities
Forestry	Federal	All	All	All
Fish and wildlife	Federal/States	All	All	All
Endangered Species	Federal	All	All	All
Solid Waste	All	Municipalities	All	All
Hazardous Waste Disposal	Federal	Federal/States	Federal/States	Federal/States
Pollution Prevention	States	States	Federal/States	States
Public Health	All	All	All	All
Climate change/global warming	All	All	All	All
Migratory Birds	Federal	Federal	Federal	Federal
Biodiversity	All	All	All	All

Source Federal Constitution

7.4 The System in Practice

7.4.1 De Facto Assignment of Responsibilities

There are differences between the legal and de facto assignments. These differences stem not only from the confusion established by overlapping functions across tiers of government, a result of the lack of clarity of Article 23 (VI), but also of the centralizing behavior of the federal environmental agency (IBAMA). This behavior has led to clashes with the States, especially with respect to the competence to grant environmental licenses. The majority of municipalities have also difficulties to fulfill their legal assignments because of low tax revenues and lack of institutional capacity to formulate and implement policies and to carry out environmental management. In fact, many municipalities do not want to take constitutional assigned environmental responsibilities because it overburdens them with tasks for which they do not have the necessary resources to comply with. This has shifted some attributions to the states sometimes by means of agreements or fiscal instruments.

The basic pillars of Sisnama are well founded. However, the system produces vertical incoherent and inconsistent actions that lead to waste of resources because of overlapping functions and confusion of responsibilities among the three levels of government. The Constitution's Article 23 sets concurrent responsibilities with respect to many environmental issues but it is still unclear about institutional assignments. Since environmental protection is a common goal for the three levels of government, it is essential to establish a system that assigns functions across the federative entities as well as the means for coordination among them so as to avoid overlapping, inefficiency, waste of resources, discoordination, and system disarticulation.

Congress approved a Law establishing responsibilities and assignments across the three tiers of government in 2011[8] (Brasil, 2011). It lays down rules, pursuant to paragraphs III, VI and VII of the caput and the sole paragraph of Article 23 of the Federal Constitution. It states the cooperation between the Union, the States, the Federal District, and the Municipalities in administrative actions arising from the exercise of

---

[8] It is the Complementary Law No. 140/2011 originally proposed by the former Ministry of the Environment and Congress Representative José Sarney Filho.

common competence relating to the protection of notable natural landscapes, the protection of the environment, combating pollution in any form and preserving forests, fauna and flora; and amends Law No. 6,938 of August 31, 1981 (Brasil, 1981).

On this matter, legislation relies both on the principles of the prevalence of interests and subsidiarity. In addition, it has to respect the political and administrative autonomy of the federal entities when they exercise their common material competence.[9] Environmental licensing, for instance, is a clear case of common material competence and a source of many conflicts between the Federal and States Governments.

> **Box: 7.2 Environmental Licensing**
> Environmental licensing is a tool of the National Environmental Policy. It was instituted by the Act that created the National Environment System in 1981 (Law 6938) and regulated by Decree No 99274 and by Directive 237 from Conama that typifies activities and undertakings subjected to licensing. Projects or activities that can potentially or effectively pollute or degrade the environment require licensing or studies that assess their environmental impact. Environmental licensing is required for the location, construction, installation, expansion, and modification and functioning of undertakings and activities that use environmental resources that effectively or potentially pollute the environment as well as for activities that under any circumstances may cause environmental degradation. Licensing is an administrative procedure in which the environmental agency sets the conditions, restrictions and measures for environmental control. They encompass three modalities: previous, installation and operation. Licensing and assessments are carried out by the federal (IBAMA) and States' environmental agencies and, in some instances, for the State Environmental Council or even by the municipality, depending on the location, size and nature of the activity or project. Since licensing is conditional for the granting of fiscal incentives or for credit financing, practically all projects and undertakings that pose environmental risks request it. A Previous Environmental Impact Assessment (EIA) is required for all undertakings that may harm the environment. The Report on Environmental Impact Assessment (RIMA) is just a conclusive summary of the study about the

---

[9] A common material competence is defined as the administrative cooperation among Union, States, and Municipalities for the concurrent and continuous exercise of functions concerning issues listed in Article 23.

> environmental impact and of its recommendations. Lack of clarity, despite Directive 237 from Conama, with respect to assignments of responsibilities across tiers of government as well as overlapping and multiple criteria and procedures have led to conflicts, deadlocks and delays in the licensing process.

A forum where vertical coordination across levels of government could take place is Conama. Instead of coordination, there have been disagreements between federal and state environmental authorities with respect to some issues, especially licensing.

In an attempt to resolve conflicts of competence regarding environmental licensing, Conama issued Directive 237/97. This norm sets competence for issuing licenses by the different levels of government. Article 7 of this Directive specifies that licensing will be issued for one single level of competence in accordance with the principles stated in the norm. Directive 237 is based on the principle of the prevalence of interest since it assigns competence to issue environmental licenses according to the amplitude of the environmental impacts (across international borders, national, regional, across municipalities and local). Notwithstanding this principle, this norm is faulty. First, Conama does not have the competence to regulate licensing across tiers of government. Second, the appropriate piece of legislation for doing so is a Complementary Law. Third, Directive 237 attributes some types of licensing to the Federal Government on the principle that some natural assets (territorial sea, continental platform, etc.) is under its responsibility (entitlement). However, critics assert that this criterion ignores the common competence of the three tiers of government and the diffused nature of environmental goods (Marçal, 2005).

Despite this Directive, there have been conflicts of competence between IBAMA and the States. In fact, States have expressed deep dissatisfaction with what they consider as undue administrative intervention of the environmental federal agency (IBAMA) over licensing in their territorial jurisdictions. In this case, there is not only lack of coordination but also a conflict of competence that disrespects Directive 237/97, Articles 4, 5, and 6. A recent intervention of IBAMA in the licensing of activities potentially harmful to the states' environment was considered unlawful and as such was subjected to a formal protest of the States' Governments. Concerns IBAMA, according to Directive 237, the licensing of undertakings or activities that have significant national or regional impact provided that: (i) it is located in the country or jointly developed with a neighboring country; in the territorial sea; in the continental undersea platform; in exclusive economic zones; in indigenous lands and in conservation units under the responsibility of the Federal Government; (ii) it is located or developed in two or more States; (iii) spillover effects reach other countries or other States; (iv) deals with radioactive materials or that use nuclear energy; and (v) are related to military installations. The rest, which encompasses the majority of the cases, is assigned to States and municipalities. Oates (2001) considers that if spillovers effects are present in two or more States, the best alternative to a centralized decision is some sort of a concerted regional cooperation that involves the joint interests of the affected federative entities. This possibility was never practiced in Brazil and it is not even considered in the environmental legislation.

The Brazilian Association of States' Environmental Entities (ABEMA) that comprises about fifty Secretariats and Executive Agencies has often manifested a disappointment with the setback regarding the constitutional principle of the federative pact that it is based on autonomy and respect for each other's jurisdiction. ABEMA has stated that IBAMA must observe and respect its different responsibilities, attributions, and rights in comparison with the corresponding States' entities. "Along the same line of reasoning, ABEMA has claimed clearer definitions, under the current legislation, of the role played by the environmental federal agency offices in the States" (IBAMA's Executive Agencies in the States) that, in some scenarios, seems not to have a clear view of their constrained competence. Further, ABEMA suggests that an agreement be signed between the Ministry of the Environment and the States' entities in order to remove all conflicts of competence with respect to licensing. Vertical coordination in this case means to align specific responsibilities and regulating tools across the different tiers of Government to achieve the aims of environmental policy (Oates, 2001).

Therefore, one reason why there is a difference between the legal assignment and the facto assignment of responsibilities is the attempt by the federal environmental agency-IBAMA-of having a hegemonic and centralized role with respect to some environmental issues, especially licensing.

However, the undue intervention of IBAMA occurs sometimes by "invitation." Eventually, whenever a conflict over licensing bursts between an undertaking and the States' Government, the former calls upon

IBAMA to settle it. As an example, a large Brazilian firm appealed to IBAMA to grant licensing for the operation of one of its units in the State of Pará. The invitation was made despite the fact that IBAMA did not have legal competence for issuing the requested license. It was argued that the State of Pará did not certify the undertaking because the firm refused a request by the State's Government to build thirty thousand houses as an environmental compensation (sic).

However, there are instances of vertical coordination that have worked perfectly well. This is the case of the program of Prevention and Quick Response to Emergencies with Hazardous Chemicals. The Ministry of the Environment and the State's Government have worked together to elaborate the plan in which the priorities were defined by the States.

Another difference between the legal and de facto assignment stems from the difficulties faced by Brazilian Municipalities to implement their environmental functions. The majority of the local governments in Brazil has a poor tax base, is heavily dependent upon Federal and States' transfer and is unprepared to carry out the environmental duties defined by the Constitution and complementary legislation. Thus, scarcity of resources, weak institutional capacity, and unskilled staff create difficulties for the municipalities to fulfill their functions.

Despite these difficulties—and also as a result of them—the vertical coordination between states and municipalities is more frequent than between the two higher levels of government. This is so because states and municipalities tend to be closer as a result of the local demands that usually have a political expression in the Executive and Legislative branches of the State Government. Furthermore, the scarcity of human and financial resources brings the municipalities closer to the State's Government mediated by the partisan circumstances of the States' political power. States are the most important partners of the local governments.

One of the instances of vertical coordination between states and municipalities emerged as a result of the adoption of the Ecological Value Added Tax (VAT). The states transfer to the municipalities a quarter of the VAT collected every month. A quarter of this transfer is dependent upon criteria established by the State's Government. Through a State's Law the amount to be transferred is conditional on the performance, measured by appropriate indicators, of the municipalities environmental policies with regard to conservation units, creation, and protection of water basins, treatment of sewerage, disposal of solid waste, and construction of landfills (Jatobá, 2003). This mechanism triggers an expenditure competition among the municipalities although is not binding the allocation of the additional resources on environmental investments. The municipalities are free to do whatever they want with the surplus they get because of their environmental achievements. However, they know that their municipalities will be more than compensated for the investments they did to improve the environment. Ecological VAT is the major source of environmental funds for the municipalities. It is a fiscal instrument designed for environmental management that brings closer the fiscal authorities—the ones that transfer the additional VAT—and the environmental authority—the ones that measure the outcomes. Thus, this case—and it is a rare one—is emblematic not only of a good coordination between fiscal and environmental policies but also of a vertical coordination between two tiers of government in a federal republic.[10]

7.4.2 How Well the System Works to Protect the Environment?

The United Nations Environmental Report for Latin American and the Caribbean (PNUMA) states that environmental degradation in the region has worsened in the last three decades. Brazil is not an exception to this evidence that it is manifested in the burning of forests, in the loss of biodiversity, in the deterioration in the quality of the soil and of the water, in the advances of desertification, in the increasing contamination of urban life by noise and air pollution, and in the high degree of environmental vulnerability in many of the country's natural resources. These results adversely affect the quality of life of the population especially the ones who live in more sensitive urban settings like the large metropolitan areas, or in the Amazon or yet in the semi-arid areas of the Northeast. Table 7.2 provides some environmental indicators for Brazil in comparison with Latin America and the Caribbean, and Upper Middle Income Countries. It should not go unnoticed that the percentage of territorial protected areas (28.4%) is above the averages and that the country is performing badly with respect to rural sanitation as compared to upper middle income countries and the Latin America and Caribbean region.

In spite of these problems Brazil has, along the region, made substantial advances in terms of setting goals, of strengthening institutional capacity, of formulating and implementing policies and in devising

---

[10] Many states have adopted the Ecological VAT as an economic tool for environmental management. The criteria have evolved from merely ecological objectives to social targets, especially in health and education (Jatobá, 2003).

Table 7.2 Environmental indications for Brazil compared to Latin America and the Caribbean region and upper middle income countries

	Brazil	Latin America and Caribbean Countries	Upper middle income group
GNI per capita, World Bank Atlas method ($)	9,990	8,968	8,263
Adjusted net national income per capita ($)	7,650	7,249	6,302
Urban population (% of total)	85.7	79.9	64.1
Agriculture			
Agriculture land (% of land area)	34	38	35
Agricultural irrigated land (% of total agricultural land)	.	.	.
Forests			
Forest area (% of total land area)	59.0	46.3	34.9
Deforestation (avg. annual %, 2000–2015)	0.4	0.4	0.0
Threatened species, higher plants	521	5,108	6,808
Biodiversity			
Terrestrial protected areas (% of total land area)	28.4	23.3	15.2
Energy			
Energy use per capita (kg oil equivalent)	1,471	1,337	2,192
Eletric power consumption per capita (kWh)	2,578	2,122	3,495
Electricity generated using fossil fuel (% of total)	24.3	43.1	71.1
Energy depletion (% of GNI)	0.7	0.9	1.1
Emissions and Pollution			
CO_2 emissions per capita (metric tons)	2.5	3.0	6.6

(continued)

Table 7.2 (continued)

	Brazil	Latin America and Caribbean Countries	Upper middle income group
Water and Sanitation			
Freshwater resources per capita (m^3)	27,470	22,160	8,261
Access to improved water source (% of rural pop.)	87	84	91
Access to improved water source (% of urban pop.)	100	97	97
Access to sanitation rural (% of rural pop.)	52	64	67
Access to sanitation urban (% of urban pop.)	88	88	87

Source The Little Green Data Book 2017

management and economic instruments for improving the environment. How, then, can we explain this paradox? Is it apparent or real? First, there is poor coordination either horizontally (across sectors in the same level of government) or vertically (across tiers of government) in the formulation and implementation of environmental policy. The former reveals lack of mainstream in the formulation and implementation of environmental policy whereas the latter provides evidence of failures in the federative system that may be having difficulties to work in a cooperative way. Second, there is insufficient use of market-based mechanisms for improving the management of the environment that relies too heavily on commands and controls emanated from the National and States' Environmental Councils.

The gap between institutional advancements and environmental quality may be the result of the inefficiency and ineffectiveness of public interventions (institutions, organizations, and public policies) designed to prevent or to correct the adverse effects on the environment stemming from natural disasters or market failures. In order to improve efficiency and effectiveness, which are government aims, it is necessary to improve coordination and cooperation across sectors and levels of government. One way of improving coordination is to lessen federative conflicts and to promote mainstreaming. Lerda et al. (2004), highlights that uncoordination across sector or tiers of government can generate negative incentives and aggravate the current environmental problems. Negative externalities may arise from the political, economic, and social costs stemming from the lack of or from failed coordination that might be of sufficient relevance to explain the paradoxical cleavage between aspirations and outcomes as far as sustainable development is concerned.

The description of the main environmental issues faced by Brazil highlights the contrast between progress in the institutional front and persistent problems in the environment situation.

The most important environmental issues in Brazil are: deforestation and protection of the Amazon rain forest, air, water, and soil pollution, the protection of the remaining Atlantic Forest and the deterioration of the Brazilian Savannas. The continental size of Brazil and its endowment of natural resources spread in an area of 8.5 million km^2 make it an especial challenge for environmental policy. Almost half of the Brazilian territory is covered by the Amazon rain forest, one of the most valuable planet assets. Further, around one quarter of the country's territory holds, in its heartland, the world's largest savanna. The remaining Atlantic Forest still holds a valuable biodiversity. The environmental challenges are big because the country is huge and diverse.

Recent news on the advance of deforestation in the Amazon provides a picture of a situation that is getting worse year after year. The causes of deforestation rely, among others, in the expansion of cattle ranching, of commercial crop agriculture, especially soy, of predatory logging and in the squatting of public lands.

The Cerrado Savannas are subject to a similar process of degradation as a result of cattle ranching and from the expansion of export crop agriculture. Soil degradation is a major problem in a biome in which less than 20%, by now, has been left undisturbed. The Savannas are also plagued by pollution from agrochemicals.

Less than 4% of Brazil's territory is protected.[11] However, since the country is large, such percentage, although small, represents a large area. The destruction of natural vegetation in the Amazon and elsewhere advanced at a slow but steady pace of 0.4% per year during the nineties

---

[11] This figure does not include the land occupied by indigenous people.

and up to 2015. Since then, the situation is getting worse year after year. Recent evidence shows that the rhythm of devastation is increasing.

In Jair Bolsonaro's Government (2019–2022), there were substantive increases in deforestation in Brazil. The lack of priority for environmental issues, denialism in relation to climate change, flexibilities in the regulatory framework, failures in supervision and insufficient cooperation between the Federal Executive and State and Municipal governments have led to a situation unparalleled in the country's recent history. This is shown by data from the Deforestation Report in Brazil (RAD in Portuguese) produced by "Mapbiomas Alerta"[12] for the year of 2021. A summary of the main numbers is specified below:

1. An area of 16,557 km² was deforested in 2021, a 20% increase over 2020. In the period 2019–2021, 42,517 km² of forest were felled;
2. The number of deforestation alerts was 69,796 in 2021;
3. Almost 90% of deforestation (89.2%) occurred in the Amazon biomes (59%; 977,733 ha) and the Cerrado Savannas (30.2%; 500,557 ha);
4. Five states (Pará, Amazonas, Mato Grosso, Maranhão and Bahia) accounted for 67% of deforestation;
5. Ten municipalities accounted for 23% of deforestation: five in Pará, three in Amazonas, one in Mato Grosso and one in Rondônia;
6. Almost 50% (49.4%) of deforestation occurred in 50 municipalities, most of them (15) in the State of Pará;
7. Deforestation has been a rapid process; 189 hectares are deforested per hour;
8. Deforestation has also occurred in conservation units. Some 167,000 hectares were devastated by 2021, an increase of 24% compared to 2020;
9. About 40.4% of the indigenous lands had some deforestation;
10. Agriculture is the main vector of deforestation (97%);
11. Almost all deforestation has indications of illegality (a little over 98%);
12. Only 5.2% of the deforested area had some action from IBAMA;
13. About 75% of the deforestation that occurred between 2019 and 2021 did not have Ibama inspection actions.

---

[12] Available on http://alerta.mapbiomas.org.

Further, 813 km² of public States forests were wiped out in 2021, an increase of 26% over 2020. In State's Conservation Units, 690 km² of forests were devastated, an increase of 24% in relation to 2020 that had also been the worst year over a decade.

Brazil is well endowed in the availability of freshwater per capita but this asset is unevenly distributed among its regions. The Amazon basin holds 70% of the stock of freshwater whereas the northeastern semi-arid region, that absorbs 28% of the country's population, gets only 5% of Brazil's water resources. In this region, plagued by poverty and social exclusion, desertification is progressing. The South and Southeast regions, holding 60% of the country's population, have water but most of the rivers crossing big cities like São Paulo are heavily polluted. The people most affected by it are the poor that live by the riverside and in the periphery of metropolitan areas. This phenomenon is not exclusive of cities located in the South and Southeast. A city like Recife situated in the Northeast, face the same problem or worse.

Brazil has had a fast urbanization. Nowadays, around 85% of the country's population lives in urban areas according to the definition of urban areas made by IBGE, the Brazilian Statistics Institute. This has attracted and geared poverty. In fact, more than half of the Brazilian poor live in towns and cities, especially the metropolitan ones that absorb about one third of Brazil's population. This means that urban Brazil is suffering the same environmental damage found in similar countries, that is, air, water, and soil pollution along with lack of sanitation and collection of solid waste. These last two issues together with water supply affect the majority of the small municipalities that hardly have resources to face them. The worst problem is not the supply of clean water but the removal of dirty water. The poor population is the one who suffers the most with the lack of sanitation and with the ill-treated wastewater.

There has been progress in the institutional setting designed to formulate and implement environmental policies as well as to manage environmental assets and issues. The Word Bank considers that Brazil has one of the most advanced environmental management systems among emerging economies. Further, it acknowledges that the country has made steady and substantial improvements over the last three decades. However, the environmental problems are growing and, in some dimensions, like in the Amazon region it seems to be getting worse. Coming back to the original question stated at the beginning of this section. How to reconcile the two contrasting evidence? The explanation lies in the inefficiency and ineffectiveness in the conduction of environmental policies. The institutions are functioning without integration, coherence, and coordination. On the one hand, the Ministry of the Environment has not been able of mainstreaming environmental policy, that is, of coordinating it across sectors (horizontal coordination).[13] On the other, there has been a lack of vertical coordination across tiers of government in the implementation of environmental policies. Federative conflicts render even more difficult such coordination. Thus, the lack of clear assignments of functions across levels of government contributes not only to federative conflicts but also to poor or failed vertical coordination of environmental policies.

7.4.3 Is There a "Race to the Bottom"?

There is no evidence that the States lax environmental standards for attracting private investments. This is a common practice among States as a result of a "fiscal war" whereas tax incentives are provided to attract new investments. Thus, the bait used by the States to allure new investors is not a lax environmental legislation but tax abatements.

There are situations, however, that may be considered in the threshold of starting a race to the bottom competition. This is the case of shrimp farming where many states, especially in the Northeast, have adopted different criteria for licensing. All States, as it was affirmed before, are free to legislate if the norms, criteria, and guidelines are more stringent than the ones emanated from the National Environment Council. Therefore, States may apply different licensing criteria but all must be harder than the national norm. Investors being aware of these differences have taken as criterion for locating new undertakings, the State's legislation concerning licensing for shrimp farming. Thus, they prefer to invest, *"Ceteris Paribus,"* where the requirements for licensing are more lax. Licensing costs and renewal periods are among some of the requirements that are bargained by entrepreneurs. This has motivated the loosing States either to lower its standards, constrained by the general norm, for the functioning of shrimp farming or to claim that a homogeneous criterion be established for all of them. A trend toward homogenization is being established backed by the States.

---

[13] It must be also affirmed that weak or even inexistent horizontal coordination is also found at the State level.

Water charges and tax on the exploration of wood products may entail predatory competition among States. Thus far, these cases are not triggering yet a race to the bottom. However, the severity of IBAMA concerning deforestation licensing as compared with lax legislation on the part of some States for the same end has led entrepreneurs to locate in areas under the jurisdiction of the latter. Thus, a lax legislation may entail the growth of family farming at the cost of increasing deforestation. Other States may retaliate by easing on the legislation, thus triggering a race to the bottom competition.

7.5 The Judicial and Legal System

7.5.1 Role of Courts

The three branches of government are involved in environmental management in Brazil. The Executive branch is responsible for the conception and implementation of public policies, guidelines, and Directives as well as for its enforcement, evaluation, and monitoring. Congress sets the legislation and decides on budget allocation. The Judiciary reviews actions of the Executive and it is the locus where people can appeal against decisions of the other two branches of government and launch law suits against environmental offenders. The three branches of government exist for the Union and for all states as independent powers of a federal republic.

There is no specialization in the Judiciary for dealing with environmental litigation.[14]

If the Government fails in the protection of the natural capital, it is the duty of citizens to provoke the Judiciary in order to restore the environment by punishing the offender even if it is the State itself. The Brazilian 1988 Constitution embodies the principles of: (i) sustainable development; (ii) polluter-payer; (iii) civil responsibility; and (iv) environmental compensation[15] (Nalini, 2003). Unfortunately, not all legislation is enforced. As a matter of fact, law enforcement is a major weakness

---

[14] Brazil has a separate Judiciary to deal only with labor issues. The Labor Court System replicates the standard Judiciary (Supreme and lower courts that function in a decentralized way).

[15] The principle of objective civil responsibility widens the compensation possibilities since it exempts the prosecutor's office of proving the damage as a requirement to consider the offending partner as guilty. Environmental compensation might take the form of financial reparation or of restoring the environment to its original state.

of the Brazilian legal system as a whole and the environmental legislation is not an exception to the general rule. Few environmental criminal investigations are open in the Police Offices. Even fewer are charged with environmental offenses and punished accordingly. However, judges state that they can only act when they are provoked. Therefore, government and society are responsible for bringing the environmental offenders to Court.

Access to Courts is, de jure, open to all litigants. The problems faced by the Judicial System in Brazil are its slowness, inefficiency, and ineffectiveness that contrast to its high cost for society. The Judiciary takes long time to decide. Recurrent, almost endless appeals, bureaucracy, poor management, and lack of information technology are some of the reasons for the long delays. This has contributed for a public disbelief in the functioning of the Brazilian Judiciary and even in its capacity to be fair and just. However, there is another power in the Brazilian legal system that facilitates the access to courts by public litigants.

It is the Public Prosecutor's Office (Ministério Público) which plays a significant role in the enforcement of the environmental and water resources legislation. It is an independent and decentralized institution that operates both at the federal and states' tiers of government. Among its duties are the protection of peoples' environmental rights and the capacity to bring civil law suits and criminal charges against those people or entities (firms, public institutions) that have caused or may cause harm to the environment.[16] Any person or organization can incite the Public Prosecutor's Office to action. The Public Prosecutor's Office also brings the investigated party to sign legally binding court agreements to adjust behavior (TAC) that is considered offensive to the environment. This is a preventive measure that intends to correct environmental misbehavior. A grace period is given to the violator to achieve full compliance. If this tool fails, then the Public Prosecutor's Office can bring administrative, civil, and criminal charges to the offender.

---

[16] Article 129 (III) of the Federal Constitution (1988). The Public Prosecutor's Office is the guardian of the Law. Its attributions are broadly defined in Article 127 of the Constitution and are not restricted to the enforcement of environmental legislation.

7.5.2 Protections Laid Out in the Constitution

A healthy environment for the present and future generations is a fundamental right of all Brazilians and residents in Brazil. This means that the preservation of the Brazilian natural capital is an unalienable right that cannot be disposed of. The duty to preserve the environment is both of the State and of the common citizen. The Government does not have the discretionary power to protect the environment. The Brazilian Constitution imposes that obligation. It is not a matter of choice but a mandate that the Government is compulsory required to exercise.

It is also an obligation of the Government to foster the ecological management of species and ecosystems, to protect biodiversity, to define the territories to be specially protected (conservation units, biological reserves, etc.), and to control those entities responsible for research and manipulation of genetic material.

Chapter VI of the Federal Constitution addresses only the environment. It has been laid out in Article 225, clause VII, which states that it is mandate of the Government:

> To protect the fauna and flora, being prohibited as stated in Law, any practice that endangers its ecological function, lead to the extinction of species or that may submit animals to cruelty.

This same clause determines the elaboration of environmental impact assessments establishes the "polluter-pays principle" and strengthens conservation by declaring the Amazon rain forest, the Pantanal (Marshlands), the Atlantic Forest, Serra do Mar, and the Coastal Zone as national biomes deserving especial protection. However, it does not grant special constitutional protection to the Savannas of Central Brazil (Cerrados) and to the Caatinga (Freire et al., 2018), located in the semi-arid area of the Northeast, two important biomes of the Brazilian eco-system.

The Constitution also requires miners to recover the environment degraded by predatory exploration. It also poses a three-fold responsibility: administrative, civil, and criminal for all persons and corporations that have committed environmental offenses. Bringing sanctions to corporations was an advance with respect to previous understanding that corporations could not be indicted for environmental crimes (Nalini, 2003).

Two major environmental disasters have occurred in Brazil in the last ten years caused by Brazilian and foreign companies. Two dams with iron mining waste (Mariana in November 2015 and Brumadinho in January 2019) in the State of Minas Gerais collapsed killing hundreds of people and destroying in their course heritage, natural, and material, and seriously polluting several and important watercourses. The social, economic, and judicial implications of these two disasters continue to be felt to this day, including in international courts. It was also a test for the coordination of the federal, state, and municipal environmental systems because two states and dozens of cities were hit.

7.6 Political Issues and Impact on Environmental Standards

7.6.1 Influence of Political Party Competition

Environment is a component of every Brazilian political party program and the theme is, de jure, above partisan disputes. After the Constitution of 1988, the Brazilian Congress continued to approve important legislation to protect the environment. If the environmental agenda is classified into green (forests, biodiversity), brown (noise, air pollution, etc., in urban areas), and blue (water resources), the Brazilian Congress approved legislation in all those areas to improve the environment (Gabeira, 2003). The principle across most of these measures is the mainstreaming of environmental policy.

Nevertheless, Brazil has a political party devoted entirely to the environmental cause (The Green Party-Partido Verde, PV). Further, the Brazilian Congress is not exempted from pressures stemming from environmentalists and business interests. In one extreme are rural proprietors that fight thru their representation in Congress against any measure that may create difficulties or make unfeasible the expansion of the agrobusiness. On the other are the environmentalists that use the Green Party and other means to pressure Congress for the adoption of more severe measures to protect the environment.

7.6.2 Impact of Societal Consensus on Environmental Standards

Brazil is one of the developing countries in which organized civil society has had a key role in the shaping of the environmental agenda. Non-governmental organizations have also played an active role in the formulation of public policies and in the generation of initiatives designed to improve the environment. Nevertheless, there are still many institutional weaknesses in the Brazilian environmental movement, which is not monolithic since it presents different approaches and perspectives concerning environmental problems and policies (Born, 2003). The Brazilian environmental movement has performed eight roles or functions. The first one is public denunciation and dissemination of information concerning environmental problems as well as the calling upon of environmental authorities to tackle them. The second function is related to education and training of these groups so that they can strengthen awareness of environmental problems in society and in the press. The third one is advocacy of rights and public policies for the environment and sustainable development. The fourth role concerns research development, and generation and dissemination of knowledge focused on the quality and integrity of environmental management. As a result of the previous functions comes out a fifth one that performs monitoring and evaluation. Reacting to criticisms that NGOs points problems but do not solve them, they have been performing a sixth role related to the conception and implementation of pilot projects as a way to induce their adoption by governments and the private sector. A natural outcome of the previous function is the development and provision of advisory services and the dissemination and replication of good practices and ideas. The eighth and last function concerns staff training which is aimed at strengthening professional skills and abilities to undertake the previous functions (Born, 2003).

In their struggle against environmental degradation, the Brazilian environmental movement has contributed to the institutionalization of social control mechanisms such as the public hearings which are held to assess the potential environmental impact of large undertakings. The publication of licensing requests, the freedom to access information and the participation of NGOs in councils and boards were classical demands of the Brazilian environmental movement. Meeting these demands upgrades the democratic instruments for dealing with issues of public interest. However, in the current government there has been important setback in this process.

The environmental movement exerts influence in terms of shaping public opinion and of placing demands for the Legislative and Executive branches of Government concerning environmental issues and policies. They have instances and tools for doing so. There are many forums of environmental NGOs such as the Permanent Councils of Environmental Entities for the Defense of the Environment (Apedemas) and the Social Movement for the Environment and Development (FBOMS). These groups were important for the design of the new 1988 Constitution's environmental chapter and articles and have had an active role in the Rio-92 environmental summit and in the 2002 Johannesburg Rio+10 conference. In addition, NGOs participate in environmental councils, at the national, state, and local levels. NGOs have also been important partners of local, state, and federal governments and of international organizations as well. During the Jair Bolsonaro Government (2019–2022), there was a substantial reduction in the participation of civil society in various forums, especially environmental ones. When not reduced, it was completely eliminated (Brasil, 2019). This setback was important especially in the Amazon, which even lost resources conditional on the participation of NGOs or members of civil society.

There is a trend for environmental NGOs or a network of them to represent the interests of specific biomes and regions, especially the Amazon rain forest, the Savannas of Central Brazil (Cerrados), and the Pantanal.[17] Other examples are the Network for the Atlantic Forest (RMA) and the Secretariat of Northeastern Environmental Entities (SEAN). They also get together around some themes such as climate (Network Climate Brazil) or rivers (Live Rivers Coalition).

7.7 International Agreements and Implications for Environmental Federalism

The Brazilian Constitution embodies many principles contained in international treaties and conventions. According to Nalini (2003), there is an important "open clause" that takes into the Federal Constitution protective norms established in international treaties and conventions that have been ratified by Brazil. These treaties and conventions cover a wide

---

[17] NGOs, academics, States' Governments and the local population (traditional residents) will also encompass the new National Council of the Biosphere (Atlantic Forest, Cerrado, Caatinga and Amazon) which is sponsored by UNESCO.

diversity of themes: big environmental spaces; workers' protection; toxic material regulation; protection to fauna, flora, and biodiversity, in general; water basin; international fishing; combat desertification, maritime and ocean spaces, international rivers, and lakes; protection of the ozone layer; atmosphere and climate change, peaceful use of nuclear energy; protection of natural, cultural and social assets, environmental compensation and civil responsibility. Among those, Brazil has signed four important conventions: biodiversity, desertification, climate change, and Agenda 21. Three of them have implications for the federative entities.

7.7.1 The Convention on Biological Diversity (CBD)

Biological diversity comprises a variety of life on Earth including plants, animals, and microorganisms, as well as the ecosystems of which they are part. Currently, biodiversity is being lost as a result of human activities that degrade natural habitats and increase all forms of pollution, contributing as a consequence to climate change. The Convention on Biological Diversity address this problem, being an agreement among nations aimed at preserving the diversity of life forms through conservation and sustainable use. The Convention was opened for signature at the United Nations Conference on Environment and Development in Rio de Janeiro (1992) and entered into force on December 29, 1993. Currently, 187 countries and the European Community are participants of the Convention. The Parties to the Convention seek to achieve its objectives through agreed policies and practices.

The importance of this Convention for Brazil stems from the fact that the country holds between 15 and 20% of the world's biodiversity. Further, it has the largest number of endemic species in the world. Unfortunately, Brazilian biomes present severe losses in biodiversity. Between 5 and 7% of the Atlantic Forest remains barely intact and the rhythm of devastation of the Amazon is increasing sharply. Forests cover about two thirds of the Brazilian territory, most of them native. They are relevant both in economic and environmental terms since they contribute to Biodiversity and Climate and constitute the resource base for the production of a variety of products (timber, paper, and cellulose, among many others including non-timber products).

Therefore, the Brazilian Government approved the Convention on February 4, 1994, and on the 30th of December of the same year created the National Program of Biological Diversity (PRONABIO). In 2002, the

Brazilian Government established the National Biodiversity Policy, which is the legal instrument that provides the principles and guidelines for the protection of the Brazilian biological diversity and more recently created the National Commission on Biodiversity (CONABIO), the Coordination Committee of the National Forest Plan (CONAFLOR) and launched the Plan for the Prevention and Control of Deforestation in the Legal Amazon (PPCD). The Institute Chico for the Conservation of Biodiversity (ICMBio) created in 2007 is co-responsible with IBAMA for the implementation of these policies. One of the strategies for the conservation of biodiversity is the creation of protected areas. The National System of Conservation Units (SNUC) was established by Law in 2000 after seven years of debate with the States and other stakeholders. It provides standardized criteria for the creation and management of Conservation Units. The SNUC's Law placed the Federal Government, States, and Municipalities under the same legal framework. In spite of this regulation, the Federal Government has created Conservation Units within many States without discussing it with the local environmental authorities.

The Convention on Biodiversity does not have any global funding as other conventions do. This has forced the States to invest their own financial resources in the knowledge and strengthening of its biodiversity, especially the revision and reclassification of Conservation Units, the elaboration of maps (locating the States' biodiversity assets), the elaboration of the Red Book describing the local endangered species (Fauna and Flora) and in the setting up of an infrastructure to support inspection and the implementation of the Conservation Unit's management plan. This has influenced the States' green agenda and mobilized financial and human resources for that end. Some States have elaborated their plans to protect forests and to conserve biodiversity (SECTMA, 2001).

The Convention on Biodiversity has advanced slowly in the States' environmental policy because it lacks a flow of sustainable resources for its successful implementation. In addition, this policy requires the inclusion of special minority groups like the isolated communities of Afro-Brazilians (Quilombolas) and the indigenous populations.

7.7.2 *The Convention on Climate Change*

The United Nations Framework Convention on Climate Change is the foundation of global efforts to combat planet warming. It was opened for signature at the Rio Earth Summit. Its main objective is the stabilization of greenhouse gas concentrations in the atmosphere at a level that would prevent human induced interference with the climate system. The Kyoto Protocol, adopted in 1997 to the Convention on Climate Change will strengthen the international response to climate change and recently entered into force despite resistance of some countries and its no ratification by the United States.

The Convention on Climate Change is under the coordination of the Federal Government, which is responsible for formulating, implementing, financing, and monitoring its evolvement. The emphasis has been on clean technologies and in the possibilities open as a result of the Mechanism for Clean Development (MDL) that allows Brazil to trade carbon monoxide emissions with industrialized countries.

7.7.3 The Convention to Combat Desertification

The United Nations Convention to Combat Desertification was adopted in 1994 and entered into force in 1996 as the only international legally binding instrument to address the threat of desertification. It stems from a recommendation of Agenda 21 created in the 1992 Rio Earth Summit. The Convention's objective is to combat desertification and mitigate the effects of drought.

In the Brazilian case, desertification is under way in the semi-arid region of the Northeast, comprising eight States and the north of the State of Minas Gerais, which are subjected to periodic droughts. It is estimated that about 100 thousand km^2 out of area of one thousand is severely affected by desertification. This process gravely hits about 80 thousand km^2 while almost 400 thousand km^2 are moderately affected by it.

Brazil ratified the Convention in 1997. The instrument to combat desertification in Brazil is the National Action Plan to Combat Desertification (PAN) that was launched in August, 2004. A workgroup comprised of Ministries, States, and civil society was created with the objective of drafting it. The Convention is under the supervision of the Water Resources Secretariat of the Ministry of Regional Development, which has allocated too few resources for the States to implement the plan. Nevertheless, even before this mechanism emerged, many States have taken the lead with respect to the federal government to decelerate the pace of desertification, especially in the Brazilian Northeast.

This convention is synergic with the previous two. Further, it has a specific source of financing based on the Global Environmental Facility (GEF) that was established in 2003 as a financial mechanism to combat desertification.

7.7.4 Agenda 21

Agenda 21 was created in the 1992 Rio Earth Summit with the objective of integrating the economic, social, and environmental dimensions of sustainable development. The Brazilian Agenda 21 was elaborated according to the guidelines emanated from the Rio Conference on Environment and Development. It was launched in 2001. The states and municipalities are supposed to elaborate their specific Agenda 21 based on the guidelines set by the Brazilian Agenda.

Among the States, Pernambuco (2002) and Santa Catarina were the only ones to conclude the Agenda 21 in accordance with the international and Brazilian guidelines.

Agenda 21 is an instrument for the promotion of sustainable development. It is elaborated with the participation of all social actors involved directly or indirectly with environmental issues. Therefore, it is a concerted pact between stakeholders for the improvement of the environment.

The Brazilian Agenda 21 has stated the following propositions, among a range of other themes: the adoption of the Ecological VAT, the creation of a green tax to inhibit environmental degradation and the compensatory financing by federal financial institutions of undertakings designed to offset environmental damages caused by projects subsidized by those institutions' loans (The green protocol).

7.7.5 Sustainable Development Goals

In the past decade, important events have occurred with positive implications on environmental federalism. The year 2015 was a milestone in this process. In the 2015 Millennium Charter, respect for nature was one of the pillars of the values and principles of sustainable development in the context of an integral strategy for human progress. The transition from the Millennium Goals to the Sustainable Development Goals (SDGs) was established, among others, to ensure the sustainability of the environment (Objective 17).

Thus, the year 2015 hosted four important conferences: the Third World Conference on Disaster Risk Reduction in Myiagi, Japan; the Third International Conference on Financing for Development in Adis Abeba, Ethiopia; the United Nations Summit on Sustainable Development when the 2030 Agenda and the Sustainable Development Goals were established in New York (USA), and; the United Nations Climate Change Conference in Paris (21st Conference of the Parties on Climate Change). In the latter, it was agreed: (1) that rising temperatures must remain well below 2 degrees centigrade and that the countries that signed the agreement make efforts to limit the temperature increase by 1.5 degrees compared to the pre-industrial era; (2) countries commit to setting their national targets to reduce greenhouse gas emissions every five years; (3) that rich countries continue to provide financial support to poor countries to help them reduce their emissions and adapt to the effects of climate change, even if no mention is made of specific figures.

The institutions to implement the 2030 Agenda are governments, including subnational ones, civil society and public-private partnerships. In Brazil, it was created in 2016 during the Government of President Michel Temer (Brasil, 2016) and extinguished in 2019 (Brasil, 2019), in the Government of President Bolsonaro, the National Commission for Sustainable Development Goals (CNODS), that had equal number of members between governments and civil society. Six ministries, eight representatives of civil society, a representative of the states and the federal district and a representative of municipal governments, comprise it. Decree No. 9,980 of August 20, 2019, reduced the status and institutional functions of the CNODS to the Special Secretariat of Social Articulation of the Secretariat of Government of the Presidency of the Republic with the following objectives set out in Article 15:

1. Assist the Minister of State on issues related to the Sustainable Development Goals;
2. Articulate, within the federal government, together with the Special Secretariat of Federative Affairs, with the federative entities, the actions of internalization of the 2030 Agenda for Sustainable Development of the United Nations; and
3. Request and consolidate information on the implementation of the Sustainable Development Goals provided by government agencies.

At the Conference of the Parties on Climate Change (COP 26) held in Glasgow in 2021, Scotland, progress has been assessed since 2015 and new agreements and targets were established especially for reducing the emission of greenhouse gases and methane. Brazil, in the first case, opted for compensation via carbon credits instead of direct actions to reduce them. In this Conference, in addition to the participation of the Federal Government, states and municipalities were present, presenting plans and proposals. Despite the precarious coordination between the Federal Government and the other federative entities observed since 2019, the joint participation of the three levels of government represented an advance in the contribution of environmental federalism to sustainable development. It should be highlighted, contrary to what has been observed in Brazil, a strong presence of civil society in the discussions for funding policies designed to sustainable development.

7.8 Has Environmental Federalism Contributed to Policies for Sustainable Management of the Environment and Natural Resources?

Thirty years after the United Nations Conference on Environment and Development (The Rio Earth Summit, 1992) and fifty years after the Stockholm Conference (1972), environmental federalism in Brazil has contributed, despite the still persisting intergovernmental conflicts, to the growing participation of subnational entities in the formulation and implementation of sustainable management of the environment and natural systems. States, the Federal District, and Municipalities have created and expanded the competence of their formulation bodies for the implementation and supervision of environmental policies. These entities have increasingly participated in national and international meetings to evaluate the advances made in relation to the preservation of the environment and biodiversity and have adhered to multilateral agreements under the sponsorship of international institutions, especially those linked to the United Nations system. The states of Pernambuco and São Paulo, for example, among others, adhered to the Edinburgh Declaration of August 31, 2020, when it was recognized that none of the 20 biodiversity targets of the Conference of the Parties of Aichi, Japan, in 2010, had been fully met. In fact, of the 20 goals set, only two came close to being achieved. It was therefore recognized that convergence between multilateral environmental agreements would be advancing at too slow a pace.

The Edinburgh Declaration, preceded in February 2020 by the Charter of São Paulo, is linked to the Post-2020 Convention on the Conservation of Biodiversity. The Edinburgh Declaration was signed by subnational entities to create the means necessary for international cooperation programs to be implemented with the participation of regions, states, and municipalities in order to put the planet on the visionary route for 2050 to live in harmony with nature. The conventions, such as that on Biodiversity Conservation are signed by the countries, but those who execute are the subnational entities that can implement viable goals, develop good projects, capture resources, and establish solid partnerships through a roadmap for action.

Five states, including Pernambuco and São Paulo, have also developed local plans to combat climate change through actions to decarbonize their economic environments. In Brazil, seventeen of the twenty-seven states, including the Federal District, have committed to initiatives that will zero greenhouse gas emissions by 2050 (race to zero). Pernambuco has developed a carbon neutral plan that aims to create a roadmap to neutralize greenhouse gas emissions by 2050, promoting sustainable development. This was possible because the State developed the first Greenhouse Gas Inventory that analyzes CO_2 emissions in the State, from 2015 to 2018. Launched on November 6, 2019, during the Brazilian Climate Change Conference, in Recife it was prepared by the State Secretariat of Environment and Sustainability, in partnership with the international organization Under2, promoter of the Climate Footprint project, in several countries.

A few years ago, the actions of subnational entities were timid, subjected to initiatives that came from the federal government when they were not inhibited by conflicts of competence arising from the lack of clarity of the Brazilian Constitution of 1988, which still remains in some dimensions, but that do not limit the participation of subnational entities in the formulation and execution of sustainable development. In the past, only environmental education initiatives were frequent in most states and municipalities. Today, they also include environmental services, programs, and projects, among other actions, which are shared with other federative entities.

A consensus is forming in the country that policies to combat climate change and biodiversity conservation require differentiated governance with close cooperation between the Central Government, the states, the

Federal District, and municipalities and, within each state, between state and municipal governments. This means that the strengthening of environmental federalism is increasing for the successful implementation of environmental management policies, nature preservation, and renewable systems.

7.9 Conclusions

Brazil has made substantial progress in the institutionalization of environmental policy. The country built a network of institutions and conceived a set of environmental policy tools that, de jure, makes it capable of facing its huge environmental challenges. In addition, the conception of environmental policy and the exercise of many of its instruments are democratic. In fact, there are environmental councils at all levels of governments. In these councils, civil society is widely represented although in recent years there has been severe reversals in this process. As a result, environmental awareness is increasing in the country, keeping pace with the dynamism of the Brazilian environmental movement and with the maturing of environmental government institutions. Environmentalists are organized in NGOs from where they press the legislative and executive branches of all levels of government to advance with the environmental agenda. Access to information is improving, a system of environmental statistics is being shaped and the legal system is instrumented to punish environmental crimes. Further, the State and Federal Public Prosecutor's Offices have been keen observers of unlawful environmental practices and very attentive to any threats to environmental quality.

Despite all this, the country faces severe environmental problems as illegal deforestation, water, soil, and air pollution, desertification and other environmental plagues strike all regions. In the Amazon and in the Central Savannahs, soy expansion, cattle ranching, and illegal logging are wiping out native forests at increasing rates; in the Northeast, desertification is advancing; in the large metropolitan areas water, noise, and air pollution worsens the quality of life, especially of the poor population. In fact, this group is the most stricken by the deterioration of the environment since they live in the periphery of the big cities, on the side of polluted rivers, in the rural areas of the semi-arid Northeast and in the violent and devastated agricultural frontier of North and Central Brazil.

The contrast between institutional advances and the severity of the environmental problems poses an apparent contradiction. This paper attempted to reconcile this paradox by stressing the discoordination of environmental policies and practices across sectors and tiers of government. Of special concern for this paper is the lack of coordination and harmonization across levels of government which reveals many weaknesses in the Brazilian Federation.

Brazil assures environmental protection in its Constitution. The government is mandated to protect the Brazilian natural resources. However, the Constitutional Charter is still ambiguous, confusing and overlapping with respect to the assignment of environmental functions across tiers of government despite the advances made by Complementary Law 140 of 2011. The lack of clarity concerning the competence of each level of government has brought about federative conflicts and impaired the efficiency and efficacy of environmental policy. Mainstreaming the environment in sectoral policy and improving the degree of coordination and harmonization in the conception and implementation of policies and programs across federative units would lead to gains in efficiency and welfare. Recent experience has moved toward these goals.

The Brazilian environmental system is also vulnerable to frauds and corruption. There are big economic interests at stake because of the richness of the Brazilian natural resources. To reconcile economic interests with environmental protection is at the heart of sustainable development. Brazil has an immense natural capital to preserve and also to use as productive resources for its development. The achievement of sustainable, inclusive, and equitable development in harmony with Brazilian natural endowment is a challenge that government and the environmental movement will face for many years ahead.

Bibliography

Born, Rubens H. (2003). Terceiro Setor in: Trigueiro, André. *Meio Ambiente no Século 21.* Ed. Sextante: Rio de Janeiro.

Brasil (1981). Lei No. 6.938 de 31 de Agosto de 1981 (The National Environment System Act).

Brasil (1997). Resolução Conama 237 de 19 de Dezembro de 1997. Brasília: Conselho Nacional de Meio Ambiente.

Brasil (2007). Lei 11.516 de 28 de agosto de 2007 (The Chico Mendes Institute for the Conservation of Biodiversity—ICMBio).

Brasil (2011). Lei Complementar No. 140 de 08 de dezembro de 2011 (Regulates Law No. 6938).
Brasil (2016). Leis e Decretos. Decreto No. 8.892 de 27 de outubro de 2016.
Brasil (2019). Leis e Decretos. Decreto No. 9.759 de 11 de abril de 2019.
Carvalho, José Carlos (2003). *Poder Executivo* in Trigueiro, André. Meio Ambiente no Século 21. Ed. Sextante: Rio de Janeiro.
Constituição da República Federativa do Brasil (2001). São Paulo: Editora Atlas, 17ª Edition.
Freire, Neison et al. (2018). *Atlas of the Caatingas*. Recife. Fundação Joaquim Nabuco. Editora Massagana.
Gabeira, Fernando (2003). *Poder Legislativo* in Trigueiro, André. Meio Ambiente no Século 21. Ed. Sextante: Rio de Janeiro.
Instituto Brasileiro de Geografia e Estatística (2005). *Perfil dos Municípios Brasileiros:Meio Ambiente 2002.* Pesquisa de Informações Básicas Municipais. Rio de Janeiro.
Jatobá, Jorge (2003). "O ICMS como instrumento econômico para a gestão ambiental". Divisão de Meio Ambiente e de Assentamentos Humanos da Comissão Econômica para a América Latina e o Caribe-CEPAL. Santiago de Chile.
Jatobá, Jorge (2005). *A Coordenação entre as políticas fiscal e ambiental no Brasil: a perspectiva dos governos estaduais.* Serie Médio Ambiente y Desarollo No. 92. Santiago: CEPAL: División de Desarollo Sostenible y Asentamientos Humanos.
Lerda, J.C.; Acquatella, J.; Gómez, J.J. (2004). "Coordinácion de Políticas Públicas: Desafios y oportunidades para una agenda fiscal-ambiental". Versão preliminar do trabalho apresentado no *II Taller sobre Política Fiscal y Medio Ambiente*, realizado em Santiago de Chile em 27 January 2004.
Marçal, Cláudia (2005). *Análise da distribuição de competências no licenciamento ambiental: necessidade de estabelecimento de regras claras*, in Benjamin, Antonio H. Paisagem, Natureza e Direito Anais do Congresso Internacional de Direito Ambiental. São Paulo: Instituto "O Direito por um planeta verde". Vol. 1.
Martins de Castro, Deborah I.; Fernandes, Rodrigo (2005). *O papel do ente municipal para promover o desenvolvimento sustentável* in Benjamin, Antonio H. Paisagem. Natureza e Direito Anais do Congresso Internacional de Direito Ambiental. São Paulo: Instituto "O Direito por um planeta verde". Vol. 2.
Nalini, Renato (2003). *Poder Judiciário* in Trigueiro, André. Meio Ambiente no Século 21. Ed. Sextante: Rio de Janeiro.
Oates, Wallace (1997). "On the Welfare Gains from Fiscal Decentralization". *Journal of Public Finance and Public Choice* 2–3: 83–92.

Oates, Wallace (1998). "Environmental Federalism in the United States: Principles, Problems and Prospects". *Resources for the Future*. Washington, DC.

Oates, Wallace (2001). "A Reconsideration of Environmental Federalism". *Resources for the Future*. Discussion Paper 01-54. Washington, DC.

Oates, Wallace and Portney, Paul R. (2001). "The Political Economy of Environmental Policy". *Resources for the Future*. Discussion Paper 01-55. Washington, DC.

Pernambuco. Secretaria de Ciência, Tecnologia e Meio Ambiente-SECTMA (2001). *Desenvolvimento Florestal e da Conservação da Biodiversidade de Pernambuco*. Recife.

Pernambuco (2002). *Agenda 21 do Estado de Pernambuco*. Recife: Secretaria de Ciência, Tecnologia e Meio Ambiente.

Programa das Nações Unidas para o Meio Ambiente (2004). GEO América Latina y el Caribe: perspectivas del medio ambiente (2003). México, D.F.

Rio Grande do Sul. Resolução CONSEMA No. 004 de 28 de Abril de 2000. Porto Alegre: Conselho Estadual de Meio Ambiente.

Vogel, David; Toffel, Michael; Post, Diahanna (2003). *Environmemtal Federalism in the European Union and the United States*. Paper prepared for an international conference on Globalization and National Environmental Policy. Veldhover, The Netherlands, September.

PART IV

Combating Global Climate Change

CHAPTER 8

Carbon Tax as a Tool for Tax Reform and Protecting Local and Global Environments

Bjorn Larsen and Anwar Shah

8.1 Introduction

The last few decades have witnessed a dramatic growth in worldwide concern over global climate change and a proliferation of proposals to limit or reverse global environmental damage. Carbon taxes and tradeable permits figure prominently in proposed economic policy responses. During the past decade, more than 30 countries have introduced carbon taxes (see World Bank, 2021). A carbon tax is a fee imposed on the carbon content of fossil fuels such as coal, oil and gas at the point

B. Larsen
Environmental Health and Resource Economics Global Consultancy Services, Asker, Norway

A. Shah (✉)
Brookings Institution, Washington, DC, USA
e-mail: shah.anwar@gmail.com

© The Author(s), under exclusive license to Springer Nature Switzerland AG 2023
A. Shah (ed.), *Taxing Choices for Managing Natural Resources, the Environment, and Global Climate Change,*
https://doi.org/10.1007/978-3-031-22606-9_8

of production or use to provide a relative disincentive for the use of these fuels consistent with their relative contributions to the emission of greenhouse gases that are known to contribute to global climate change. Figure 8.1 shows that coal releases the most CO_2 and natural gas the least. Table 8.1 illustrates carbon content and implied tax rates for various fossil fuels. The purpose of a tradeable permit regime to limit aggregate emissions to a targeted level. Despite this policy interest in combating global climate change, empirical work relevant to developing countries is scant. Even for developed countries, research on carbon taxes was initiated only in the early 1990s (see e.g., Jorgenson and Wilcoxen, 1990; Poterba, 1991; Pearce, 1991; Goulder, 1991; Fisher et al., 1995), and still largely in progress. A careful analysis of carbon taxes in terms of their impacts on efficiency, equity, economic growth, government revenues, and environmental protection is needed for an informed debate on policy development (see Summers, 1991; Shah, Larsen and Whalley, 1991). This chapter reproduces, with minor updates, the first rigorous attempt in this direction by Shah and Larsen (1992a, b) to quantify the efficiency and equity implications of carbon taxes in the international context using a combined sample of industrial and developing countries. The methodology presented in this chapter would be of interest to future studies on this subject. Policy conclusions drawn here remain valid for the sample countries and in the global context even though the data used in these calculations is dated.

The chapter is organized into seven sections. The remainder of Sect. 1 outlines the global warming issue and suggested policy responses. Section 2 briefly outlines the potentials and perils of global carbon true regimes. Section 3 deals with the economics of a national carbon tax. Calculations on the revenue potential and differential incidence of carbon taxes are presented for India, Indonesia, Pakistan, the United States, and Japan. For Pakistan, detailed calculations on the distributional implications are also presented. Section 4 provides estimates of the impact of carbon taxes on greenhouse gases and focal pollutants for the sample countries. Impacts of carbon taxes on industrial performance for selected industries in Pakistan are traced in Sect. 5, using dynamic production structure empirical models. Section 6 evaluates the use of tradeable permits as an alternative to carbon taxes. A final section presents a summary of the conclusions. The chapter concludes that whereas a global carbon tax may be a more distant policy option, national carbon taxes—if introduced in revenue-neutral fashion by reducing corporate

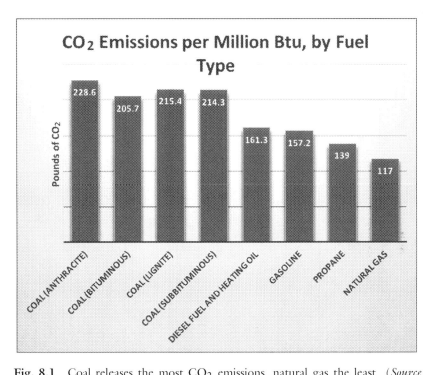

Fig. 8.1 Coal releases the most CO_2 emissions, natural gas the least (*Source* U.S. Energy Information Administration; Carbon Tax Center, USA)

Table 8.1 CO_2 content and carbon tax rates for fossil fuels @US$49/metric ton of Carbon Dioxide equivalent (metric ton CO_2)

Fuel	CO_2 content	Tax@$49/mt CO_2	Tax @US$49 ton of Carbon (not CO_2) equivalency in terms of Per million Btu
Coal (Anthracite)	2579 kg/short ton	$126/short ton	$1.40
Crude Oil	432 kg/barrel	$21/barrel	$1.12
Gasoline	9 kg/gallon	$0.44/gallon	$1.07
Diesel Fuel	10 kg/gallon	$0.50/gallon	$1.18
Natural Gas	53 kg/mcf	$2.60/mcf	$0.80

Source https://www.carbontax.org

income taxes—offer significant potential in combatting global change and local pollution as well as reforming the tax system. Further, a conservative evaluation of the benefits of reducing local externalities overwhelms the negative output effects of carbon taxes. Thus, even ignoring global externalities, a case for carbon taxes for most countries can be made purely on own national economic interest or simply based upon tax reform considerations.

8.1.1 The Problem and the Status of Policy Discussions

In recent years, worldwide concern about the atmospheric accumulation of so-called greenhouse trace gases—carbon dioxide (CO_2), methane (CH_4), nitrous oxides (N_2O), tropospheric ozone (O_3), and chlorofluorocarbons (CFCs)—has been mounting. By trapping some of the sun's heat in the atmosphere, these gases permit the existence of life on earth. Their rapid accumulation, however, can contribute to a rise in the earth's temperature (commonly termed the "greenhouse effect" or "global warming"). CO_2 is estimated to contribute 80.3% of total warming potential (Nordhaus, 1991a, b). Scientists fear that if the current pace of accumulation continues unchecked into the twenty-first century, a point might be reached when the absorptive capacity of the earth's atmosphere would become exhausted, and a natural disaster of unprecedented proportions would consequently ensue. Even without this point being reached, significant warming of the earth's surface is expected to have major economic consequences (see Churchill and Saunders, 1991, for an overview of the scientific and economic issues of relevance to developing countries). Developing countries with agrarian economies and/or coastlines would be particularly vulnerable to natural calamities associated with global warming. It must be emphasized that there is considerable uncertainty at the present time regarding global climate change, its magnitude, its regional manifestations, and its consequences. While much scientific work remains to be done, a broader consensus among the scientific community regarding the contributions of greenhouse gases has emerged in recent years. The uncertain state of our present knowledge of global warming coupled with the potentially large and irreversible damages that might result, call for public policy responses that are both flexible and reversible. The possible use of carbon taxes and tradeable permits to deal with global climate change has initiated a controversial debate.

These debates reflect a wide spectrum of views on this issue (see Cullenwood and Victor 2020). Some argue that in view of the uncertainties regarding climate change, inaction would be the best policy (Eckaus, 1991). At the other extreme, some environmentalists argue that we may have already missed the boat and immediate economic policy responses that may forsake growth are needed (see Postel and Flavin, 1991). A majority, however, take a middle view. Energy economists argue that energy policy options consistent with restraining the greenhouse effect also make good economic sense. Churchill and Saunders (1991), for example, exhort developing countries to seize the initiative and "increase incentives for sustainable energy use, shift to cleaner alternative fuels and technologies, and improve efficiency in energy production, distribution and end use" (p. 28). Some public finance economists espouse the same middle-of-the-road view by presenting a somewhat different perspective that emphasizes reliance on flexible and less distortionary tools to deal with an uncertain but potentially serious problem. Summers (1991) has argued that corrective taxes, e.g., taxes on carbon contents of fossil fuels, can raise significant amounts of revenue at a relatively small deadweight loss while furthering global and local environmental protection and discouraging "bads", and therefore represent "what we pay to preserve civilization". It has also been argued that in developing countries, carbon taxes offer a potential for enhancing the environment as well as financing developmental expenditures and could therefore serve to enrich civilization.

It is interesting to note the large energy subsidies that prevail in a handful of large carbon-emitting countries. Getting energy prices right would prima facie represent a first-order priority in any economic policy response designed to curtail greenhouse gas emissions. Larsen and Shah (1992a and Chapter 9, this volume) examine energy pricing practices around the world. In determining the level of subsidy, they use border prices of fossil fuels as reference prices (as proxies for marginal opportunity costs of production). Total world energy subsidies in 2019 are estimated to be more than US$1.38 trillion. Oil, natural gas, and coal account for 52% of these subsidies and, in revenue terms, are equivalent to a negative carbon tax of US$250 per ton of carbon (Chapter 9, this volume). The removal of such subsidies could reduce global carbon emissions by 11% and would translate into a 15% reduction in carbon emissions in the subsidizing countries. To achieve an equivalent reduction in tons of emissions in the OECD countries, a carbon tax in the range of

US$151–$341 per ton would need to be imposed in the OECD countries (Chapter 9, this volume). This would result in a total annual cost (in terms of foregone output, adjustment costs, etc.) of US$31 billion for OECD countries. This amount would then represent the upper bound for OECD compensatory transfers to the subsidizing countries. It is also worth noting that very large (37–68%) reductions in global carbon emissions could be achieved, were Japanese or German standards of energy efficiency to be universally adopted.

While this debate continues to rage, some countries have already moved to adopt tax policies that, intentionally or otherwise, bear on the issue of global warming. Among European countries, Finland took the lead in introducing the world's first carbon tax at a rate of $6.10 per ton of carbon on all fossil fuels in January 1990. Netherlands and Sweden ($45/ton tax) have followed suit in February 1990 and January 1991, respectively. During the last decade (2010–2022), about 30 countries have implemented carbon taxes or pilot projects as a prelude to the introduction of carbon taxes (World Bank, 2021). Sweden has imposed the highest rate (US$119/ton) followed by Switzerland and Liechenstein ($99/ton) and Finland ($68/ton), Norway ($53/ton), and France ($49/ton). Korea, Iceland, Ireland, British Columbia, Canada, and UK impose taxes in the range of $21 to $33 per ton and about a dozen countries and several Canadian provinces have imposed taxes that range from $1 to $20 per ton (World Bank, 2021). Despite these efforts, skeptics are not convinced about the seriousness of these efforts. For example, Jenkins (2021) notes that all these. Efforts in aggregate so far amount to just one greenhouse gas being priced at or above $15 per metric ton of CO_2 equivalent—at a largely token level (see also Carbon Tax Center, 2022).

At the international level, momentum has steadily built behind the proposition that global warming and other aspects of climate change are of major consequence and require a concerted global policy response. The discussion in these international fora has centered on both domestic and global policy options to combat global climate change. These have included: immediate term options such as a global carbon tax or permits (tradeable or otherwise) and emission limits; intermediate term measures such as increased energy efficiency, afforestation, biomass, nuclear energy, and population control; and long-term measures such as backstop technologies that use solar, solar-hydrogen, and other environmentally safe sources. Developing countries are fully involved in the debate on these issues. One argument often advanced is that the greenhouse effect results from the accumulation over a long period of trace gases contributed primarily by industrial activity in developed countries, and consequently that developing countries should not be asked to sacrifice their current developmental goals in order to address a problem created by past policies of developed countries. In fact, if one were to construct an index of "global warming debt" by level of development, this argument would have some empirical validity. It is also frequently asserted that any global attempt to limit environmentally harmful emissions would ultimately slow the economic development of LDCs. Attempts to develop energy-intensive manufacturing capability in the early to mid-stages of development would be more costly, and hence more difficult. Also, as importers of energy-intensive manufactures (primarily capital goods), developing countries would end up bearing the burden of the policy response applied to emission-generating activities. In general, it is commonly perceived that, unless accompanied by compensatory transfers, the relative costs of action are likely to be higher for developing countries, given that their relative contribution to the accumulation of these gases is expected to grow faster than that of the OECD countries over the next century. The available literature offers little guidance in determining the validity of these arguments. Nevertheless, the importance of policy action to combat global climate change is now widely recognized. In 2015, 196 countries adopted the *Paris Agreement* that came into effect in November 2016 to contain global warming to well below 2 degrees Celsius (see Chapter 7, this volume for a detailed history of international agreements and conventions on climate change). The COP26 (The Conference of Parties 26) in December 2021 adopted the Glasgow Climate Pact aiming to turn the 2020s into a decade of climate action and support. The package of decisions consists of strengthenef efforts to build resilience to climate change, to curb greenhouse gas emissions and to provide the necessary finance for both. Nations reaffirmed their duty to fulfill the pledge of providing 100 billion dollars annually from developed to developing countries (Note that as of December 2022, this pledge remains to be fulfilled). And they collectively agreed to work to reduce the gap between existing emission reduction plans and what is required to reduce emissions, so that the rise in the global average temperature can be limited to 1.5 degrees celsius. For the first time, participating nations are also called upon to phase down unabated coal power and inefficient subsidies for fossil fuels. More recently, the United Nations Climate Change

Conference COP 27 held in November 2022 at Sharm el-Sheikh, Egypt, reaffirmed commitments to limit global temperature rise to 1.5 degrees Celsius above pre-industrial levels. Further, it reached an agreement to set up a special fund to provide "loss and damage" funding for vulnerable countries hit hard by climate disasters. The following section provides preliminary and tentative guidance on these questions.

8.2 Global Carbon Taxes: Potentials and Perils

Taxes on the carbon content of fossil fuels have been advocated in recent years as part of a proposed concerted international effort to combat global climate change. While both the need for and the mechanics of such taxes remain unsettled issues, a general consensus is emerging that, if adopted globally, such taxes would represent a flexible, reversible, and lower cost alternative to regulatory responses, including the widely discussed notion of equal percentage reductions in greenhouse emissions by all countries. The latter measure is unlikely to lead to the equalization of marginal emission reduction costs from all sources and would not, therefor, result in a cost-efficient outcome for the world (see Hoel, 1991). Tietenberg (1985) reports that cost savings associated with moving from equal percentage reductions to a market-based instrument such as a carbon tax, could be substantial (exceeding 40% of total costs). Maler (1989) also reports that a uniform percentage reduction strategy for greenhouse gas emissions would capture only one-third of the total potential gains from optimal allocation. A uniform level carbon tax (i.e., tax per unit of carbon emissions equal for all gases and all countries), if imposed by a global agreement, *would* equalize the marginal costs of emission reductions (by fossil fuel and by location), and would therefore be cost-efficient. Several alternative designs for such an agreement are possible, with each presenting its own shortcomings. Consider the case of a domestic carbon tax that is imposed by an international agreement. Since perspectives on global warming vary among countries, national commitment to impose such taxes will also vary. If a country has signed such agreement under international pressure, that country can make the carbon tax an ineffective instrument by reducing existing energy taxes, by taxing close substitutes of fossil fuels (e.g., hydroelectricity), by providing subsidies to complements or products that are fossil fuel energy intensive, and by lax enforcement of the agreed-upon carbon tax (see Hoel, 1991).

Thus, by following a suitable strategy, a free ride becomes possible. A global carbon tax imposed by an international agency, on the other hand, would impinge on national sovereignty and therefore would not likely be accepted internationally. A third alternative would have globally imposed but nationally administered and collected carbon taxes; countries would make a positive or negative net transfer to an international agency based upon an agreed revenue disposition scheme. Basic criteria for such redistribution would be population and GDP, or a combination of these factors. Additionally, a small fraction of the revenue pool could be distributed on the basis of special considerations, e.g., to provide an inducement to countries which might view global warming as beneficial (such as Russia, Canada and Nepal) to join an international agreement. Tables 8.2 and 8.3 provide illustrations of net transfers involved based on the three revenue redistribution schemes outlined above, using either standard GDP or GDP adjusted by purchasing power parity—so-called Penn GDP. It must be acknowledged that these tables as presented by Shah and Larsen (1992a, b) are dated and are based on the world political map of 1987 but are useful in illustrating the arguments presented here. From these tables, it is apparent that a revenue redistribution alternative based on population alone would be unacceptable to most industrialized countries, whereas one based solely on GDP would not be agreeable to developing countries. Note that under the formula that uses population as the sole factor, net transfers to developing countries would dwarf current official development assistance. It is possible that a formula that uses a combination of both factors and therefore redistributes only a very small fraction of total carbon tax revenues, might find acceptance by most countries.

A recent proposal by Norway would have mandatory greenhouse reduction targets imposed on industrial countries; such targets could be exceeded only by financing the transfer and/or adoption of green technology in developing countries. If such a proposal is received well in OECD countries, some of these countries might well choose to adopt carbon taxes to achieve the agreed-upon targets, then partly use the proceeds from carbon taxes to finance technology transfer to developing countries. In general, the prognosis for the acceptance of a global carbon tax regime is quite pessimistic. The degree of scientific uncertainty that surrounds global warming makes it unlikely that most countries would agree to an international convention that is seen to forsake their current growth. The critical question then is that if one ignores the important yet

Table 8.2 Net transfer by country under alternate carbon tax regimes (using UN national Accounts GDP)

	GDP per capita (US $) 1987	% of world emissions	Carbon emissions to GDP (kg/$)	Carbon emissions per capita (kg)	Tax revenues (tax $10/ton) million US$	Tax revenue to GDP (%)	Tax regime 1 World tax revenue distributed by population (Million US$)
Bangladesh	166	0.06	0.179	30	32	0.18	1160
Nigeria	229	0.16	0.366	84	90	0.37	1165
China	286	10.38	1.868	533	5699	1.87	11,882
India	322	2.65	0.567	182	1454	0.57	8868
Pakistan	325	0.24	0.394	128	132	0.39	1140
Indonesia	443	0.48	0.346	155	263	0.35	1906
Zimbabwe	598	0.08	0.774	463	42	0.77	100
Egypt, Arab republic	709	0.35	0.536	380	190	0.54	558
Korea, Dem People's	889	0.71	2.063	1834	382	2.06	238
Total		15.10			8292		
Average	318		1.073	341			
Ave w/o China & Korea			0.491				
Mexico	1715	1.41	0.550	943	772	0.55	810
Brazil	2145	0.92	0.166	356	503	.17	1573
South Africa	2493	1.38	0.919	2292	759	0.92	364
Venezuela	2629	0.42	0.485	1276	233	0.49	203
Korea Republic of	3221	0.82	0.432	1067	449	0.34	468
Total		4.95			2716		

Average	2228		0.385	857		0.38	
Ave w/o South Africa			0.314			
Poland	1697	2.29	1.967	338	1257	1.97	419
Former Yugoslavia	2848	0.6	0.492	403	328	0.49	260
USSR (former)	8325	18.45	0.430	378	10,129	0.43	3148
Former Czech	9242	1.17	0.444	4110	640	0.44	173
Germany, east (former)	11,261	1.63	0.477	5369	894	0.48	183
Total		24.13			13,248		
Average	7489		0.470	3520		0.47	
Ave w/o Poland			0.433			
Australia	11,364	1.16	0.346	3926	638	0.35	181
Canada	16.36	1.99	0.263	4221	1091	0.26	287
Germany, West (former)	18,249	3.23	0.159	2898	1773	0.16	680
US	18,434	22.69	0.277	5112	12,461	0.28	2711
Japan	19,437	4.32	0.100	1942	2371	0.10	1358
Total		33.39			18,334		
Average	18,295		0.214	3908		0.21	

(continued)

Table 8.2 (continued)

	GDP per capita (US $) 1987	% of world emissions	Carbon emissions to GDP (kg/$)	Carbon emissions per capita (kg)	Tax revenues (tax $10/ton) million US$	Tax revenue to GDP (%)	Tax regime 1 World tax revenue distributed by population (Million US$)
Sample total		77.56					
Africa	665	2.80	0.405	261	1540	0.40	6561
South America	1951	2.46	0.249	486	1350	0.25	3091
Europe	11,136	21.13	0.210	2343	11,600	0.21	5504
North and central	12,897	26.04	0.281	2620	14,300	0.28	4392
Oceania	9372	1.31	0.319	2992	718	0.32	267
Asia	1333	24.40	0.350	466	1340	0.35	31,947
Former USSR	8328	18.39	0.429	3369	10,100	0.42	3147
World	3635	100	0.306	1112	54,910	0.31	54,910

	Net revenue per capita (US$)	Net revenues % of GDP	Tax regime 2 World tax revenue distributed by GDP (million US$)	Net revenue per capita (US$)	Net revenue % of GDP	Tax regime 3 World revenue distributed half and half by population (million US$)	Net revenues per capita (US$)	Net revenues % of GDP
Bangladesh	1.82	6.50	54	0.21	0.13	617	5.52	3.31
Nigeria	10.28	4.48	75	−0.14	−.06	630	5.07	2.21
China	5.79	2.03	934	−4.44	−1.56	6408	0.66	0.23

India	8.30	2.89	785	−0.84	−0.26	4827	4.23	1.32
Pakistan	8.84	3.02	102	−0.29	−0.09	621	4.77	1.47
Indonesia	8.59	2.16	232	−0.18	−0.04	1069	4.71	1.06
Zimbabwe	6.49	1.09	16	−2.60	−0.47	58	1.85	0.31
Egypt, Arab republic	7.32	1.03	108	−1.63	−0.23	333	2.85	0.40
Korea, Dem People's	−7.22	−0.81	58	−15.42	−1.76	148	−11.42	−1.28
Total								
Average	7.87	2.45		−2.30	−0.72		2.74	0.86
Ave w/o China & Korea								
Mexico	1.69	0.10	430	−4018	−0.24	670	−1.25	−0.07
Brazil	7.56	0.35	928	3.01	0.14	1251	5.29	0.25
South Africa	−11.80	−0.47	255	−15.29	−0.61	310	−13.55	−0.54
Venezuela	−1.64	−0.06	147	−4.71	−0.18	175	−3.17	−0.12
Korea Republic of	0.43	0.01	402	−1.12	−0.04	435	−0.33	−0.01
Total								

(continued)

Table 8.2 (continued)

	Net revenue per capita (US$)	Net revenues % of GDP	Tax regime 2 World tax revenue distributed by GDP (million US$)	Net revenue per capita (US$)	Net revenue % of GDP	Tax regime 3 World revenue distributed half and half by population (million US$)	Net revenues per capita (US$)	Net revenues % of GDP
Average	2.33	0.11		−1.76	−0.08		0.39	0.02
Ave w/o South Africa								
Poland	−22.26	−1.31	196	−28.18	−1.66	307	−25.22	−1.49
Former Yugoslavia	−2.91	−0.10	204	−5.31	−0.19	232	−4.11	−0.14
USSR (former)	−24.66	−0.30	7215	−10.30	−1.12	5180	−17.48	−0.21
Former Czech	−29.98	−0.32	440	−12.82	−0.14	307	−21.40	−0.23
Germany, east (former)	−42.57	−0.38	574	−19.23	−0.17	379	−30.90	−0.27
Total Average								
Ave w/o Poland	−24.08			−12.28	−0.16		−18.18	−0.24

Australia	−28.14	−0.25	565	−4.48	−0.04	373	−16.31	−0.14
Canada	−31.09	−0.19	1270	6.93	0.04	779	−12.08	−0.08
Germany, West (former)	−17.86	−0.10	3414	26.87	0.15	2048	4.50	0.02
US	−40.00	−0.22	13,753	5.30	0.03	4232	−17.35	−0.09
Japan	−8.30	−0.04	7265	40.07	0.21	4310	15.88	0.08
Total average Sample total	−27.96	−0.15		16.91	0.09		−5.53	−0.03
Africa	8.31	1.32	1164	−0.64	−0.10	3862	3.94	0.61
South America	6.26	0.32	1660	1.12	0.06	2376	3.69	0.19
Europe	−12.31	−0.11	16,870	10.65	0.10	111,187	−0.83	−0.01
North and central	−25.08	−0.19	15,591	3.27	0.03	9992	−10.91	−0.08
Oceana	−18.80	−0.20	688	−1.23	−0.01	478	−10.02	−0.11
Asia	6.46	0.48	11,723	−0.58	−0.04	21,835	2.94	0.22
Former USSR	−24.57	−0.30	7213	−10.20	−0.12	5180	−17.39	−0.21
World	0.00	0.00	54,910	0.00	0.00	54,910	0.00	0.00

Note Carbon emissions are from fossil fuel combustion only. Emissions from deforestation are not included. Note also that country names represent the political status as of 1987

Source Shah and Larsen (1992a)

Table 8.3 Net transfer by country under alternate global carbon tax regimes (using PENN GDP)

	Penn GDP per capita (US$) 1987	% of world emissions	Carbon emission to GDP (kg/$)	Carbon emission per capita (kg)	Tax revenues (tax $10/ton) million US$	Tax revenue to GDP (%)	Tax Regime 1:1 World tax revenue distributed by population (Million US$)
Nigeria	248	0.16	0.393	84	90	0.34	1185
Bangladesh	831	0.06	0.036	30	32	0.04	1180
India	947	2.65	0.191	182	1454	0.19	8868
Zimbabwe	1047	0.08	0.442	463	42	0.44	100
Indonesia	1269	0.48	0.121	153	263	0.12	1906
Pakistan	1391	0.24	0.092	128	132	0.09	1140
Egypt	1454	0.35	0.261	380	190	0.26	558
China	2233	10.38	0.239	533	5699	0.24	11,882
Total		14.39			7900		
Average	1534		0.213	328		0.21	
Venezuela	3422	0.42	0.373	1276	233	0.37	203
Mexico	3780	1.41	0.249	943	772	0.25	910
Korea republic	4236	0.82	0.252	1067	449	0.25	468
Brazil	4613	0.92	0.077	356	503	0.08	1573
South Africa	5921	1.38	0.387	2292	759	0.39	368
Total		4.95			2714		
Average	4414		0.194	859		0.19	
Former Yugoslavia	5000	0.60	0.281	1403	328	0.28	260
Poland	5579	2.29	0.598	3338	1257	0.60	419

USSR	6791	18.45	0.527	3379	10,129	0.33	3148
Former Czechoslovakia	8521	1.17	0.482	4110	640	0.48	173
Former Germany east	8767	1.63	0.612	5369	894	0.61	183
Total		24.13			13,248		
Average	6171		0.524	3520		0.52	
Australia	11,782	1.16	0.333	3926	638	0.33	181
Japan	12,506	4.32	0.135	1942	2371	0.16	1358
Canada	15,730	1.99	0.268	4221	1091	0.27	287
Former Germany west	16,893	3.23	0.172	2898	1773	0.17	680
US	17,503	22.69	0.292	5112	12,461	0.29	2711
Total		33.38			18,334		
Average	15,828		0.247	3908		0.25	
Sample total % of World total		76.85					
Africa	1065	2.80	0.245	261	1540	0.25	6561
South America	3976	2.46	0.122	486	1350	0.12	3091
Europe	11,146	21.13	0.210	2343	11,600	0.21	5504
North and Central America	12,828	26.40	0.282	3620	14,300	0.28	4392
Oceania	9824	1.31	0.305	2992	718	0.30	267
Asia	2126	24.40	0.219	466	13,400	0.22	31,947

(continued)

Table 8.3 (continued)

	Penn GDP per capita (US$) 1987	% of world emissions	Carbon emission to GDP (kg/$)	Carbon emission per capita (kg)	Tax revenues (tax $10/ton) million US$	Tax revenue to GDP (%)	Tax Regime 1:1 World tax revenue distributed by population (Million US$)	Net revenues % of Penn GDP
Former USSR	6793	10.39	0.525	3369	10,100	0.53	3147	
World	4169	100	0.267	1112	54,910	0.27	34,910	

	Net revenue per capita (US$)	Net revenues % of GDP	Tax Regime 2 World tax revenue distributed by Penn GDP (million US$)	Net revenue per capita (US$)	Net revenue % of Penn GDP	Tax Regime 3 World revenue distributed half and half by Penn GDP population (million US$)	Net revenues per capita (US$)	Net revenues % of Penn GDP
Nigeria	10.28	4.14	71	−0.18	−0.07	628	5.05	2.03
Bangladesh	10.82	1.30	235	1.92	0.23	708	6.37	0.77
India	9.30	0.98	2014	0.70	0.07	3441	3.00	0.53
Zimbabwe	6.49	0.62	25	−1.84	−0.18	63	2.33	0.22
Indonesia	9.59	0.76	580	1.85	0.15	1243	3.72	0.45
Pakistan	9.84	0.71	380	2.43	0.17	760	6.13	0.44
Egypt	7.32	0.50	195	0.08	0.01	376	3.70	0.25
China	5.79	0.26	6370	0.63	0.03	9126	3.21	0.14
Total								
Average	7.84	0.51		0.82	0.05		4.33	0.28
Venezuela	−1.64	−0.05	167	−3.63	−0.11	185	−7.63	−0.08
Mexico	1.69	0.04	823	0.65	0.02	868	1.17	0.03
Korea republic	0.45	0.01	474	0.60	0.01	471	0.53	0.01
Brazil	7.56	0.16	1740	8.75	0.19	1657	8.16	0.16

South Africa	−11.80	−0.20	523	−7.13	−0.12	466	−9.47	−0.16
Total Average	2.55						2.87	0.07
Former Yugoslavia	−2.91	−0.06	312	−0.69	−0.01	286	−1.80	−0.04
Poland	−22.26	−0.40	560	−18.50	−0.33	490	−20.38	−0.37
USSR	−24.66	−0.36	5128	−17.66	−0.26	4138	−21.16	−0.31
Former Czechoslovakia	−29.98	−0.35	354	−18.38	−0.22	284	−16.16	−0.28
Former Germany east	−42.57	−0.49	389	−30.31	−0.55	287	−36.44	−0.42
Total Average	−24.08	−0.36		−17.28	−0.26		−20.62	−0.31
Australia	−27.18	−0.24	511	−7.83	−0.07	346	−17.99	−0.15
Japan	−8.30	−0.07	4073	13.94	0.11	2713	2.82	−0.02
Canada	−31.09	−0.20	1085	−0.25	0.00	686	−15.67	−0.10

(continued)

Table 8.3 (continued)

	Net revenue per capita (US$)	Net revenues % of GDP	Tax Regime 2 World tax revenue distributed by Penn GDP (million US$)	Net revenue per capita (US$)	Net revenue % of Penn GDP	Tax Regime 3 World revenue distributed half and half by Penn GDP population (million US$)	Net revenues per capita (US$)	Net revenues % of Penn GDP
Former Germany west	−17.86	−0.11	2756	16.08	0.10	1718	−0.89	−0.01
US	−40.00	−0.23	11,383	−4.42	−0.03	7047	−22.21	−0.13
Total Average Sample total % of World total	−27.96	−0.18		3.14	0.02		−12.41	−0.083
Africa	8.51	0.80	1677	0.23	0.02	4119	4.37	0.41
South America	6.26	0.16	2949	5.75	0.14	3020	6.01	0.15
Europe	−12.31	−0.11	14,718	6.30	0.06	10,111	−3.01	−0.03
North and Central America	−25.08	−0.20	13,516	−1.98	−0.02	8954	−13.53	0.11
Oceania	−18.80	−0.19	629	−3.71	−0.04	448	−11.25	−0.11
Asia	6.46	0.30	16,294	1.01	0.05	24,120	3.73	0.16
Former USSR	−24.57	−0.36	5128	−17.57	−0.26	4138	−21.07	−0.31
World	0.00	0.00	51,910	0.00	0.00	54,910	0.00	0.00

Note Carbon emissions are from fossil fuel combustion only. Emissions from deforestation are not included
Source Shah and Larsen (1992a)

uncertain phenomenon of global warming, is there a case for the adoption of national carbon taxes on other grounds, such as tax reform or a reduction in environmental externalities? The following sub-sections present a benefit-cost calculus of carbon taxes based on these latter considerations.

8.3 Economics of a National Carbon Tax

As discussed earlier, taxes on the carbon content of fossil fuels to combat global climate change have been lately widely advocated (Metcalf, 2021; Rabe, 2020; Urban Institute and Brookings Institution, Tax Policy Center, 2015; World Resources Institute, 2015), and implemented in selected countries (World Bank, 2021). In the following section, the case for carbon taxes is examined in terms of their revenue potential, efficiency, and distributional implications, and impacts on global and local externalities. For these calculations, a small fossil fuel carbon tax of the order of $10/ton ($25 in 2020 current prices) of carbon contents is selected. Such a tax results in 2% and 8.6% increases in the aggregate price of fossil fuels, and 1.0% and 5.6% reductions in consumption of fossil fuels, in Japan and India, respectively (Table 8.4). Partial equilibrium calculations presented in this paper, offer reasonable and defensible approximation of the impact of small carbon taxes; the same confidence could not be asserted for those taxes of $100/ton or higher which are frequently discussed in global models.

8.3.1 *Revenue Potential of Carbon Taxes*

The revenue potential of carbon taxes is extremely large. For example, a $10/ton carbon tax (equivalent to US$25 in 2020 prices), individually imposed by all nations of the world could raise $55.5 billion (US$140 billion in 2020 prices) in the very first year of its operation (see Table 8.4). For some countries, like China and Poland, such revenues would amount to about 2% of GDP and would be sufficient to wipe out central government's budgetary deficit. On the average, countries having a 2020 per capita GDP of less than US$2000 could raise revenues exceeding one percent of GDP and 5.7% of government revenue. For the OECD countries, comparable figures would be 0.21% of GDP and 1.0% of government revenue. Carbon taxes in general are easier to administer than personal and corporate taxes and thereby less prone to tax avoidance and evasion. Due to tax evasion, the latter taxes raise revenues that

Table 8.4 Revenue potential of a US $10/ton domestic carbon tax (using UN National Accounts GDP)

	Population (mill) (1987)	GDP per capita (US$) 1987	Carbon emission to GDP (kg/$)	Carbon emission per capita (kg)	Tax revenues (tax $10/ton) mill US$	Tax revenue to GDP (%)	Govt Rev to GDP (%)	Carbon tax to total Govt revenue (%)	Carbon tax revenue to Govt deficit (%)
Bangladesh	106.1	166	0.179	30	32	0.18	9.12	1.96	4.18
Nigeria	106.6	229	0.366	84	90	0.37	15.71	2.33	262.31
China	1068.5	286	1.868	533	5699	1.87	21.19	8.81	6.65
India	797.53	322	0.567	182	1454	0.57	14.73	3.85	4.63
Pakistan	102.48	325	0.394	128	132	0.39	17.29	2.28	21.97
Indonesia	171.44	443	0.346	153	263	0.3	21.33	1.62	7.03
Zimbabwe	8.99	598	0.744	463	42	0.77	33.10	2.34	9.03
Egypt Arab rep	50.14	709	0.536	380	192	0.54	38.07	1.41	
Korea, Dem People's	21.37	889	2.063	1834	392	2.06			
Total					8292				
Average						1.07	18.78	5.71	25.20
Mexico	81.86	1715	0.550	943	772	0.55	17.41	3.16	4.06
Brazil	141.43	2145	0.166	356	503	0.17	33.29	0.50	1.42
South Africa	33.11	2493	0.919	2292	759	0.92	23.02	3.99	16.11
Venezuela	18.27	2629	0.485	1276	233	0.49	21.61	2.25	27.25
Korea Republic	42.08	3121	0.342	1067	449	0.34	17.27	1.98	−77.26
Total					2716				
Average						0.38	25.16	1.53	4.56
Poland	37.66	1697	1.967	3338	1257	1.97	38.78	5.07	137.92
Former Yugoslavia	23.41	2848	0.492	1403	328	0.49	6.86	7.18	−1288.78

Former USSR	283.1	8325	0.432	3578	10,129	0.43	48.35	0.92	527.11
Former Czechoslovakia	15.57	8242	0.445	4110	640	0.44			
Former Germany East	16.65	11,261	0.477	5369	894	0.48			
Total					13,248				
Average						0.47			
Australia	16.25	11,364	0.346	3926	638	0.35	26.53	1.30	28.73
Canada	25.85	16,056	0.263	4221	1091	0.26	20.29	1.30	−10.32
Former Germany west	61.17	18,249	0.159	2898	1773	0.16	29.34	0.54	15.01
US	234.77	18,434	0.277	5112	12,461	0.28	20.23	1.37	8.45
Japan	122.09	19,437	0.100	1942	2371	0.10	13.77	0.73	2.82
Total					18,334				
Average						0.21	19.77	1.08	7.81

Note Carbon emissions are from fossil fuel combustion only. Emissions from deforestation are not included
Source Shah and Larsen (1992a)

are considerably less than their potential yield. Carbon taxes therefore present an attractive alternative to income taxes in developing countries. But how do such taxes fare in terms of equity and efficiency?

8.3.2 Distributional Implications of Carbon Taxes

The existing literature on industrialized countries typically portrays carbon taxes as regressive charges. This is because expenditures on fossil fuel consumption as a proportion of current annual income, falls with income. Poterba (1991) relates carbon taxes to annual consumption expenditures—a proxy for permanent income—and still finds a regressive incidence, although one considerably less pronounced than with respect to annual income. These results nevertheless cannot be generalized to developing countries, where the incidence of carbon taxes would be affected by institutional factors. Some important factors that may have a bearing on the tax-shifting are: market power, price controls, import quotas, rationed foreign exchange, the presence of black markets, tax evasion, and urban–rural migration.

Case (a): Full Forward Shifting. The degree of tax-shifting depends upon the relative elasticities of supply and demand for the taxed commodity. For example, carbon taxes on production or use of fossil fuels can be fully forward shifted in the short run if the firms in the industry have full market power, or the demand for the taxed commodity is perfectly inelastic, or the supply is perfectly elastic. In Table 8.5, columns (a) and (b) present carbon tax ($10/ton) incidence calculations for Pakistan using data from the 1984/85 Household Income and Expenditure Survey and employing two alternative concepts of household income. Column (a) relates carbon tax payments to household current income by income class and column (b) to household expenditure by income class. In either case, the carbon tax burden falls with income, thereby yielding a regressive pattern of incidence. Such regressively is nevertheless less pronounced with respect to household expenditures, thereby confirming the same conclusions reached by Poterba (1991) for the US.

Case (b): Complete Absence of Forward Shifting. Under a variety of circumstances, the burden of carbon taxes can fall entirely on capital owners. This can happen if price controls apply and legal pass-forward of the tax is disallowed, or if supply is completely price inelastic. The carbon tax will then be fully borne by fixed factors of production. With binding import quotas or rationed foreign exchange, carbon taxes will reduce rents received by quota recipients, rather than affect prices paid by consumers. Under the assumption of zero forward shifting, the burden of a carbon tax is attributed to capital income alone. The allocation of tax by capital income is then related to household income and household expenditures. Both these calculations yield a progressive distribution of the carbon tax burden (see Table 8.5: columns (c) and (d)).

Case (c): Partial Forward Shifting. Clearly, (a) and (b) above are polar cases and are unlikely to be fully satisfied for energy products in any country. There are only a handful of empirical studies which examine shifting assumptions for developing countries. One such study was carried out for excise taxes in Pakistan by Jeetun (1978). He finds 31% forward shifting of excises in Pakistan. Given than a tax on the carbon content of fossil fuels at their production stage is by its very nature an excise tax, it would be reasonable to use this assumption for assessing the distribution of the carbon tax burden. In Table 8.5, columns (e) and (f), 31% of the carbon tax is attributed to final consumption and 69% to general capital income; these series are then related to household incomes and expenditures by income class. This results in a roughly proportional incidence of carbon taxes under the former series and a progressive incidence pattern under the latter series.

8.3.3 Comparison with the Incidence of Personal and Corporate Income Taxes

The above analysis suggests that the regressivity of carbon taxes should be less of a concern in developing countries than in developed countries. This conclusion is further reinforced when one examines the incidence of personal income tax in a typical developing country. Personal income tax may not necessarily turn out to be a progressive element in the overall tax system, given both tax evasion and urban–rural migration effects, and their significance in lower to middle income countries. With respect to tax evasion, Shah and Whalley (1991) argue that, if the bribe rate is high and tax compliance low, the redistributive impact of the bribe system is likely to dominate the direct redistributive effects of income taxes. The relevant issue then is who receives the bribes. If public service is dominated by a seniority system, then high officials with higher income and wealth receive a large portion (or the majority) of the bribe, along

Table 8.5 Carbon tax ($10/ton) incidence—Pakistan (Carbon taxes [TAX as percent of monthly income (Y) or expenditure (EXP)])

Monthly income (Rupees)	Full forward shifting TAX/Y (a)	TAX/EXP (b)	Capital owners TAX/Y (c)	TAX/EXP (d)	Capital owners (0.69) TAX/Y (e)	Consumption (0.31) TAX/EXP (f)
– 600	1.49	1.19	0.66	0.53	0.92	0.74
600–700	0.89	0.83	0.62	0.58	0.71	0.66
701–800	0.91	0.86	0.64	0.60	0.72	0.68
801–1000	0.80	0.77	0.68	0.66	0.72	0.69
1001–1500	0.81	1.81	0.72	0.72	0.75	0.75
1501–2000	0.81	0.85	0.76	0.79	0.78	0.81
2001–2500	0.82	0.87	0.74	0.79	0.77	0.82
2501–3000	0.74	0.80	0.77	0.83	0.76	0.82
3001–3500	0.76	0.83	0.75	0.81	0.75	0.82
3501–4000	0.78	0.83	0.77	0.83	0.77	0.83
4001–4500	0.68	0.78	0.78	0.90	0.75	0.86
4500	0.51	0.67	0.80	1.06	0.71	0.94
	Regressive	Regressive	Progressive	Progressive	Proportional	Progressive

Source Authors' calculations

with professionals (accountants) who often act as "middlemen" in this process. Increasing income tax can thus trigger a reverse distributional process from middle class businessmen and others to wealthy elites, an entirely opposite conclusion to that commonly reached. Thus, tax evasion either reduces or offsets the progressivity of the tax system. The perceived progressivity of personal income tax is further clouded by the operation of the Harris-Todaro effect. In developing countries, personal income tax is imposed on urban sector incomes only. Under such circumstances, if expected wages are equalized across modern and traditional sectors through rural–urban migration effects, some of the burden of the (urban) tax is shifted to the rural sector through intersectoral wage effects. Thus, rural workers, although they face no legal liability to pay the tax, bear part of the burden of the tax through reduced wages. The potential importance of this effect is illustrated by Shah and Whalley (1991) using 1984–1985 data for Pakistan. They find that incorporation of the Harris-Todaro effect in incidence calculations clouds the progressivity of the personal income tax in Pakistan (see Table 8.6). Shah and Whalley (1991) also present calculations establishing the progressivity of corporate income taxes that consider complications introduced by foreign and public ownership of the corporate sector in Pakistan.

The above analysis suggests that concerns over the regressivity of carbon taxes may be overstated. If the lowest income group is protected from the regressive impact of carbon taxes by direct subsidies or alternate measures, then the regressivity of carbon taxes may not pose a serious policy concern. Further, if carbon taxes are used to reduce personal income taxes, traditional concerns that such a tax change would represent a move to a less progressive tax structure are not fully justified. Thus, a commonly perceived and widely accepted case against carbon taxes, based on equity grounds, does not hold up under a closer scrutiny.

8.3.4   *Efficiency Costs of Carbon Taxes*

By design, carbon taxes distort production, investment, and consumption decisions and thereby internalize the social costs of global and local externalities. For every dollar of carbon tax revenues raised, consumers lose more than a dollar in direct and indirect costs. It is the indirect or hidden costs of carbon taxes relative to other forms of taxation that are of interest to policy makers. The literature commonly refers to these costs as marginal welfare costs of taxation. In evaluating the potential

Table 8.6 Incidence of personal and corporate income taxes in Pakistan under alternate approaches

(a) Incidence of Personal Income Taxes in Pakistan under alternate approaches
(tax as a percentage of total income)[a]

Annual household income (rupees)	No personal income tax in rural sector			Form of income tax shift to rural sector					
				Reduced wages for low-income rural households			Reduced rural wages overall		
	Urban	Rural	Total	Urban	Rural	Total	Urban	Rural	Total
Under 7,200	0	0	0	0	0.74	0.58	0	0.54	0.42
7,200–8,400	0	0	0	0	0.83	0.63	0	0.60	0.45
2,400–9,600	0	0	0	0	0.88	0.63	0	0.64	0.46
9,600–12,000	0	0	0	0	0.73	0.46	0	0.52	0.34
12,000–18,000	0.02	0	0.01	0.01	0.57	0.35	0.01	0.41	0.25
28,000–24,000	0.04	0	0.02	0.02	0.70	0.38	0.02	0.32	0.18
24,000–30,000	0.02	0	0.02	0.01	0.01	0.01	0.01	0.29	0.13
30,000–36,000	0.20	0	0.13	0.09	0.02	0.06	0.09	0.26	0.16
36,000–42,000	0.22	0	0.16	0.10	0.02	0.07	0.10	0.31	0.17
42,000–48,000	0.40	0	0.29	0.18	0.03	0.13	0.18	0.19	0.18
48,000–54,000	0.77	0	0.50	0.35	0.01	0.23	0.35	0.18	0.29
Above 54,000	1.33	0	1.04	0.61	0.11	0.47	0.61	0.13	0.48

(b) Incidence of Corporate Taxes in Pakistan under alternative approaches
(tax as a percentage of total Income)

Annual household income (rupees)	Income category subject to tax burden			Tax incidence excluding taxes paid by state and foreign enterprises
	Capital	Capital and consumption	Capital and labor	
Under 7,200	1.18	1.71	1.56	0.85
7,200–8,400	1.06	1.55	1.64	0.77
2,400–9,600	1.04	1.53	1.70	0.76
9,600–12,000	1.26	1.62	1.69	0.91
12,000–18,000	1.46	1.70	1.69	1.06
28,000–24,000	1.70	1.79	1.68	1.24
24,000–30,000	1.68	1.76	1.69	1.22
30,000–36,000	1.75	1.78	1.68	1.28
36,000–42,000	1.77	1.78	1.66	1.29
42,000–48,000	1.81	1.79	1.65	1.32
48,000–54,000	1.89	1.76	1.63	1.34
Above 54,000	2.01	1.74	1.64	1.46

Note [a]Calculations under "no personal income tax in rural sector" are based on actual tax collections by income class as reported in Pakistan, Government (1985). All figures from this survey are adjusted to bring the total in line with data from Pakistan, Government (1988). Income tax collections on household income derived from urban sources or from graduated surcharges on land revenue are effectively zero
Source Shah and Walley (1991)

of carbon taxes, one needs to determine what will be the impact on economic efficiency if the same revenues were to be raised by carbon taxes rather than by existing (and distortionary) taxes on income. The empirical literature on this question is regrettably sparse. Poterba (1991), for example, provides estimates of average and marginal deadweight loss associated with carbon taxes relative to a no-tax scenario. Such calculations are interesting, yet, as the following analysis demonstrates, pre-existing taxes have a major bearing on welfare costs. Further, it is the differential (relative to other taxes), rather than the absolute incidence of carbon taxes, that offers useful policy insights. Goulder (1991) has pursued this line of inquiry for the United States using a computable general equilibrium model. Browning (1987) has argued that a properly specified partial equilibrium model of taxation's welfare costs offers superior insights on the measurement of welfare costs since, in such an analysis, the contribution made by key parameters to the final estimate remains transparent, whereas it is obscured in CGE models. He further demonstrates that almost all the differences in welfare costs of taxation for the United States can be traced to different assumptions regarding key parameters, rather than differences in the nature of models (i.e., partial vs general equilibrium). In the following, two measures for the differential costs of carbon taxation and a measure for the absolute burden of carbon taxes are presented. All these measures explicitly recognize existing taxes. Derivations of these expressions are laid out more fully in Appendix.

Case A: Welfare Costs Under a Revenue-Neutral Change That Displaces Equal Yield Personal Income Taxes by a $10/ton Carbon Tax. An evaluation of the welfare costs of carbon taxation is carried out here by using a frequently employed concept of applied welfare economics known as the Hicksian compensating variations. According to this measure, welfare loss is defined as the additional income required to maintain the consumer's original utility level, given the vector of new consumer and producer prices resulting from the policy change. Thus, it is the additional income that would make the consumer indifferent to the new vector of consumer prices. A Taylor-series approximation of the expenditure function yields the following expression for the welfare cost of the tax system under the equal yield scenario mentioned above.

$$\begin{aligned}L^N = L - L' = &-\frac{1}{2}\,\epsilon_{xp}\left[\frac{(T_1^x)^2 - (T_1^x + T_2^x)^2}{p_1^2}\right]p_1 x_1 \\ &- \frac{1}{2}\,\epsilon_{xw}\left[\frac{T_1^x T_1^H - (T_1^x + T_2^x)(T_1^H + T_2^H)}{p_1 W_1}\right]p_1 x_1 \cdot \theta \\ &+ \frac{1}{2}\,\epsilon_{Hp}\left[\frac{T_1^x T_1^H - (T_1^x + T_2^x)(T_1^H + T_2^H)}{p_1 W_1}\right]W_1 H_1 \cdot \theta \\ &+ \frac{1}{2}\,\epsilon_{Hw}\left[\frac{(T_1^H)^2 - (T_1^H + T_2^H)^2}{W_1^2}\right]W_1 H_1 \end{aligned} \quad (8.1)$$

where

ε_{xp} = own-price elasticity of fossil fuel demand.
ε_{xw} = cross-price elasticity of fossil fuel demand with respect to after-tax wages.
ε_{Hp} = elasticity of labor supply with respect to prices of fossil fuels.
ε_{HW} = elasticity of labor supply with respect to after-tax wages.
p_1 = composite price of fossil fuels before carbon tax.
X_1 = quantity of annual consumption of fossil fuels before carbon tax.
W_1 = after-tax hourly wages before revenue-neutral labor income tax change.
H_1 = man-hours of labor per year.
θ = share of fossil fuel expenditures to total expenditures.
T_1^x = pre-existing unit taxes on fossil fuels.
T_2^x = unit carbon tax (US$10/ton).
T_1^H = pre-existing labor income taxes per man-hour.
T_2^H = reduction in per man-hour labor income taxes.

The first term in the expression above captures the direct effect of higher fossil fuel prices on fossil fuel consumption. The two middle terms are the indirect effects (cross effects) of higher after-tax wages on fossil fuel consumption and higher fossil fuel prices (lower real wages) on labor supply. The fourth term captures the direct effect of higher after-tax wages on labor supply. The key parameters needed for the evaluation of this expression are hours worked per year; current labor income tax rate;

prices of energy products; quantity of energy consumption; current tax rate on energy; carbon tax rate (per unit of energy); elasticity of labor supply; and elasticity of energy demand. The data required to calculate these parameters for India, Indonesia, Pakistan, the United States, and Japan were collected from a variety of sources. Table 8.6 presents' data on carbon emissions, carbon prices, and energy taxes for the sample countries, and Table 8.7 reports a summary of results on welfare effects based on the above model. These calculations suggests that replacement of personal income tax by an equal yield $10/ton carbon tax represents a welfare-deteriorating proposition in the sample countries. Estimates of the welfare loss (compensating variations) range from a low of 1.5 cents per dollar of carbon tax revenues in Indonesia to a high of 17.5 cents per dollar in Pakistan. On economic efficiency considerations alone, therefore, carbon taxes cannot be supported as a replacement for personal income taxes. The difference in the welfare costs of a US$10 carbon tax arises primarily from variations in elasticity values (quite similar for our sample countries), pre-existing fossil fuel taxes, labor income taxes, carbon prices (i.e., market value of total fossil fuel consumption divided by carbon emissions), and energy price changes from the carbon tax. The price of carbon, a key parameter in the welfare cost calculations, is a function not just of fossil fuel prices, but also of the types and mix of fossil fuels consumed. A country that is a large consumer of coal will have a low price of carbon relative to a country that is a large consumer of natural gas or oil, even if the latter has the same level of fossil fuel prices. The relatively low welfare loss indicated for Indonesia is primarily attributable to lower levels of energy taxation in Indonesia, and the relatively large loss for Pakistan is due to high pre-existing energy taxes. In the case of Japan, the welfare loss is substantially lower than for Pakistan despite high pre-existing energy taxes. This results from the high price of carbon in Japan, which implies the percentage increase in energy prices due to the US$10 carbon tax will be low. The welfare loss for India compares well with that for the United States, even though pre-existing taxes in India are much lower (see Tables 8.7 and 8.8). This is because the price of carbon in India is only half of the price in the United States—the result of India's high consumption of coal.

The welfare gain associated with the direct effect of lower labor income tax on labor supply is very small for India, Indonesia, and Pakistan (at most 0.5% of the total welfare loss), because labor income taxes are low relative to wage income in these countries. Higher labor income taxes

Table 8.7 Carbon emissions, Carbon prices and energy taxes in selected countries

Country	Carbon emissions (million tons)	Carbon price ($/ton)	Energy taxes ($/ton)
India	148.2	117	10.69
Indonesia	26.6	200	0.00
Pakistan	13.2	253	65.13
USA	1246	198	26.64
Japan	237.1	538	104.80

Note The above calculations are based on 1987 data
Data Sources
Carbon Emissions—World Resources Institute (1990)
Carbon Price—Authors calculations based on data from Asian Development Bank and Energy Information Administration
Energy Taxes in 1987—Authors calculations based on data from International Energy Agency
Source Shah and Larsen (1992a)

make the equivalent effect substantially larger in the U.S and Japan (20% and 15%, respectively). For the first three countries, the indirect effects are small but positive, which indicates that the positive effect of higher real wages on energy consumption that results from the lowering of labor income taxes, dominates the negative effect of higher energy prices on labor supply. Again, this is caused by low initial labor income taxes. In absolute terms, the indirect effects are negative and larger for the United States and Japan. This is because the negative effect of higher energy prices on labor supply dominates the positive effects on energy consumption of higher real wages associated with income tax reductions, and because the initial effective taxation of labor income is higher in the United States and Japan than in developing countries. These results imply that analyses which ignore pre-existing taxes will be in error and could consequently result in possibly quite misleading policy advice. The difference in measured welfare costs can be substantial if pre-existing taxes are high, as is the case for Pakistan. If these pre-existing taxes were to be ignored, one would obtain for Pakistan low estimates for the welfare costs of carbon taxes, like those for Indonesia. For India, the welfare costs of carbon taxes in a no-tax case scenario would then be twice the level of Pakistan and Indonesia (since, due to the Indian use of inexpensive coal, carbon prices in India are nearly half of those in Pakistan and Indonesia).

Table 8.8 Summary of welfare effects of a $10/ton Carbon tax

	Carbon tax revenues	Welfare loss (−) or Gain (+)			
	Million USD	Million USD	% of Carbon Tax Revenues	% of Total Revenues	%GDP
A. Revenue Neutral Change by Equal Yield Reduction in Personal Income Tax					
India	1482	−129	−8.7	−0.39	−0.06
Indonesia	266	−4	−1.5	−0.03	−0.005
Pakistan	132	−23	−17.5	−0.39	−0.07
USA	12,461	−1049	−8.4	−0.11	−0.02
Japan	2371	−269	−11.4	−0.07	−0.008
B. Revenue Neutral Change by Equal Yield Reductions in Corporate Income Tax					
India	1482	+250	+16.9	0.8	0.11
Indonesia	266	+23	+8.7	0.2	0.03
Pakistan	132	+12	+9	0.2	0.04
USA	12,461	−773	−6.2	−0.08	−0.017
Japan	2371	+213	+9	0.06	0.007
C. Raising Additional Revenues with NO Change in Existing Taxes					
India	1482	−130	−8.8	−0.40	−0.06
Indonesia	266	−4	−1.5	−0.03	−0.005
Pakistan	132	−23	−17.7	−0.40	−0.07
USA	12,461	−1269	−10.2	−0.14	−0.03
Japan	2371	−291	−12.3	−0.08	−0.009
D. Raising Additional Revenues with NO Change in Existing Taxes but Accounting for Subsidies					
India	1482	0	0	0	0
Indonesia	266	+1	0.4	+0.01	0.005
Pakistan	132	−23	−17.7	−0.40	−0.07
USA	12,461	−1269	−10.2	−0.14	0.03
Japan	2371	−291	−12.3	−0.08	−0.009

Note Calculations based on the models presented in Appendix
Source Shah and Larsen (1992a)

Case B: Revenue Neutral Introduction of a $10/Ton Carbon Tax by Equal Yield Reductions in the Corporate Income Tax. Feldstein's (1978) model is adapted to derive the following expression for the welfare costs of taxation (see Appendix, Case B for details):

$$L^N = -\frac{1}{2}\epsilon_{xp}\left[\frac{(T_1^x)^2 - (T_1^x + T_2^x)^2}{(p_1)^2}\right]p_1 X_1$$

$$- (\eta_{sr}/r_1 T + 1 - \sigma)\left[\frac{T_1^x T_1^R - (T_1^x + T_2^x)(T_1^R + T_2^R)}{(p_1 p_1^R)}\right]p_1^R R_1 \cdot \theta$$

$$+ \frac{1}{2}(\eta_{sr}/r_1 T + 1 - \sigma)\left[\frac{(T_1^R)^2 - (T_1^R + T_2^R)}{(p_1^R)^2}\right]p_1^R R_1 \qquad (8.2)$$

η_{sr} = elasticity of corporate savings with respect to after-tax rate of return.
r_1 = after-corporate tax rate of return on corporate savings.
T = number of years from time of savings to dis-saving.
σ = marginal propensity to save.
T_1^R = pre-existing unit tax on consumption in the period of dis-saving; i.e., unit tax on return on corporate savings.
T_2^R = reduction in unit tax on return on corporate savings.
p_1^R = after-corporate tax discounted price of consumption in period of dis-saving, i.e., after-tax price of savings.
R_1 = savings in real terms such that $p_1^R R_1$ = nominal after-tax value of savings.

The first term portrays welfare loss associated with a $10/ton carbon tax, and is equivalent to the corresponding term in Case A. The second term captures the interaction of reductions in corporate income taxes and a simultaneous increase in carbon taxes. An increase in the after-tax price of energy products is likely to affect the consumption of energy products, and a reduction in after-tax return on savings is likely to affect savings decisions. This term could be either positive or negative. The third term represents the welfare gain associated with a reduction in corporate income taxes. Corporate income may be considered as a return on savings, i.e., on a firm's total assets or shareholders' equity. Thus, the third term captures the welfare effects of changes in after-tax rate of return on savings in the corporate sector. Corporate income taxes induce intertemporal inefficiencies by reducing savings and increasing current consumption. Key parameters needed for the evaluation of this expression include energy and retirement (future) consumption expenditures and prices, taxes on energy and retirement consumption, savings, and marginal propensity to save out of exogenous income, uncompensated elasticity of savings with respect to after-tax rate of return, and price elasticity of energy demand. These parameter values are obtained from a variety of sources. The model's results, presented in Table 8.9, suggest that, with the major exception of the United States, an equal yield introduction of carbon taxes in part replacement of corporate income tax would uniformly represent a welfare-improving proposition for the sample countries. The estimated net welfare gain varies from a high of 0.11% of GDP for India, to a low of 0.007% of GDP for Japan. These positive net welfare effects lend support to the widely supported view that corporate income taxes are far more distortionary than labor income taxes.

For the United States, the revenue-neutral introduction of a $10/ton carbon tax to replace corporate tax revenues is, in contrast to the above, a welfare-deteriorating proposition. The welfare loss is estimated to equal 6.2% of carbon tax revenues or 0.017% of GDP. The effect is due to lower marginal taxation of corporate income in the United States in comparison with other sample countries.

Case C: Raising Additional Revenues from Carbon Taxes with No Change in Existing Taxes. The following expression for the evaluation of net welfare captures the direct effect of carbon taxes on energy demand through price increases, and also their indirect effect through reduced wages—the latter being associated with an increase in consumption taxation.

$$L^N = -\frac{1}{2} \epsilon_{xp} \left[\frac{(T_1^x)^2 - (T_1^x + T_2^x)^2}{(p_1)^2} \right] p_1 X_1$$

Table 8.9 Costs and benefits of carbon taxes for selected countries

	Pakistan	Indonesia	India	United States	Japan
Fossil fuel consumption (million local currency in 1987)	58,209	8,793,837	222,744	246,502	15,759,000
Carbon (C) emissions (million tons)	13.2	26.6	148.2	1246.1	237.1
Price of carbon (per ton): Local Currency	4409	330,595	1503	198	66,465
US$	253	200	117	198	538
Energy taxes (US$/ton of carbon)	65.13	0.0	10.69	26.64	104.80
Carbon tax (US$/ton)	10	10	10	10	10
Carbon tax (local currency/ton)	174	16,500	129	10	1235
Elasticity of energy demand	−0.64	−0.6	−0.651	−0.6	−0.55
Price increase (from carbon tax) of:					
Coal	37.8%	17.5%	26.2%	18.3%	8 7%
Petroleum products	3.2%	5.8%	2.3%	3.4%	0.15%
Natural gas	2.6%	4.4%	3.0%	4.3%	1.4%
Emissions of (000 tons):					
PM	44	87	1192	6478	463
SO_2	321	337	2207	17,900	1600
NO_x	203	434	2090	17,400	1400
Emission reductions (%):					
C	−4.5%	−3.9%	−13.3%	−5.3%	−1.6%
PM	−11.6%	−5.0%	−15.3%	−7.8%	−0.6%
SO_2	−19.1%	−4.6%	−15.9%	−10.0%	−2.3%
NO_x	−3.8%	−3.8%	−11.9%	−5.6%	−1.2%
(1) Welfare cost of a US$10 per ton carbon tax					

(continued)

Table 8.9 (continued)

	Pakistan	Indonesia	India	United States	Japan
(Revenue increasing tax) million US$ (Table 8.8)	−23	−4	−130	−1270	−292
(2) Cost of carbon (C) reductions (US$/ton) (1) divided by tons of C reductions	38.7	3.9	6.6	13.8	78.9
(3) Price level (GDP/Penn GDP 1987)	0.23	0.35	0.34	1	1.55
(4) Benefit-Cost ratio[a] High (SO_2 + NOx+ PM)	1.8	17.9	9.5	11.2	1.3
Medium (SO_2 + NOxt PM)	1.6	12.9	7.5	8.7	1.0
Low (SO_2 + NOx + PM)	0.5	2.2	1.9	2.1	0.2

Note All calculations based on data for 1987. A US$10/ton carbon tax would be equivalent to a US$25/ton tax in 2022 prices
[a] 'High' is based on Glomsrod et al. (1990); 'Medium' is based on Bernow and Marron (1990); 'Low' is based on EPA/Energy and Resource Consultants, Inc. Referenced in Repetto (1990)
Source Shah and Larsen (1992a)

$$- \in_{HW} \left[\frac{-T_2^x T_1^H}{p_1 W_1} \right] W_1 H_1 \cdot \theta \tag{8.3}$$

The key elasticity parameters required for the evaluation of the above expression are the demand elasticities for fossil fuels and supply elasticity for labor. The results presented in Table 8.9 suggest that although the welfare costs of carbon taxes are significant, they represent only a small fraction of carbon tax revenues. Estimates for the sample countries range from a low of 1.5 cents per dollar for Indonesia (0.005% of GDP), to a high of 17.7 cents per dollar for Pakistan (0.07% of GDP). The welfare losses for India, Indonesia, and Pakistan are only slightly higher than those obtained in case A. This is because, given very ineffective preexisting labor income taxes and substantial tax evasion, the direct welfare effect of labor income tax reductions is very small for these countries. The difference in the two cases is larger for the U.S and Japan because of higher pre-existing labor income taxes and levels of tax compliance. Poterba (1991) finds a much lower welfare loss for the United States (average welfare costs of 3 cents per dollar of carbon tax revenues, or about 0.01% of GDP) in the revenue increase scenario by assuming no pre-existing taxes and no wage effects from carbon taxes. Thus levels of pre-existing taxes (on energy, income etc.) are critical in the estimation of the overall welfare effects associated with tax changes. Calculations that ignore these effects will understate the welfare cost of tax policy changes.

Case D: Raising Additional Revenues from Carbon Taxes with No Change in Existing Taxes but Accounting for Subsidies. The efficiency costs of carbon taxes will be overstated if, as in Cases A through C, subsidies are ignored. An efficient energy pricing policy calls for price to equal long-run marginal cost (in the case of no externalities). Thus, it is interesting to re-evaluate this welfare calculation by recognizing *existing* subsidies (Larsen and Shah, 1992a, b and Chapter 9, this volume). For the sake of simplicity, only the welfare cost of the carbon tax's direct effect on fossil fuel consumption is calculated, and the indirect effect on labor supply of higher fossil fuel prices is ignored. This is justified because the indirect effect on labor supply is less than 1% of total welfare costs. To calculate the welfare cost, petroleum products, natural gas, and coal are considered separately, and the same own-price elasticity of demand is applied to all product groups. Furthermore, the welfare calculation ignores the substitution effect between coal and petroleum products in Cases A–C, thus overstating true welfare costs.

Significant fossil fuel subsidies exist in India and Indonesia. The price of coal in India was only 85% of long-run marginal cost in 1990 (Bates and Moore, 1991), implying a 15% subsidy. By (conservatively) assuming a similar level of subsidy in 1987, the year used here for welfare calculations, a US$10 carbon tax leads to an approximately 26% increase in the price of coal at 1987 prices. Thus a large proportion of the tax acts to remove the subsidy and should be considered a welfare gain. The welfare cost of the carbon tax on petroleum products and natural gas is estimated to be equal to the welfare gain of the subsidy removal on coal. The overall welfare effect of a US$10 carbon tax is therefore approximately zero, rather than the −8.8% of carbon tax revenues in Case C. Similarly, petroleum products in Indonesia are priced significantly below world prices—approximately 35% lower in 1987. Following the same approach as for India, the carbon tax on petroleum products in Indonesia represents a welfare gain, although it is too small to eliminate the subsidies completely. The welfare gain is larger than the welfare costs of the carbon tax on coal and natural gas. Thus, in comparison with Case C's welfare loss of −1.5%, the net effect is a small welfare gain of 0.4% of carbon tax revenues. This section illustrates that pre-existing taxes and subsidies are critical in estimating the welfare effects of carbon taxes. Calculations that ignore pre-existing subsidies will overstate the welfare costs of tax policy changes.

In conclusion, the case for carbon taxes on efficiency considerations alone depends on whether they are introduced in a revenue-neutral manner, whether they replace corporate income taxes, and whether fossil fuel subsidies exist. According to the calculations presented here, such taxes do not fare so well against personal income taxes, at least for countries with pre-existing energy taxes and no subsidies. Clearly, however, an overall assessment of carbon taxes must therefore consider their impact on greenhouse gases and local pollutants, as well as on industrial performance and economic growth. These issues are taken up next.

8.4 THE IMPACT OF CARBON TAXES ON GREENHOUSE GASES AND LOCAL POLLUTANTS

Through their impact on aggregate use and composition of fossil fuel consumption, carbon taxes may reduce the emissions of local and regional pollutants such as nitrous oxides (NOx), carbon monoxides (CO), particulates (PM), sulfur dioxides (SO_2) as well as carbon emissions. This section deals with the impact of carbon taxes on NOx, SO_2, and PM emissions. These extent of these latter three emission types depends on technology, combustion processes, and sulfur content of fossil fuels; emission coefficients therefore vary greatly across sectors and countries. The data on emissions are derived here from available sectoral emission coefficients and sectoral fossil fuel consumption (OECD, 1989; Radian

Corporation, 1990). Table 8.9 illustrates the impact of a US$10 carbon tax on fossil fuel prices, and on CO_2, SO_2, NOx, and PM emissions for selected countries.

The impact of the carbon tax on CO_2, SO_2, NOx, and PM depends on the percentage increase in the end-user price of each fuel, in addition to the price elasticity of demand and emission coefficients. It is calculated as follows:

$$Z = \Sigma_{ij} e_{ij}^z \delta Q_{ij} = \Sigma_{ij} e_{ij}^z Q_{ij} \in_{ij} \delta p_{ij} / p_{ij} \qquad (8.4)$$

where Z is tons of reductions in CO_2, SO_2, or NOx; i are sectors; j are fuels (coal, natural gas, and petroleum products); e_{ij}^z is the emission coefficient of Z for fuel j in sector i; Q_{ij} consumption of fuel j in sector i; $\in_{ij}$ the own price elasticity for fuel j in sector i; and $\delta p_{ij}/p_{ij}$ is the percentage increase in price of fuel j in sector i from the carbon tax. Inter-fuel substitutions are ignored.

The elasticity of energy demand, being fairly similar across all the sample countries, does not contribute to the cross-country differences in emission reductions. The price of coal shows the largest increases primarily because of the low price of coal per ton. The increases for petroleum products and natural gas are only marginal in comparison because of their much higher current prices per ton. India shows the highest estimated emission reductions principally because coal is the predominant fossil fuel in consumption; it experiences relatively large reductions due to the high price increase induced by the carbon tax. Reductions are lowest in Japan because of high pre-existing energy prices that induce very low price increases from the carbon tax and thus low reductions in fossil fuel consumption. SO_2 emission reductions are highest in Pakistan because most such emissions are from high sulfur (5–6%) coal. SO_2 emission reductions are also quite high in the United States because of the large share of coal in consumption. Because of low coal use, Indonesia experiences relatively modest emission reductions. In all sample countries, percentage PM reductions tend to follow percentage reductions in the other pollutants.

A benefit-cost analysis of a US$10 carbon tax can now be made by comparing the welfare loses (Table 8.10) of a revenue-increasing carbon tax (with no reductions in either labor income taxes or corporate income tax) with the benefits of emission reductions. Welfare cost calculations are for the case which does not account for subsidies. Thus, welfare costs are substantially overstated for both India and Indonesia. Benefits are estimated given only SO_2, NOx, and PM emission reductions; no attempt is made to estimate the benefits of reductions in emissions of CO_2, CO, lead, and ground level ozone. The monetary value of emission reductions for any of these gases will be highly uncertain, in part because the damage emissions cause depends on: the aggregate level of emissions, climatic and topographic conditions, population density around emission sources, and on concentration levels of the pollutant. The main monetary benefits per reduced ton of SO2, NOx, and PM emissions, come from improvements in health and reduced corrosion (see Table 8.10, for results from three independent studies). Glomsrod et al. (1990) and Bernow and Marron (1990) report the highest estimates based on their studies for Norway and the United States, respectively. EPA/Energy and Resource Consultants Inc. report (for the United States) significantly lower benefits, for NOx. This low benefit estimate for NOx may result from excluding chronic health effects. Benefit figures are adjusted by Penn GDP relative Purchasing Power Parity indices (Summers and Heston, 1991) for each sample country, thereby allowing more meaningful cross-country comparisons. Note that this procedure assumes a degree of transferability for different countries' externality measures that is unlikely to be satisfied in practice; estimates of such measures are therefore likely to be crude at best.

Notwithstanding the above caveat, the comparison of costs and benefits (Table 8.10) suggests that, on local environmental grounds alone, Indonesia, India, and the United States can benefit substantially from a carbon tax. Benefits exceed costs by a ratio of more than 7 in two cases, and approximately 2 in the case of the lowest benefit estimates. In the case of Pakistan and Japan, because of high pre-existing energy taxes and thus

Table 8.10 Marginal benefits of NOx, SO_2 and PM Reductions (US$/ton)

	NO_x	SO_2	PM
Glomsrod et al.[a] ("High")	10,300	1400	3300
Bernow and Marron ("Medium")	6500	1500	4000
EPA/Energy and Resource Consultants ("Low")	230	637	2550

[a] The first study is for Norway and the last two for the United States
Source Glomsrod et al. (1990), Bernow and Marron (1990), and Repetto (1990)

high a welfare cost for carbon taxes, the benefit-cost ratio is significantly lower, although still greater than one.

It is important to note that, although the monetary benefits of emission reductions are uncertain, there emission reductions have additional benefits that are not accounted for here as already mentioned. Furthermore, welfare losses are based on the worst-case scenario of a revenue-increasing carbon tax not compensated for by a reduction in other taxes. Last, but not least, significant energy subsidies in India and Indonesia are not incorporated in the welfare calculations, which consequently overstate welfare losses.

Note also that these benefit-cost ratios do not depend on the price elasticity of demand for fossil fuels, which is assumed identical for each fuel. Both the welfare costs of carbon taxes and the quantity of emission reductions are proportional to that elasticity parameter, which is therefore canceled out in the ratio of benefits and costs. The latter depends primarily on preexisting taxes on fossil fuels (which affects welfare costs) and on the valuation of emission reductions of SO_2 and NOx in both relative and absolute terms. Furthermore, the calculations presented here do not attempt to identify least-cost policies for local pollutant reduction. They merely quantify various additional benefits from carbon taxes that are frequently ignored in the literature.

One means of accounting for the non-uniformity of emission externality costs across countries is to adjust the benefits of emission reductions for variations in population density and rural/urban population ratio. Here, an equal weight is applied to population density and urbanization. In consequence, benefits are larger by an average factor of two for Pakistan, Indonesia, and India. Thus, the benefit-cost ratio is larger than one for Pakistan even when lowest benefit estimates are used. For Japan, benefits are as much as twelve times higher. This is the result of a very high population density, which brings the ratio to 2.4 in the case of lowest benefit estimates and to as much as 14 in the case of highest benefit estimates. In this circumstance, Japan would benefit even more from a carbon tax than the United States.

A cost analysis of carbon reductions is also illustrated in Table 8.10. The cost of carbon reductions is stated in terms of the welfare costs of a revenue-increasing US$10/ton carbon tax, divided by tons of carbon reductions. The large cost differences across countries are caused mostly by differences in pre-existing energy taxes (high pre-existing energy taxes implying high welfare costs) and percentage carbon emission reductions. To illustrate this point, the cost of carbon reductions may be stated as follows:

$$C = (W/R)(R/E) = (W/R)(t * C_e/E) = W/E \qquad (8.5)$$

where C is the cost per ton of carbon reductions; W is the total welfare cost of the carbon tax; R is total carbon tax revenues; E is tons of carbon emission reductions; t is the carbon tax rate (US$10/ton); and C_e is the total tons of carbon emissions (thus C_e/E is the reciprocal of percentage carbon emission reductions). Equation (8.5) reveals that C is high if welfare cost per tax revenue dollar (W/R) is high (Table 8.7c), and/or if percentage carbon emission reduction is low (Table 8.4). The cost per ton of carbon emission reduction is lowest for Indonesia, even though percentage emission reduction is low. This is because of virtually non-existent energy taxes, which imply very low welfare costs per tax revenue dollar. Cost per ton is highest in Japan because of the combination of high welfare costs per tax revenue dollar and very low percentage emission reductions. The results in Table 8.10 also suggest that optimal carbon taxes are not uniform across countries because of different levels of pre-existing energy taxes and impact on local pollutants.

The preceding analyses of fossil fuel consumption and emission reductions considers only aggregate fuel reductions and not inter-fuel substitution. But since a carbon tax may induce significant inter-fuel substitutions, it is to be expected that the estimated emission reductions in Table 8.10 are overstated, given own-price elasticities. However, allowing for inter-fuel substitution would reduce the welfare costs of the carbon tax, such that the overall ratio of benefits to costs would most probably be only marginally affected.

In conclusion, the above analysis suggests that a carbon tax has significant benefits in terms of both local pollutant and CO_2 reductions. A monetary benefit-cost analysis indicates that, for countries with low or non-existent energy taxes, a carbon tax can be justified on local environmental grounds alone, even ignoring its benefits from a public finance viewpoint.

8.5 CARBON TAXES, INDUSTRIAL PERFORMANCE AND ECONOMIC GROWTH

Carbon taxes by changing the relative prices of inputs can impact on the production, financing, and investment decisions of firms. In this section, the Bernstein-Shah dynamic model of production structure (Bernstein and Shah, 1991) is used to examine the impact of carbon taxes on the economic performance of Pakistan's apparel and leather products industries (1966–1984). Several features of this dynamic model are noteworthy.

The costs of adjustment are treated as internal to the firm and are explicitly modeled. These capital adjustment costs imply that capital input does not necessarily attain its long-run desired level within any one contemporaneous period. The model formulation allows for estimation of this speed of adjustment. Investment in capital results in some foregone output in the short run. The model distinguishes short-run, intermediate-run, and long-run effects of tax policy initiatives. These effects are influenced by the varying degree of capital adjustment. The model also treats the determination of output supplies, variable, and quasi-fixed input demands simultaneously. Thus, both the direct and indirect effects of tax policy changes are captured in the model. Moreover, the dynamic nature of the model allows for direct and indirect effects to be estimated in all three runs of production. In addition to the explicit modeling of adjustment costs, the Bernstein-Shah model incorporates several features of producer behavior which are absent from the Jorgenson-Wilcoxon framework. Output supply is endogenous and not solely a function of factor demand or of investment. Furthermore, product markets are not assumed purely competitive and the nature of firm interdependence, as measured by the conjectural elasticity parameter, governs the structure of product markets. Finally, the model recognizes financial capital market imperfections as firms are constrained by the rate of return that can be earned on their financial capital. Rates of return on equity and debt capital are treated as exogenous to firm's behavior and cannot therefore be influenced by shareholders. Under such circumstances, the interest of owners is best served by maximizing the expected present value of the flow of funds to shareholders and bondholders. In other words, the firm's objective function is to maximize the expected present value of financial capital. The above-mentioned product and financial market imperfections are germane to most developing countries.

Accounting for them thus adds a sense of realism to the analysis of producer responses in these countries.

The estimation model is characterized by an after-tax normalized shadow variable profit function, output supply and input demand, and capital input demand equations. The model fits the data quite well. Furthermore, estimated parameters satisfy the conditions that the after-tax shadow variable profit function be concave in capital and net investment and convex in prices.

Table 8.11 provides estimates of carbon tax elasticities with respect to input demands and output supply in the short, intermediate, and long runs. These tax elasticities are then used to calculate carbon tax effects at mean sample values. A US$10/ton carbon tax on the apparel and leather industries leads to reductions in output and input demands in all periods, with the leather industry experiencing slightly higher reductions in output than the apparel industry. This difference is primarily attributable to the slightly higher energy intensity of the leather industry. Long-run output impacts are (−) 0.09% for apparel and (−) 0.11% for leather goods, both of which are higher than intermediate and short run impacts. Higher adverse effects in the long run arise because the model estimation suggests energy inputs serve as complements to both labor and capital in the two industries.

To examine the same effects for manufacturing industries in Pakistan overall, a flexible accelerator type dynamic factor demand model developed by Shah and Baffes (1991) is implemented using time series data for the period 1956 to 1985. This model employs a flexible and non-restrictive technology and captures the short-run divergence of fixed factors from their equilibrium values as well as the speed of such adjustment. Parameter estimates from the model suggest some pairwise substitutability among energy (materials) inputs and capital and labor. The model results suggest that the imposition of a $10 per ton carbon tax on Pakistani manufacturing industry will result in an output loss of 0.21% in the short run (see Table 8.11). The primary reason for larger output losses in aggregate manufacturing than in the apparel and leather industries is the substantial impact of the carbon tax on the price of coal. Coal is used primarily in industries other than apparel and leather. Thus, energy prices for aggregate manufacturing increase substantially more than for the apparel and leather industries.

Table 8.11 Impact of Carbon taxes on Pakistan industries application of Bernstein-Shah Dynamic Variable Profits Model and Shah-Baffes Dynamic Flexible Accelerator Model

		Own price elasticities			Carbon tax elasticities		Impact of a US$10/ton			Carbon emissions reductions	
		Apparel	Leather	Aggregate manufacturing	Apparel	Leather	Apparel (%)	Leather (%)	Aggregate manufacturing (%)	Apparel & Leather (%)	Aggregate manufacturing (%)
Short run	Y				−0.00081	−0.00098	−0.032	−0.039	0.205	−4.8	−7.6
	L	−1.521	−0.979	−0.514	−0.00086	−0.00090	−0.034	−0.036	−0.0137		
	M				−0.00193	−0.00133	−0.076	−0.053	−0.482		
	K				−0.00052	−0.00072	−0.021	−0.028	0.499		
Intermediate run	Y				−0.00111	−0.00145	−0.044	−0.057		−4.9	—
	L	−1.402	−1.135	—	−0.00108	−0.00147	−0.043	−0.058			
	M				−0.00178	−0.00154	−0.070	−0.061			
	K				0.00085	−0.00121	−0.034	−0.048			
Long run	Y				−0.00220	−0.00272	−0.087	−0.107		−10.4	—
	L	−2.461	−2.879	—	−0.00198	−0.00307	−0.079	−0.121			
	M				−0.00313	−0.00392	−0.124	−0.155			
	K				0.00201	−0.00255	−0.079	−0.101			

Notations
Y = Output
L = Man Years worked
M = Intermediate Inputs
K = Capital stock
Source Model Results as reported in Shah and Larsen (1992a)

Table 8.12 Impact of a US$10 Carbon Tax on manufacturing value added and local externalities

Pakistan	Pakistani Apparel & Leather Industries			Pakistan aggregate manufacturing industries
	Short run	Intermediate run	Long run	Short run
Output effect (%)	−0.035%	−0.051%	−0.098%	−0.205%
Output effect (in 000 US$)	−102	−148	−284	−20,900
Value added effect (in 000 US$)	−19.12	−27.3	−52.3	−6650
Emission Reductions (%)[a]				
NOX	4.9%	−5.0%	−10.5%	−7.4%
SO_2	−4.7%	−4.8%	−10.2%	−18.4%
PM	−4.8%	−4.9%	−10.3%	−12.9%
CO_2	−4.8%	−4.9%	−10.4%	−7.6%
Cost of CO_2 reductions—US$/ton (loss of value added divided by tons of CO_2 reductions)	44.2	61.9	55.9	14.5
Benefit-Cost Ratios associated with the impact of a US$10 carbon tax on value added and local pollutant[b]				
High	2.5	1.8	1.9	3.9
Medium	1.6	1.1	1.2	3.5
Low	0.09	0.06	0.07	1.1

[a] Emission reductions are percentage of emissions from the Apparel and Leather industries or from total manufacturing industries
[b] Includes sulfur dioxides (SO_2), nitrous oxides (NO_X) and particulate matter (PM). 'High" is based on Glomsrod et al. (1990); 'Medium is based on Bernow and Marron (1990); 'Low" is based on EPA/Energy and Resource Consultants, Inc. The last study does not include chronic health effect of NO_X emissions

Source Model based calculations as reported in Shah and Larsen (1992a)

A comparison of value-added losses with the health benefits of reductions in local externalities throws some (albeit limited) light on the cost-benefit calculus of carbon taxes. Table 8.12 reports estimates of costs associated with carbon truces, as well as benefits arising from a reduction in local externalities. Data limitations restrict the analysis to NOx, SO_2, and particulate matter (PM) only. The dollar values on local externalities are based on the same three studies used in Sect. 4, adjusted for purchasing power parity. Benefits to cost ratios are higher for aggregate manufacturing than for the apparel and leather industries because relatively large emission reductions from reduced consumption of high sulfur coal more than offsets the higher loss of value added in manufacturing industries. Ratios are larger than one except in the case with the lowest benefit estimates for the apparel and leather industries. These tentative calculations suggest that losses of value added are offset by health benefits associated with NOx, SO_2, and PM emission reductions, even if the reduction in global externalities associated with curtailing CO_2 emissions is completely ignored. Table 8.11 also reports estimates for the average cost of carbon reductions associated with a US$10/ton carbon tax in terms of US$/ton of carbon reductions, ignoring the benefits of reductions in local externalities. Calculations suggest that such costs are higher in the apparel and leather industries. The is primarily because total carbon emissions relative to value added in these two industries are much lower than in the overall manufacturing sector, while model results suggest that the elasticity of output or value added with respect to energy prices is only slightly lower. Thus, losses of value added relative to carbon emission reductions are higher in the apparel and leather industries.

8.6 Tradeable Permits

Tradeable emission permits represent an alternative instrument that can ensure marginal costs of emission reductions are equalized across domestic sources and across countries. Given both perfectly competitive markets and certainty, permits are equivalent to emissions taxes (see Hoel, 1990). Tradeable permits afford direct control over quantities of emissions as opposed to a carbon tax regime's indirect influence through prices. They are also easier to implement as an initial allocation of such permits reduces the resistance of existing emitters. Furthermore, tradeable permits in terms of their regulatory effects are more transparent to policymakers and administrators (see Oates and Portney, 1990). Tradeable permits have also been cited for their potential as a hedging instrument against risk and a vehicle for international technology transfer.

Epstein and Gupta (1990) have argued that tradeable permits could serve as an instrument to reduce the risk of investing in backstop technology R&D. They argue that agents or nations that invest in R&D are exposed to a high probability of failure, although also to high profits in the event of success. If R&D investments turn out a successful technology that significantly reduces the costs of carbon emission reductions, the price of emissions permits will fall. If the investments yield no return, the price of permits is expected to rise. This means that risk averse investors can purchase futures on emission permits as a hedging against risk. In this case, total investments in R&D can be expected to be higher than if a market for emission permits did not exist. One could further argue that carbon taxes would also induce higher levels of investment if tax revenues were pooled (fully, or in part) in an R&D fund or used to subsidize R&D. A closer analysis of the effectiveness of these alternatives seems appropriate given potential gains from the development of backstop technology.

Emissions permits will induce international technology transfers if initial emissions allocations are such that industrialized countries will purchase emissions permits from developing countries. If this is the case, developing countries may purchase more energy efficient technology from industrialized countries until the marginal benefit is equal to the permit price. This transfer could potentially be quite substantial and significant for developing countries. Its magnitude depends on the costs of emission reductions and initial permit allocations (Larsen and Shah, 1992b, 1994b). If costs of emission reductions are high (after some smaller initial reductions) in industrialized countries, then industrial countries will wish to purchase more emission permits from developing countries than if costs are low. This would imply larger revenue accumulations in developing countries which could be used to purchase more energy efficient technology. Technology transfers may turn out to be significant for developing countries because, in addition to reducing energy dependency, new capital embodies technological progress and thus contributes to increased total factor productivity. Total factor productivity gains are considered an important component for economic growth and improved international competitiveness.

In practice, tradeable permits are subject to important limitations. These include the "thinness" of permit markets, the presence of large buyers and sellers, and lack of any mechanism to deal with overshooting the mark. In the United States, it is observed that the main reason the permit markets are not as well functioning as envisioned is the "thinness" of the market, especially on the supply side, that is largely due to trading restrictions and unclear definitions of property rights. When permits are infrequently traded, clear price signals are absent, thereby impairing the functioning of the permit system. On the other hand, a carbon tax is a clear measure of the cost of emissions.

To avoid ill-functioning permit markets, the number of potential traders should be sufficiently large. In the case of carbon emission permits, an insufficient number of traders may be avoided by integrating international (inter-country) and domestic markets. Market power is then eliminated, and sufficient liquidity provided, especially if the market is open to outside parties as well as "emitters". In this case, any agent— a producer or consumer—obtains emission permits at a price quoted at trading boards, in much the same way as foreign exchange is traded and rates are quoted in international markets.

There are alternative market arrangements, although an international (inter-country) market seems a minimum requirement because the costs of emission reduction can be expected to differ substantially across countries. Emission permits, traded internationally, allow marginal costs of reduction to be equalized across nations. Permits may be traded independently within nations so that marginal costs are equalized across domestic sources. It is also possible that permits will only be traded internationally and that carbon taxes will be used domestically. Alternatively, some countries may use emissions permits to reduce domestic emissions while other countries use taxes. In the latter case, there may be separate international and domestic permit markets. Any market arrangement that reduces the number of traders below that in a globally integrated market is exposed to the danger of market inefficiencies (market power, illiquidity). However, the transactions costs of such markets may be too high to justify the establishment of a market that involves all "emitters" of carbon gases, from large industrial firms through to the individual household using fossil fuels. A carbon tax avoids these transactions costs. In global trading of permits, large countries can influence prices. For a large seller, it is optimal to have higher emissions than the level indicated by the marginal cost of reductions (the market price for quota), and the opposite holds true for a large buyer (see Hoel, 1990).

A potential problem with permit markets is that the supply of permits is by no means guaranteed to be intertemporal fixed. New information about the costs of emission reductions and of global warming will induce policymakers to change the total supply. Furthermore, such changes cannot be preannounced at the initial time since the changes are a function of the new information in future periods. New information is therefore like random shocks. These exposes permit holders to the risk of permit price changes that cannot be ignored. Two ways of getting around this problem are to establish a futures market, or to let permits expire at the end of each period to issue the new supply at market determined prices. Clearly, additional transactions costs will be unavoidable, thereby making tradeable permits less of an attractive instrument.

It is not clear whether there will be a regional or global policy response to the greenhouse effect. In the event of such a response, the most talked about scenario is to set a target of a certain percentage global emissions reduction below their current (or some future) level, or to stabilize the current (or some future) global stock of emissions. The most frequently discussed target is a 20% reduction below current levels by a specific year, although a 50% reduction is considered necessary to stabilize the stock of global emissions at current levels (World Resources Institute, 1990). What is the optimal policy instrument to achieve this objective? A carbon tax will result in some uncertainty about the magnitude of reductions but less uncertainty about the cost of reductions. Under a regime with tradeable permits the magnitude of emission reductions will be known, but there may be great uncertainty about the total cost of reductions. This is an important distinction between the two instruments in the case of global warming. Oates and Portney (1990) make this distinction when comparing a carbon tax with tradeable permits. If there is great uncertainty regarding the costs of emission reductions, a tax is preferred to avoid potentially large, unexpected costs. (This is particularly important if the marginal costs of reduction are rising steeply after some initial reductions have been achieved.) However, if the costs of global warming are believed to be unacceptably high or there is a threshold effect, it becomes very important to limit total emissions to an upper bound. In this case, tradeable permits are preferred to a tax even though there will be great uncertainty regarding the costs of emission reductions. At this point in time, we do not know whether there is a threshold with respect to the stock of carbon emissions beyond which temperatures would rise exponentially. Furthermore, we know little about the economic costs and environmental consequences of global warming. Given present ignorance regarding the global warming phenomenon, one might currently argue for a carbon tax to limit unexpected costs of emission reductions. When, or if, future research reveals more about possible threshold effects and the costs of warming, tradeable permits may become the appropriate instrument.

A global tradeable permit (or carbon tax) regime poses an additional problem in terms of initial permit allocations (or redistribution of tax revenues). Larsen and Shah (1992b, 1994b) evaluate alternative allocations, such as allocations relative to GDP or population, and conclude that neither of the two are likely to induce participation from significant groups of countries. They propose an alternative allocation, based on willingness to pay for carbon reductions that may induce broader participation in an international treaty.

8.7 Summary and Conclusions

This chapter has evaluated the case for carbon taxes on national interest grounds. As a background to this discussion, it has also reviewed current energy pricing regimes in both developed and developing countries and their implications for greenhouse gas emissions (Larsen and Shah, 1992a, b, 1994a, 1995, also Chapter 9, this volume). The following conclusions emerge from the analysis:

- A global carbon tax raises difficult issues of tax administration, compliance, and international resource transfer, and is therefore unlikely to be implemented in the near future.
- National carbon taxes can raise significant amounts of revenue in a cost-effective manner and, in developing countries, are not likely to have as regressive an impact as commonly perceived. Such a tax also fares quite well in efficiency terms if introduced in a revenue-neutral manner as a partial replacement for corporate income taxes. In general, the welfare costs of carbon taxes vary directly with the existing level of energy taxes and therefore a carbon tax should be the instrument of choice for countries with no or low levels of energy taxation, such as Indonesia and India.

- A carbon tax also has significant benefits in terms of local pollutant reductions in addition to CO_2 reductions. The cost-benefit analysis for selected countries presented in this paper suggests that countries with low or non-existent energy taxes can receive substantial net gains from a carbon tax, not just in efficiency terms, but on grounds of local environmental considerations alone.
- A carbon tax of US$10/ton in 1990 constant prices (US$25 in 2022 current prices) results in very small output losses for the Pakistani industries analyzed in this paper. The estimated effects are somewhat lower than comparable estimates for the United States obtained by Jorgenson and Wilcoxen (1990). The value-added losses are, however, offset by the health benefits associated with reductions in NOx, SO_2, and particulate matter (PM) emissions, even if reductions of global externalities associated with the curtailment of CO_2 emissions are ignored.
- Tradeable permits represent a preferred alternative to carbon taxes should there be a known critical threshold in the stock of carbon emissions beyond which temperatures would rise exponentially. Given our current lack of knowledge about the costs of carbon emission reductions and the threshold effect, a carbon tax appears to be a superior and more flexible instrument that avoids potentially large and unexpected costs.

Thus, while a universal case for national carbon taxes cannot be made, even ignoring global externalities, such taxes make eminent sense for many countries in terms of efficiency, equity, and local environmental externality considerations.

8.8 Appendix: Measurement of Differential Welfare Costs of Carbon Taxes

Welfare costs L of a tax system $T_1 = (T_1^1, T_1^2 \ldots T_1^n)$ introduced at a non-distorted equilibrium with prices $p_0 = (p_0^1, p_0^1, \ldots, p_0^n)$ is defined as the difference in the expenditure level E necessary to maintain a utility level $\bar{U}$ in the presence of T and the expenditure level required to sustain $\bar{U}$ in the absence of T, minus the tax revenues R:

$$L(p_1, p_0, \overline{U}) = E(p_1, \overline{U}) - E(p_0, \overline{U}) - R(p_1, p_0, \overline{U}) \qquad (8.6)$$

with $p_1 = p_0 + T_1$.

The expenditure functions can be approximated by a second order Taylor expansion in prices. Thus, in general, the welfare cost of taxes introduced at a non-distorted initial equilibrium is

$$L = -\frac{1}{2} \sum \sum S_1^{ij} T_1^i T_1^j \tag{8.7}$$

where $S_1^{ij} = \delta X_1^i / \delta p_1^j$, the cross-derivative of the compensated demand function and T_i is the unit tax of good i.

In the presence of existing taxes, welfare costs of changes in the tax system is not simply L. An intermediate step becomes necessary (Feldstein, 1978). Consider a revenue-neutral tax policy change such that $p_2 = p_1 + T_2$ with T_2 a vector of additional taxes. The total welfare costs of the tax system $T_1 + T_2$ is

$$L'(p_2, p_0, \overline{U}) = E(p_2, \overline{U}) - E(p_0, \overline{U}) - R(p_2, p_0, \overline{U}) \tag{8.8}$$

or in general

$$L' = -\frac{1}{2} \sum \sum S_1^{ij} \left(T_1^i + T_2^i\right)\left(T_1^j + T_2^j\right) \tag{8.9}$$

The additional welfare costs of the revenue-neutral tax change is

$$L^N = L - L' = E(p_1, \overline{U}) - E(p_2, \overline{U}) \tag{8.10}$$

Since $R(p_2, p_0, \overline{U}) = R(p_1, p_0, \overline{U})$ because of revenue neutrality.

8.9 Case A: Welfare Costs of Carbon Taxes That Displace Equal Yield Personal Income Taxes

Consider the case of two goods $(x, 1 - H)$, where x is fossil fuels and $1 - H$ is leisure (H is supply of labor). Price of fossil fuels and leisure is (P_0, W_0) an initial non-distorted equilibrium. The welfare cost of pre-existing taxes on fossil fuels and labor income (T_1^X, T_1^H), before introducing a carbon tax, is given by (8.7):

$$L = -\frac{1}{2} \frac{\partial x}{\partial p} T_1^X T_1^X - \frac{1}{2} \frac{\partial x}{\partial w} T_1^X T_1^H$$

$$-\frac{1}{2}\frac{\partial(1-H)}{\partial p}T_1^X T_1^H - \frac{1}{2}\frac{\partial(1-H)}{\partial w}T_1^H T_1^H \qquad (8.11)$$

with $T_1^H = W_1 - W_0 < 0$, $T_1^X = p_1 - p_0 > 0 =$ Writing L with compensated elasticities (8.10) becomes

$$L = -\frac{1}{2}\epsilon_{xp}\left(\frac{T_1^x}{p_1}\right)^2 p_1 x_1 - \frac{1}{2}\epsilon_{xw}\left(\frac{T_1^x}{p_1}\right)\left(\frac{T_1^H}{W_1}\right)p_1 x_1$$
$$+ \frac{1}{2}\epsilon_{Hp}\left(\frac{T_1^x}{P_1}\right)\left(\frac{T_1^H}{W_1}\right)^2 W_1 H_1$$
$$+ \frac{1}{2}\epsilon_{HW}\left(\frac{T_1^H}{W_1}\right)^2 W_1 H_1 \qquad (8.12)$$

with the elasticities evaluated at (p_1, x_1) and (w_1, H_1).

Suppose that carbon taxes are levied on fossil fuels $T_2^x = p_2 - p_1 > 0$, in addition to existing taxes T_1^x, and that labor income taxes are reduced in a revenue-neutral manner $T_2^x = W_2 - W_1 > 0$. The welfare cost of the tax system $(T_1^x + T_2^x, T_1^H + T_2^H)$ is given by (8.9):

$$L' = -\frac{1}{2}\epsilon'_{xp}\left(\frac{T_1^x + T_2^x}{p_2}\right)^2 p_2 x_2$$
$$- \frac{1}{2}\epsilon'_{xw}\left(\frac{T_1^x + T_2^x}{p_2}\right)\left(\frac{T_1^H + T_2^H}{W_2}\right)p_2 x_2$$
$$+ \frac{1}{2}\epsilon'_{HP}\left(\frac{T_1^x + T_2^x}{p_2}\right)\left(\frac{T_1^H + T_2^H}{W_2}\right)W_2 H_2$$
$$+ \frac{1}{2}\epsilon'_{HW}\left(\frac{T_1^H + T_2^H}{W_2}\right)^2 W_2 H_2 \qquad (8.12)$$

with the elasticities now evaluated at (P_2, X_2) and (W_2, H_2).

The change in welfare costs of the revenue-neutral tax change is

$$L^N = L - L' = -\frac{1}{2}\epsilon_{xp}\left[\frac{(T_1^x)^2 - (T_1^x + T_2^x)^2}{p_1^2}\right]p_1 x_1$$

$$-\frac{1}{2}\epsilon_{xw}\left(\frac{T_1^x T_1^H - (T_1^x + T_2^x)(T_1^H + T_2^H)}{P_1 W_1}\right) p_1 x_1 \cdot 0$$

$$+\frac{1}{2}\epsilon_{HP}\left(\frac{T_1^x T_1^H - (T_1^x + T_2^x)(T_1^H + T_2^H)}{P_1 W_1}\right) W_1 H_1 \cdot 0$$

$$+\frac{1}{2}\epsilon_{HW}\left[\frac{(T_1^H)^2 - (T_1^H + T_2^H)^2}{W_1^2}\right] W_1 H_1 \quad (8.13)$$

where θ = share of energy in total consumption and by noting that

$$e'_{xp}\left(\frac{T_1^x + T_2^x}{p_2}\right)^2 p_2 x_2 = \epsilon_{xp}\left(\frac{T_1^x + T_2^x}{p_1}\right)^2 p_1 x_1$$

And similarly for the other elasticities. The indirect terms are multiplied by the expenditure share of fossil fuels, IJ, to account for the fact that in reality there are more goods than leisure and fossil fuels.

The first term in (8.13) is L1e direct effect on consumption of fossil fuels of higher fossil fuel prices. The last term is the direct effect on labor supply of higher after-tax wages. The two middle terms are the indirect effects (cross effects) on fossil fuel consumption of higher after-tax wages and on labor supply of higher fossil fuel prices (lower real wages).

$L^N > 0$ would imply a welfare gain from the revenue-neutral tax change.

We would like to express the two indirect effects in terms of ϵ_{HW} which can be accomplished in two steps. The first step is to express the third term in (8.13) in terms of ϵ_{xw} by noting that

$$\frac{\partial^2 E}{\partial p \partial w} = -\frac{\partial^2 E}{\partial w \partial p} \quad (8.14)$$

From the symmetry in the two-by-two matrix in the second-order Taylor expansion of the expenditure function. The negative sign in (8.14) comes from the use of leisure as $1-H$. Given that

$$\frac{\partial E}{\partial p} = x \text{ and } \frac{\partial E}{\partial w} = H$$

Are the compensated demand functions, it follows that

$$\frac{\partial x}{\partial w} = -\frac{\partial H}{\partial p}$$

which is (8.14), and therefore

$$\epsilon_{xw}\frac{x_1}{w_1} = -\epsilon_{Hp}\frac{H_1}{p_1} \tag{8.15}$$

Thus, the two indirect terms can be expressed as

$$-\epsilon_{xW}\left[\frac{T_1^x T_1^H - (T_1^x + T_2^x)(T_1^H + T_2^H)}{P_1 W_1}\right] p_1 W_1 \cdot \theta \tag{8.16}$$

The second step is to express the compensated elasticity ϵ_{xw} in terms of ϵ_{HW}. Let

$$U = f(x, 1-H) \tag{8.17}$$

The total differential of (8.17) letting $\partial u = 0$ is

$$\frac{\partial f}{\partial x}\partial x + \frac{\partial f}{\partial(1-H)}\partial(1-H) = 0 \tag{8.18}$$

From the first order conditions of utility maximization

$$\frac{\partial f/\partial x}{\partial f/\partial(1-H)} = \frac{P}{W} \tag{8.19}$$

By (8.18) and (8.19) and dividing through by ∂W:

$$p\frac{\partial x}{\partial w} - w\frac{\partial H}{\partial W} = 0 \tag{8.20}$$

This gives

$$px\varepsilon_{xW} = WH\varepsilon_{HW} \tag{8.21}$$

To quantify L_N, T_2^H is derived from the revenue neutrality condition

$$\partial R = T_1^x X_1 - T_2^H H_1 = 0 \tag{8.22}$$

For small changes in the tax system. From (8.22), we get

$$T_2^H = T_2^x \frac{X_1}{H_2} \tag{8.23}$$

With (8.16), (8.21), and (8.23), we have

$$L^N = -\frac{1}{2} \epsilon_{xp} \left[\frac{(T_1^x)^2 - (T_1^x + T_2^x)^2}{p_1^2} \right] p_1 x_1$$

$$- \epsilon_{HW} \left[\frac{(T_1^x T_1^H - (T_1^x + T_2^x)(T_1^H + T_2^x \frac{X_1}{H_1})}{P_1 W_1} \right] W_1 H_1 \cdot \theta$$

$$+ \frac{1}{2} \epsilon_{HW} \left[\frac{(T_1^H)^2 - (T_1^H + T_2^x \frac{X_1}{H_1})^2}{W_1^2} \right] W_1 H_1 \tag{8.24}$$

Note that the elasticity values applied to (8.24) are uncompensated elasticities rather than the theoretically correct compensated elasticities. The difference in terms of welfare cost is quite small (Willig, 1976), approximately 10% with an income elasticity of 1 and $\theta = 0.05$ given our uncompensated elasticity values. This result may be derived from the Slutsky decompositions of the substitution and income effect. Thus, welfare costs are slightly overstated here.

8.10 Case B: Welfare Costs of Carbon Taxes That Displace Equal Yield Corporate Income Taxes

Corporate income may be regarded as return on savings (Feldstein, 1978), i.e., on assets or shareholders equity. Consider the case of two goods (x, R) in a two-period model where x is fossil fuel consumption in the first period and R is second period consumption of first period savings, both in real terms.

Prices of fossil fuels (x) and second period consumption (R) are (p_0, p_0^R) in an initial non-distorted equilibrium. In the presence of existing unit taxes on fossil fuels and second period consumption (T_1^x, T_1^R), welfare costs are given by (8.25):

$$L = -\frac{1}{2}\frac{\partial x}{\partial p}T_1^x T_1^x - \frac{1}{2}\frac{\partial x}{\partial p^R}T_1^x T_1^R$$
$$-\frac{1}{2}\frac{\partial R}{\partial p}T_1^x T_1^R - \frac{1}{2}\frac{\partial R}{\partial p^R}T_1^R T_1^R \qquad (8.25)$$

with $T_1^R = p_1^R - p_0^R > 0$, $T_1^x = p_1^x - p_0^x > 0$.

If r_0 is the rate of return on savings in the corporate sector, then $p_0^R = e^{-r_0 t}$ is the discounted (current) price of consumption in period $T + 1$ in the case of no tax on corporate income. Similarly $p_1^R = e^{-r_1 t}$ is the after-tax discounted price, with $r_1 = (1-t)r_0$ and t is the corporate income tax rate. Thus, corporate income taxes reduce the real value of period one savings since $p_1^R - p_0^R > 0$.

Writing L with compensated elasticities (8.25) becomes

$$L = -\frac{1}{2}\epsilon_{xp}\left(\frac{T_1^x}{p_1}\right)^2 p_1 X_1 - \frac{1}{2}\epsilon_{xp}^R\left(\frac{T_1^x}{p_1}\right)\left(\frac{T_1^R}{p_1^R}\right)p_1 X_1$$
$$-\frac{1}{2}\epsilon_{Rp}\left(\frac{T_1^x}{p_1}\right)\left(\frac{T_1^R}{p_1^R}\right)p_1^R R_1 - \frac{1}{2}\epsilon_{Rp}^R\left(\frac{T_1^R}{p_1^R}\right)^2 p_1^R R_1 \qquad (8.26)$$

with the elasticities evaluated at (p_1, x_1) and (p_1^R, R_1).

Suppose that carbon taxes are levied on fossil fuels $T_2^x = p_2 - p_1 > 0$ in addition to existing taxes T_1^x, and that corporate income taxes are reduced in a revenue-neutral manner $T_2^R = p_2^R - p_1^R < 0$. The welfare costs of the tax system $(T_1^x + T_2^x, T_1^R + T_2^R)$ is

$$L' = -\frac{1}{2}\epsilon'_{xp}\left(\frac{T_1^x + T_2^x}{p_2}\right)^2 p_2 x_2 - \frac{1}{2}\epsilon'^R_{xp}\left(\frac{T_1^x + T_2^x}{p_2}\right)\left(\frac{T_1^R + T_2^R}{p_2^R}\right)p_2 X_2$$
$$-\frac{1}{2}\epsilon'_{Rp}\left(\frac{T_1^x + T_2^x}{p_2}\right)\left(\frac{T_1^R + T_2^R}{p_2^R}\right)p_2^R R_2 - \frac{1}{2}\epsilon'^R_{Rp}\left(\frac{T_1^R + T_2^R}{p_2^R}\right)^2 p_2^R R_2 \qquad (8.27)$$

With the elasticities now evaluated at (p_2, X_2). The change in welfare cost of the revenue-neutral tax is

$$L^N = L - L' = -\frac{1}{2}\epsilon_{xp}\left[\frac{(T_1^x)^2 - (T_1^x + T_2^x)^2}{(p_1)^2}\right]p_1 X_1$$

$$-\frac{1}{2} \epsilon_{xp}^{R} \left[\frac{T_1^x T_1^R - (T_1^x + T_2^x)(T_1^R + T_2^R)}{p_1 p_1^R} \right] p_1 X_1 \cdot \theta$$

$$-\frac{1}{2} \epsilon_{Rp} \left[\frac{T_1^x T_1^R - (T_1^x + T_2^x)(T_1^R + T_2^R)}{P_1 p_1^R} \right] p_1^R R_1 \cdot \theta$$

$$-\frac{1}{2} \epsilon_{Rp}^{R} \left[\frac{(T_1^R)^2 - (T_1^R + T_2^R)^2}{(p_1^R)^2} \right] p_1^R R_1 \quad (8.28)$$

where θ = expenditures hare of fossil fuels in total consumption and by noting that

$$\epsilon_{xp}' \left(\frac{T_1^x + T_2^x}{p_2} \right)^2 p_2 x_2 = \epsilon_{xp} \left(\frac{T_1^x + T_2^x}{p_1} \right)^2 p_1 x_1$$

and similarly for the other elasticities.

The first term in (8.28) is the direct effect of higher fossil fuel prices on consumption of fossil fuels. The last term is the direct effect of lower prices on second period consumption.

The lower tax on corporate savings reduces the intertemporal inefficiency. The two middle terms are the indirect effects (cross effects) on fossil fuel consumption of lower prices on second period consumption and on savings from higher prices on fossil fuels.

$L^N > 0$ would imply a welfare gain from the revenue-neutral tax change.

We would like to express the two indirect effects in terms of ϵ_{xp}^{R} which can be accomplished in two steps. The first step is to express the third term in (8.28) in terms of ϵ_{xp}^{R} by noting that

$$\frac{\partial^2 E}{\partial p^R \partial p} = \frac{\partial^2 E}{\partial p \partial p^R} \quad (8.29)$$

From the symmetry in the two-by-two matrix in the second-order Taylor expansion of the expenditure function. Given that

$$\frac{\partial E}{\partial p} = X \text{ and } \frac{\partial E}{\partial p^R} = R$$

are the compensated demand functions, it follows that

$$\frac{\partial x}{\partial p^R} = \frac{\partial R}{\partial p}$$

which is (8.29), and therefore

$$\epsilon_{xp}^R \frac{R}{p} = \epsilon_{xp}^R \frac{X}{p^R} \qquad (8.30)$$

Thus, the two indirect terms can be expressed as

$$-\epsilon_{xp}^R \left[\frac{T_1^x T_1^R - (T_1^x + T_2^x)(T_1^R + T_2^R)}{P_1 p_1^R} \right] p_1 X_1 \cdot \theta \qquad (8.31)$$

The second step is to express the compensated elasticity ε_{xp^R} in terms of ε_{Rp^R} Let

$$U = f(x, R) \qquad (8.32)$$

The total differential of (8.32) letting $\partial U = 0$ is

$$\frac{\partial f}{\partial x} \partial x + \frac{\partial f}{\partial R} \partial R = 0 \qquad (8.33)$$

From the first-order conditions of utility maximization

$$\frac{\frac{\partial f}{\partial x}}{\frac{\partial f}{\partial R}} = \frac{p}{p^R} \qquad (8.34)$$

By (8.33) and (8.34) and dividing through by ∂p^R:

$$p \frac{\partial x}{\partial p^R} + p^R \frac{\partial R}{\partial P^R} = 0 \qquad (8.35)$$

This gives

$$pX\varepsilon_{Rp^R} = -p^r R\varepsilon_{Rp^R} \qquad (8.36)$$

By substituting (8.31) and (8.36) into (8.28):

$$L^N = -\frac{1}{2} \epsilon_{xp} \left[\frac{(T_1^x)^2 - (T_1^x + T_2^x)^2}{(p_1)^2} \right] p_1 x_1$$
$$+ \epsilon_{RP}^R \left[\frac{(T_1^x T_1^R) - (T_1^x + T_2^x)(T_1^R + T_2^R)}{p_1 p_1^R} \right] p_1^R R_1 \cdot \theta$$
$$- \frac{1}{2} \epsilon_{RP}^R \left[\frac{(T_1^R)^2 - (T_1^R + T_2^R)^2}{(p_1^R)^2} \right] p_1^R R_1 \quad (8.37)$$

It remains to express ε_{Rp^R} in terms of the elasticity of savings with respect to the after-tax rate of return for which elasticity alternatives are available.

Note that

$$\varepsilon_{Rp^R} = \eta_{Rp^R} + \sigma \quad (8.38)$$

where η is the uncompensated elasticity and σ is the marginal propensity to save out of exogenous income (Feldstein, 1978). Given that savings is $S = p^R R$, we have

$$\eta_{sp}^R = \frac{\partial S}{\partial p^R} \cdot \frac{p^R}{S} = \frac{\partial (p^R R)}{\partial p^R} \cdot \frac{p^R}{S}$$
$$= \frac{R p^R}{S} + \frac{\partial R}{\partial p^R} \frac{(p^R)^2}{S}$$
$$= 1 + \eta_{Rp}^R \quad (8.39)$$

By (8.38) and (8.39)

$$\varepsilon_{Rp^R} = \eta_{sp^R} - (1 - \sigma) \quad (8.40)$$

Recall that the discounted price of period $T+1$ consumption is $p^R = e^{-rT}$ with r the after-tax rate of return on period 1 savings. Thus,

$$\eta_{sp^R} = \frac{\partial S}{\partial p^R} \frac{p^R}{S} = \frac{\partial S}{\partial r} \frac{p^R}{S} \frac{1}{-Te^{-rT}} = \frac{\partial S}{\partial r} \frac{p^R}{S} \frac{1}{-Tp^R}$$

$$= \frac{\partial S}{\partial x} \frac{r}{S} \frac{1}{-rT} = -\eta_{sr}/rT \tag{8.41}$$

Because $\partial p^R = -Te^{-rT} \partial r$.

It follows that

$$\varepsilon_{R p^R} = -\left(\frac{\eta_{sr}}{rT} + 1 - \sigma\right) \tag{8.42}$$

To quantify L^N, T_2^R is derived from the revenue neutrality condition. With I being total tax revenues,

$$\partial I = T_2^x X_1 + T_2^R R_1 = 0 \tag{8.43}$$

for small changes in the tax system. Thus,

$$T_2^R = -T_2^x * \frac{X_1}{R_1} \tag{8.44}$$

With (8.37), (8.42) and (8.44), we have

$$L^N = -\frac{1}{2} \epsilon_{xp} \left[\frac{(T_1^x)^2 - (T_1^x + T_2^x)^2}{(p_1)^2} \right] p_1 X_1$$

$$- \left(\frac{\eta_{sr}}{r_1 T} + 1 - \sigma\right) \left[\frac{T_1^x T_1^R - (T_1^x + T_2^X)(T_1^R - T_2^X \frac{X_1}{R_1})}{p_1 p_1^R} \right] p_1^R R_1 \cdot \theta$$

$$+ \frac{1}{2}(\eta_{sr}/r_1 T + 1 - \sigma) \left[\frac{(T_1^R)^2 - (T_1^R - T_2^x \frac{X_1}{R_1})}{(p_1^R)^2} \right] p_1^R R_1 \tag{8.45}$$

8.11 Case C: Welfare Costs of a Revenue Enhancing Carbon Tax with No Change in Existing Taxes

Consider the case of two goods $(x, 1 - H)$ as in Case A. The welfare cost L^N may be derived directly from (8.26) by noting that with no other chaalgcs L^N the tax system than the carbon tax on fossil fuels,

$$T_2^H = 0 \left(\text{i.e. } T_2^x \frac{X_1}{H_1} = 0 \right)$$

in (8.26).

Thus,

$$L^N = -\frac{1}{2} \epsilon_{xp} \left[\frac{(T_1^x)^2 - (T_1^x + T_2^x)^2}{(p_1)^2} \right] p_2 x_1$$
$$- \epsilon_{Hw} \left[\frac{-T_2^x T_1^H}{p_1 W_1} \right] W_1 H_1 \theta \qquad (8.46)$$

as the last term vanishes.

The case of a revenue-increasing carbon tax may alternatively be considered by recognizing the indirect effect on corporate savings instead of the indirect effect on labor supply. In this case, L^N can be derived for (8.45) instead of (8.26). The first term will remain the same and the last term will vanish, but the indirect effect will in this case be unambiguously positive since $T^N > 0$. This is because higher prices on current period consumption induces a substitution to second period consumption, i.e., savings will increase. Thus, the welfare loss will be slightly smaller than the direct effect on fossil fuel consumption, contrary to the case previously considered with indirect effect on labor supply.

8.12 Case D: Welfare Effects of a Revenue Enhancing Carbon Tax with No Change in Existing Taxes but Accounting for Subsidies

The calculation will only include the direct effect on fossil fuel consumption from a carbon tax, i.e., the first term in (8.26) or (8.45). Fossil fuels are disaggregated as petroleum products/natural gas (x) and coal (y). Notation is the same as before. Inter-fuel substitution effects are ignored in order to be consistent with calculations in Case A–C.

The expression for welfare cost becomes

$$L^N = -\frac{1}{2} \epsilon_{xp}^x \left[\frac{(T_1^x)^2 - (T_1^x + T_2^x)^2}{(p_1^x)^2} \right] p_1^x X_1$$

$$+ \frac{1}{2} \in_{yp}^{y} \left[\frac{\left(T_1^y\right)^2 - \left(T_1^y + T_2^y\right)^2}{\left(p_1^y\right)^2} \right] p_1^y Y_1 \qquad (8.47)$$

Note that the second term is positive (welfare gain) if $T_2^y < 2 \, |T_1^y|$. This is because $T_1^y < 0$ is a subsidy.

References

Bates, R. and Moore, E. (1991). "Commercial Energy Efficiency and the Environment". Background Paper No. 5. World Development Report 1992. The World Bank. Washington, DC.

Bernow, S. and Marron, D. (1990). *Valuation of Environmental Externalities for Energy Planning and Operations, 1990 Update*. Cambridge, Massachusetts: Tellus Institute.

Bernstein, Jeffrey and Shah, Anwar (1991b). "Taxes, Incentives and Production: The case of Pakistan". Processed.

Browning, Edgar K. (1987). "On the Marginal Welfare Costs of Taxation". *The American Economic Review*, March, pp. 11–23.

Carbon Tax Center (2022). "Pricing Carbon Efficiently and Equitably". www.carbontax.org.

Churchill, A.A. and Saunders, R.J. (1991). "Global Warming and the Developing World". *Finance and Development*, June.

Cullenward, Danny and Victor, David (2020). *Making Climate Policy Work*. New York: Wiley.

Eckaus, Richard S. (1991). "Economic Issues in Greenhouse Warming". Presented at the World Bank, WDR Seminar, May 1991.

Epstein, J.M. and Gupta, R. (1990). "Controlling the Greenhouse Effect—Five Global Regimes Compared". The Brookings Institution, Brookings Occasional Papers. Washington, DC.

Feldstein, Martin (1978). "The Welfare Cost of Capital Income Taxation". *Journal of Political Economy*, 86(2):29–51.

Fisher, B., Barrett, S., Bohm, P., Kuroda, M., Mubazi, J., Shah, Anwar and Stavins, Robert (1995). "An Economic Assessment of Policy Instruments for Combating Climate Change". In James P. Bruce, Hoesung Lee and Erik F. Haites: *Climate Change 1995—Economic and Social Dimensions of Climate Change*. 1996, Chapter 11. Contributions of Working Group III to the Second Assessment Report of the Intergovernmental Panel on Climate Change. Cambridge, UK: Cambridge University Press.

Glomsrod, S., Vennemo, H. and Johnsen, T. (1990). "Stabilization of Emissions of CO2: A Computable General Equilibrium Assessment". Discussion Paper No. 48. Central Bureau of Statistics, Oslo, Norway.

Goulder, Lawrence H. (1991). "Effects of Carbon Taxes in an Economy with Prior Tax Distortions: An Intertemporal General Equilibrium Analysis for the U.S". Stanford University and NEBR. Draft.

Hoel, Michael (1990). "CO2 and the Greenhouse Effect: A Game Theoretic Exploration". University of Oslo, Preliminary Draft.

Hoel, Michael (1991). "Efficient International Agreements for Reducing Emissions of CO2". *The Energy Journal*, 12(2):93–107.

Jankins, Jesse (2021). https://www.carbontax.org/wp-content/uploads/2021/03/Jesse/Jenkins-bar-chart-via-cCullenward-Victor-23=March=201.png.

Jeetun, A. (1978). "Tax Shifting in Pakistan. A Case Study of Excise Duties, Sales Tax and Import Duties." Applied Economics Research Centre, Discussion Paper 30. University of Karachi, Pakistan.

Jorgenson, D.W. and Wilcoxen, P.L. (1990). "The Cost of Controlling U.S. Carbon Dioxide Emissions". Presented at Workshop on Economic/Energy/Environmental Modeling for Climate Policy Analysis, Washington, DC, October 22–23.

Larsen, Bjorn and Shah, Anwar (1992a). "World Energy Subsidies and Global Carbon Emissions". Policy Research Working Paper Series, WPS 1002. The World Bank, Washington, DC.

Larsen, Bjorn and Shah, Anwar (1992b). "Combating the 'Greenhouse Effect'". *Finance and Development*, 29(4):20–23.

Larsen, Bjorn and Shah, Anwar (1994a). "Energy Pricing and Taxation Options for Combatting the "Greenhouse Effect"". In Akihiro Amano, et al. (eds.): *Proceedings of the Tsukuba Workshop of the Intergovernmental Panel on Climate Change(IPCC)*, Working Group III, Tsukuba, Japan, 17–20 January, pp. 34–45. London and New York, NY: World Meteorological Organization/United Nations Environmental Programme and Cambridge University Press.

Larsen, Bjorn and Shah, Anwar (1994b). "Global Tradable Carbon Permits, Participation Incentives, and Transfers". *Oxford Economic Papers*, 46(0):841–856.

Larsen, Bjorn and Shah, Anwar (1995). "Global Climate Change, Energy Subsidies and National Carbon Taxes". In Lans Bovenberg and Sijbren Cnossen (eds.): *Public Economics and the Environment in An Imperfect World*. Boston, London, and Dordrecht: Kluwer Academic Publishers.

Maler, K.G. (1989). "The Acid Rain Game". In H. Folmer and E. van Ireland (eds): *Valuation Methods and Policy Makin& in Environmental Economics*. Amsterdam: Elsevier.

Metcalf, Gilbert E. (2021). "Carbon Taxes in Theory and Practice". *Annual Review of Resource Economics*, 13:245–265.

Nordhaus, William D. (1991a). "The Cost of Slowing Climate Change: A Survey". *The Enemy Journal*, 12(1):37–64.

Nordhaus, William D. (1991b). "To Slow or Not to Slow: The Economics of the Greenhouse Effect". *The Economic Journal*, 101(July):920–937.

Oates, Wallace and Portney, Paul R. (1990). "Policies for the Regulation of Global Carbon Emissions". Processed.

OECD (1989). Energy and The Environment: Policy Overview.

Pearce, David (1991). "The Role of Carbon Taxes in Adjusting to Global Warming". *The Economic Journal*, 101(July):938–948.

Postel, S. and Flavin, C. (1991). "Reshaping the Global Economy". In Lester R. Brown (ed.): *State of the World*, pp. 170–188.

Poterba, James M. (1991). "Tax Policy to Combat Global Warming: On Designing a carbon Tax". In R. Dornbusch and J.M. Poterba (eds.): *Global Warming: Economic Policy Response*. Cambridge, Massachusetts: MIT Press.

Rabe, Barry C. (2020). *Can We Price Carbon*. Boston, MA: MIT Press.

Radian Corporation (1990). "Emission and Cost Estimates for Globally Significant Anthropogenic Combustion Sources of NOX, N2O, CH4, CO and CO2". Prepared for Office of Research and Development, U.S. Environmental Protection Agency.

Repetto, Robert (1990). "Environmental productivity and why it is so important." *Challenge*, September-October.

Shah, Anwar and Baffes, John (1991). "Do Tax Policies Stimulate Investment in Physical and Research and Development Capital?" Policy Research Working Paper No. 689. World Bank, Washington, DC.

Shah, Anwar, Larsen, Bjorn and Whalley, John (1991). "Taxing Choices for Global Environmental Security". Research Proposal, Public Economics Division, Country Economics Department, World Bank, Washington, DC.

Shah, Anwar and Larsen, Bjorn (1992a). "Carbon Taxes, the Greenhouse Effect and Developing Countries". Policy Research Working Paper Series. World Development Report, WPS 957. World Bank, Washington, DC.

Shah, Anwar and Larsen, Bjorn (1992b). "Global Warming, Carbon Taxes and Developing Countries. Presented at the 1992 Annual Meetings of the American Economic Association, January 3, 1992, New Orleans, USA". *American Economic Review Proceedings*.

Shah, Anwar and Whalley, John (1991). "Tax Incidence Analysis of Developing Countries: An Alternative View". *The World Bank Economic Review*, 5(3):535–552.

Summers, Lawrence H. (1991). "The Case for Corrective Taxation". *The National Tax Journal*, XLIV(3, September):289–292.

Summers, Robert and Heston, Alan (1991). "The Penn World Table (Mark 5): An Expanded Set of International Comparisons, 1950–1988". *Quarterly Journal of Economics*, 106:327–368.

Tietenberg, T. (1985). "Emission Trading: An Exercise in Reforming Pollution Policy". Resources for the Future. Washington, DC.

Urban Institute and Brookings Institution, Tax Policy Center (2015). *Taxing Carbon: What, Why, and How?* Washington, DC.

Willig, Robert D. (1976). "Consumer's Surplus Without Apology." *The American Economic Review*, 66(4): 589–97

World Bank (2021). *State and Trends of Carbon Pricing.* Washington, DC: World Bank.

World Resources Institute (1990). *World Resources* 1990–91. Washington, DC: World Resources Institute.

World Resources Institute (2015). *Putting a Price on Carbon. A Handbook for US Policymakers.* Washington, DC: World Resources Institute.

CHAPTER 9

Worldwide Energy Subsidies and the Impact of Their Removal on Economic Welfare and Global Climate Change

Bjorn Larsen, Angie Nga Le, and Anwar Shah

9.1 Introduction

It has been argued that economic policies to protect local and global environments should, first and foremost, remove fossil fuel subsidies (see Summers, 1991; Churchill and Saunders, 1991; Larsen and Shah, 1992a,

---

Anwar Shah is grateful to Lawrence H. Summers, former US Treasury Secretary and Harvard President, for his encouragement for the author to undertake research on energy pricing, carbon taxes, and global climate change.

B. Larsen
Environmental Health and Resource Economics Global Consulting Services, Asker, Norway

A. N. Le (✉)
Florida International University, Miami, FL, USA
e-mail: nle@fiu.edu

© The Author(s), under exclusive license to Springer Nature Switzerland AG 2023
A. Shah (ed.), *Taxing Choices for Managing Natural Resources, the Environment, and Global Climate Change*,
https://doi.org/10.1007/978-3-031-22606-9_9

b, c, 1994a, b, 1995; Shah and Larsen, 1992a, b). In 1992, Larsen and Shah (1992c) carried out a pioneering study to document the level of worldwide subsidies on fossil fuels, the impact of their removal on world energy markets, global carbon emissions, and aggregate welfare in subsidizing and non-subsidizing countries. This chapter updates their work to assess any progress made in the removal of these subsidies over the past three decades and presents newer estimates of these subsidies based upon available data as of August 2022. Further it expands the coverage to include electricity subsidies as well. Estimated global energy subsidies in sample 87 countries as of 2019 stood at US$1.38 trillion representing a significant loss of government revenues with adverse impacts on economic efficiency, economic welfare, health care costs, and global climate change. Oil accounts for 28%, gas 18%, coal 6%, and electricity 48%, respectively, of these subsidies. It is remarkable to note that despite renewed heightened interest in combating global climate change, world has made little progress to get energy prices right over the last three decades and the use of energy subsidies worldwide has significantly increased over the past three decades.

Section 9.2 reviews existing fossil fuel pricing regimes and estimates the level of world fossil fuel subsidies. Section 9.3 develops a simple framework for estimating the impact of subsidy removal on global carbon emissions. A first estimate of carbon emission reductions is based on the assumption that world prices of fossil fuels do not change in response to the demand reduction in subsidizing countries that results from the removal of subsidies. Subsequently, world price effects and fossil fuel consumption in non-subsidizing countries are estimated using a simple model of global fossil fuel markets. Section 9.4 estimates welfare gains that result from fossil fuel subsidy removal: first, on the assumption that world prices are unchanged for both subsidizing and non-subsidizing countries; second, on the assumption that such prices do change. Section 9.5 estimates what level of OECD carbon taxes would be required to achieve world emission reductions equal to those resulting from the removal of subsidies. Section 9.6 presents a summary and conclusions.

A. Shah
Brookings Institution, Washington, DC, USA

9.2 Existing Fossil Fuel Pricing Regimes and World Energy Subsidies

Correct fossil fuel prices are prima facie first-order priority in any economic policy to curtail greenhouse gas emissions. This section explores the potential for correct fossil fuel prices by analyzing pricing practices around the world. This chapter attempts to provide as complete as possible inventory of worldwide fossil fuel subsidies, yet it may be noted that it is possible to obtain a reasonable estimate of the overall level of subsidies by studying only a small set of countries. For example, 90% of world coal production is consumed by 15 countries; 80% of world petroleum products by 28 countries; and 91% of world natural gas output by 20 countries (see Tables 9.1, 9.2, 9.3, 9.4, and 9.5). These countries are collectively responsible for 85% of fossil fuel carbon emissions. Roughly one half of coal and natural gas consumers, and one fourth of petroleum product consumers, are OECD countries with relatively insignificant subsidies.

We define fossil fuel subsidies as the difference between domestic fossil fuel prices and their opportunity cost evaluated at end-user prices. When fuels are traded internationally, border prices serve as opportunity cost: this is the case for petroleum products for all sample countries. Opportunity costs at end-user level are border prices plus a markup for distribution. US pre-tax retail prices of gasoline are used as proxies for opportunity cost at end-user level, although unit distribution costs may vary across countries to some extent. Natural gas and coal are traded less frequently than oil/petroleum products and natural gas markets are primarily regional in character. Border prices plus distribution costs are used if these fuels are imported or if there are export markets—as for the former Soviet Union in the case of natural gas, and to a lesser extent coal. For purposes of convenience, opportunity cost is henceforth referred to as world price. Exchange rates reported in the IMF's International Financial Statistics have been used to convert domestic currency figures to dollars.

Thus, total subsidies S_k for country k are given as:

$$\mathbf{S}_k = \Sigma_i \Sigma_j \left(\mathbf{p}_{ij}^w - p_{ij} e \right) \mathbf{q}_{ij} \tag{9.1}$$

where p_{ij}, is domestic end-user price of fossil fuel i in sector j, $p^w{}_{ji}$ is opportunity cost of fuel i in sector j in US dollars, e is the exchange rate in units of US dollars to domestic currency, and q_{ij} is domestic

Table 9.1 Greenhouse gas emissions from fuel combustion 2019—oil

Countries	Million tons of CO_2 eq	Percentage of world emissions	Cumulative percentage
United States	2043.09	17.70	17.70
China	1440.25	12.48	30.18
India	605.20	5.24	35.42
Japan	384.44	3.33	38.75
Saudi Arabia	322.82	2.80	41.55
Russia	318.64	2.76	44.31
Canada	286.58	2.48	46.79
Brazil	281.80	2.44	49.24
Mexico	242.14	2.10	51.33
Germany	241.29	2.09	53.43
Indonesia	222.97	1.93	55.36
Iran	191.44	1.66	57.02
France	175.49	1.52	58.54
South Korea	161.98	1.40	59.94
United Kingdom	158.80	1.38	61.32
Spain	142.30	1.23	62.55
Australia	141.05	1.22	63.77
Italy	139.69	1.21	64.98
Turkeiye	114.60	0.99	65.97
Egypt	107.11	0.93	66.90
Iraq	106.50	0.92	67.82
Thailand	103.31	0.90	68.72
Malaysia	82.86	0.72	69.44
Poland	81.12	0.70	70.14
South Africa	76.57	0.66	70.80
Argentina	66.23	0.57	71.38
Pakistan	65.58	0.57	71.95
Nigeria	62.63	0.54	72.49
Viet Nam	61.70	0.53	73.02
Taipei	60.39	0.52	73.55
Algeria	58.59	0.51	74.05
Venezuela	58.27	0.50	74.56
Philippines	56.82	0.49	75.05
Netherlands	49.91	0.43	75.48
Chile	49.28	0.43	75.91
United Arab Emirates	43.94	0.38	76.29
Belgium	42.93	0.37	76.66
Kuwait	42.50	0.37	77.03

(continued)

Table 9.1 (continued)

Countries	Million tons of CO_2 eq	Percentage of world emissions	Cumulative percentage
Colombia	40.30	0.35	77.38
Kazakhstan	38.05	0.33	77.71
Morocco	37.46	0.32	78.03
Other Africa	36.70	0.32	78.35
Ecuador	34.91	0.30	78.65
Greece	33.69	0.29	78.95
Peru	32.30	0.28	79.23
Austria	32.15	0.28	79.51
Ukraine	30.30	0.26	79.77
World[a]	11,541.87	100	
Non-OECD total	5392.53	46.72	
OECD total	4835.05	41.89	

[a] Data for over 190 countries and regions

Source International Energy Agency. (2021). Greenhouse Gas Emissions from Energy Highlight https://www.iea.org/data-and-statistics/data-product/greenhouse-gas-emissions-from-energy-highlights#overview

consumption of fuel i in sector j.[1] According to (9.1), total subsidies are the product of the price differential and quantity consumed at subsidized prices. Since the efficiency cost of subsidies is defined as the difference between total subsidies and the increase in consumer surplus, there is no need to apply price elasticities of demand in order to calculate subsidies. If total subsidies were calculated on the basis of consumption at non-subsidized prices, they would be less than the increase in consumer surplus, and welfare would therefore be higher with a subsidy.

Total subsidies by fuel and country are presented in Table 9.6, and ratios of domestic prices to world prices in Fig. 9.1. In aggregate, energy subsidies in sample countries amount to US$1.38 trillion.[2] Oil accounts for 28%, coal 6%, and gas 18% and electricity 48%, respectively, of these subsidies.

---

[1] Sectors include electricity generation, industry, transport, households, and a residual sector. Subsidies on outputs or complementary inputs to energy in any of these sectors would act as "implicit" subsidies on energy because more energy would be used than at efficient input and output prices. We do not attempt to account for such inefficiencies.

[2] For a comparative analysis of energy subsidies in the MENA region, see Boudekhdekh (2022).

Table 9.2 Greenhouse gas emissions from fuel combustion 2019—gas

Countries	Million tons of CO_2 eq	Percentage of world emissions	Cumulative percentage
United States	1662.68	22.88	22.88
Russia	855.44	11.77	34.65
China	573.78	7.90	42.55
Iran	391.74	5.39	47.94
Canada	235.58	3.24	51.18
Japan	219.87	3.03	54.20
Saudi Arabia	176.63	2.43	56.63
Germany	170.05	2.34	58.97
United Kingdom	158.12	2.18	61.15
Italy	141.32	1.94	63.09
Mexico	139.45	1.92	65.01
United Arab Emirates	128.19	1.76	66.78
South Korea	110.33	1.52	68.30
Egypt	109.42	1.51	69.80
Argentina	93.50	1.29	71.09
Uzbekistan	92.79	1.28	72.36
Indonesia	88.65	1.22	73.58
Turkeiye	84.81	1.17	74.75
France	84.80	1.17	75.92
Algeria	84.45	1.16	77.08
Thailand	84.40	1.16	78.24
India	83.06	1.14	79.38
Australia	76.63	1.05	80.44
Spain	71.12	0.98	81.42
Brazil	70.33	0.97	82.39
Netherlands	69.04	0.95	83.34
Qatar	68.85	0.95	84.28
Malaysia	63.55	0.87	85.16
Pakistan	61.67	0.85	86.01
Bangladesh	54.89	0.76	86.76
Oman	50.47	0.69	87.46
Turkmenistan	50.24	0.69	88.15
Ukraine	49.22	0.68	88.82
Kuwait	47.53	0.65	89.48
World[a]	7267.20	100	
Non-OECD total	3792.08	52.18	
OECD total	3474.66	47.81	

[a]Data for over 190 countries and regions
Source International Energy Agency. (2021). Greenhouse Gas Emissions from Energy Highlight https://www.iea.org/data-and-statistics/data-product/greenhouse-gas-emissions-from-energy-highlights#overview

Table 9.3 Greenhouse gas emissions from fuel combustion 2019—coal

Countries	Million tons of CO_2 eq	Percentage of world emissions	Cumulative percentage
China	7914.29	53.15	53.15
India	1638.76	11.00	64.15
United States	1077.84	7.24	71.39
Russia	433.15	2.91	74.30
Japan	426.15	2.86	77.16
South Africa	357.66	2.40	79.56
South Korea	300.92	2.02	81.58
Indonesia	277.52	1.86	83.45
Germany	219.28	1.47	84.92
Viet Nam	204.44	1.37	86.29
Poland	173.04	1.16	87.45
Australia	166.25	1.12	88.57
Turkeiye	164.87	1.11	89.68
World[a]	14,891.54	100	
Non-OECD total	11,912.65	80.00	
OECD total	2978.89	20.00	

[a] Data for over 190 countries and regions
Source International Energy Agency. (2021). Greenhouse Gas Emissions from Energy Highlight https://www.iea.org/data-and-statistics/data-product/greenhouse-gas-emissions-from-energy-highlights#overview

9.3 Implications for Greenhouse Gas Emissions

Removal of fossil fuel subsidies will presumably induce reductions in fossil fuel consumption and therefore carbon emissions in subsidizing countries. Conversely, consumption in non-subsidizing countries could increase if reductions in fossil fuel demand in subsidizing countries lower world prices. Furthermore, if domestic prices are below world prices because of price ceilings that are effective for producers as well as consumers, then removing such ceilings may have positive supply effects that could further reduce world prices. On the other hand, because removing producer subsidies will tend to reduce supply, we assume that the combined effect of removing of producer ceilings and subsidies is to leave supply unchanged—as far as subsidies are concerned, we therefore ignore the supply side in subsidizing countries. The extent to which reduced demand in subsidizing countries impacts on world prices and thus on increased

Table 9.4 Greenhouse gas emissions from fuel combustion 2019—total (oil, coal and natural gas)

Countries	Million tons of CO_2 eq	Percentage of world emissions	Cumulative percentage
China	9985.31	29.17	29.17
United States	4821.30	14.08	43.25
India	2371.89	6.93	50.18
Russia	1652.10	4.83	55.01
Japan	1065.85	3.11	58.12
Germany	653.90	1.91	60.03
Indonesia	596.30	1.74	61.77
South Korea	590.72	1.73	63.50
Iran	588.65	1.72	65.22
Canada	579.62	1.69	66.91
Saudi Arabia	499.45	1.46	68.37
South Africa	440.03	1.29	69.65
Brazil	428.08	1.25	70.90
Mexico	426.94	1.25	72.15
Australia	385.16	1.13	73.28
Turkeiye	370.20	1.08	74.36
United Kingdom	347.62	1.02	75.37
Italy	314.98	0.92	76.29
France	300.65	0.88	77.17
Poland	293.25	0.86	78.03
Viet Nam	285.50	0.83	78.86
Taipei	257.94	0.75	79.62
World[a]	34,233.90		
Non-OECD total	21,424.54	62.58	
OECD total	11,494.60	33.58	

[a]Data for over 190 countries and regions

Source International Energy Agency. (2021). Greenhouse Gas Emissions from Energy Highlight https://www.iea.org/data-and-statistics/data-product/greenhouse-gas-emissions-from-energy-highlights#overview

demand in non-subsidizing countries can be expected to differ for each fossil fuel.

The first part of this section estimates carbon reductions assuming no change in world prices. The last part estimates world price effects and their impact on demand for each fossil fuel in subsidizing and non-subsidizing countries.

Table 9.5 CO_2 emissions from fuel combustion by sector[a] 2019—electricity and heat production

Countries	Million tons of CO_2	Percentage of world emissions	Cumulative percentage
China	5238.13	37.23	37.23
United States	1699.24	12.08	49.31
India	1172.99	8.34	57.65
Russia	816.01	5.80	63.45
Japan	505.57	3.59	67.04
South Korea	317.05	2.25	69.30
Germany	239.06	1.70	71.00
Saudi Arabia	237.35	1.69	72.69
South Africa	230.73	1.64	74.33
Indonesia	224.79	1.60	75.92
Australia	180.58	1.28	77.21
Taipei	159.62	1.13	78.34
Iran	158.86	1.13	79.47
Viet Nam	154.42	1.10	80.57
Poland	137.43	0.98	81.55
Mexico	136.70	0.97	82.52
Turkeiye	133.75	0.95	83.47
Malaysia	116.38	0.83	84.29
Kazakhstan	104.96	0.75	85.04
Egypt	97.71	0.69	85.74
Italy	96.31	0.68	86.42
Thailand	87.84	0.62	87.04
Canada	84.66	0.60	87.65
Ukraine	80.31	0.57	88.22
United Kingdom	72.10	0.51	88.73
Philippines	71.26	0.51	89.24
United Arab Emirates	69.86	0.50	89.73
World[b]	14,068.09	100	
OECD total	4023.37	28.60	
Non-OECD total	10,044.72	71.40	

[a] Sectors include electric and heat production, other energy industry own use, manufacture industries and construction, transport, residential, commercial and public services
[b] Data for over 190 countries and regions

Source International Energy Agency. (2021). Greenhouse Gas Emissions from Energy Highlight https://www.iea.org/data-and-statistics/data-product/greenhouse-gas-emissions-from-energy-highlights#overview

Table 9.6 Fossil fuel subsidies 2019, million USD

Countries	Oil subsidies (1)	Coal subsidies (2)	Gas subsidies (3)	Electricity subsidies (4)	Total subsidies (1 + 2 + 3 + 4)
Armenia		0.11			0.11
Azerbaijan	2614.67		2185.62	285.77	5086.06
Belarus		30.76	2099.89		2130.65
Estonia		5.25	58.62		63.87
Georgia		9.92			9.92
Kazakhstan	8334.48	509.15		6182.45	15,026.08
Kyrgyzstan	483.33	32.30			515.63
Latvia		1.51	88.90		90.41
Lithuania		7.80	462.24		470.04
Moldova		4.18	55.30		59.49
Russia	33,815.96	1275.83	102,751.84	28,538.38	166,382
Tajikistan	23.24	36.48			59.72
Turkmenistan	4840.72		999.42	1416.51	6257.23
Ukraine		259.96		7712.31	8971.69
Uzbekistan		43.75		5229.09	5272.84
Argentina	4456.56		1359.13	11,862.58	17,678.27
Brazil					0
China	31,936.99	53,650.29		314,095.81	399,683.09
Czech Republic		86.06			86.06
Egypt	20,153.19	0.14		12,842.35	32,995.68
India		8646.31	18,553.45	60,040.15	87,239.91
Indonesia	46,885.00	1916.56		12,866.61	61,668.17
Mexico	1631.04			3966.64	5597.68
Poland		392.93	338.72		731.65
Saudi Arabia	47,459.28			22,346.41	69,805.69

Slovak Republic		34.31		138.96
South Africa		42.04	104.64	4459.34
Venezuela	15,392.14			15,392.14
Algeria	16,802.52	0.00	12,221.11	35,168.79
Angola	3519.99			4948.82
Australia		67.25		67.25
Austria		14.86		14.86
Bahrain	1475.61	296.07	1216.46	5044.52
Bangladesh	793.57		5537.67	11,208.72
Belgium		44.85		44.85
Bolivia	2510.27			2510.27
Brunei	534.01		169.84	962.02
Bulgaria		14.85		14.85
Canada			16,253.30	16,253.30
Chile			104.94	104.94
Colombia	9741.08		191.38	9932.46
Denmark		4.47		4.47
Ecuador	5233.50		1169.95	6403.45
Finland		23.60		23.60
France		41.14		41.14
Gabon	51.71			51.71
Germany		207.53		207.53
Ghana			948.04	948.04
Greece		8.16		8.16
Hungary		10.03	1855.36	3045.43
Iceland		4.46		667.03
Iran	59,014.39	34.12	70,611.84	162,605.19

(continued)

Table 9.6 (continued)

Countries	Oil subsidies (1)	Coal subsidies (2)	Gas subsidies (3)	Electricity subsidies (4)	Total subsidies (1 + 2 + 3 + 4)
Ireland		17.20			17.20
Israel		1.38			1.38
Italy		33.02			33.02
Japan		1941.67			1941.67
South Korea		739.65	300.74	20,277.71	21,318.09
Kuwait	8710.88			5666.03	14,376.91
Libya	8195.24		1580.04	2251.72	12,027.00
Luxembourg		1.44	181.25		182.69
Malaysia	21,543.40	147.30	6853.26	10,489.13	39,033.09
Morocco		0.00		931.73	931.73
Netherlands		29.50			29.50
New Zealand		14.79			14.79
Nigeria	14,652.62	0.08		1660.98	16,313.68
Norway		24.58		545.85	570.43
Oman	4133.80			1643.35	5777.15
Pakistan	2633.78	716.26	10,345.71	5199.62	18,895.38
Portugal		0.44			0.44
Qatar	5224.69			4307.84	9532.53
Romania		24.92			24.92
Singapore		14.35			14.35
Slovenia		2.03	79.82		81.85
Spain		19.88			19.88
Sri Lanka				1384.59	1384.59
Sweden		28.80			28.80
Switzerland		3.07			3.07
Taipei	2139.69	497.62	1411.64	7438.35	11,487.30

Thailand		610.96	5927.80	6538.76
Trinidad and Tobago	13.13		684.45	697.57
Turkiye		445.13	20,679.94	33,252.09
United Arab Emirates		73.15	5697.61	5770.76
United Kingdom		83.38		83.38
Vietnam		1273.71	11,952.10	13,225.81
Sample total[a]	384,950.47	74,501.38	666,762.19	1,379,764

[a]The sample includes data for 87 countries (the number of subsidizing countries is 59)

Source Authors' Calculations

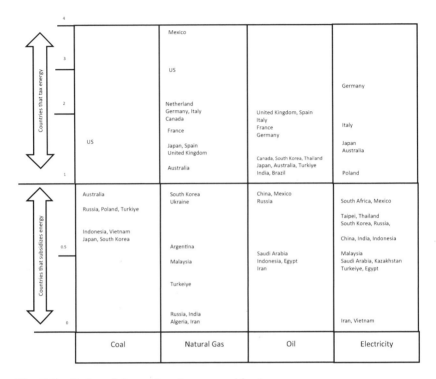

Fig. 9.1 Ratios of domestic prices to world prices

9.3.1 No World Price Effects

The magnitude of carbon reductions that result from the removal of fossil fuel subsidies clearly depends on the relevant price elasticities of demand. Bohi (1981) presents a comprehensive survey of price elasticities of energy demand. Long-run elasticities are in the range of −0.5 to −1.0 for natural gas, *−0.7* to −1.5 for petroleum products, −0.5 to −1.0 for coal, and −0.5 to −1.0 for electricity. In a cross-sectional study of OECD countries, Hoeller and Wallin (1991) estimate the long-run price elasticity of carbon demand at −1.04. These elasticity estimates are only valid for marginal price changes. In countries where subsidies are high, such as the Russian Federation, elasticity estimates for marginal price changes cannot be used to estimate emission reductions. Instead, much smaller elasticities must be considered. The elasticities used in most of the cases considered here range from −0.15 to −0.25, and to −0.6 where subsidy levels are low (see Tables 9.8, 9.9, 9.10, and 9.11).

The analysis ignores inter-fuel substitution. For the Russian Federation, where fossil fuels are subsidized across the board in almost the same proportion, this is an unproblematic assumption. However, in other countries (accounting for some 30% of carbon emissions reductions), to the extent a potential for inter-fuel substitution exists, the estimates of emission reductions presented here may be too high. Estimates of emission reductions resulting from the removal of subsidies can also be in serious error for countries where supply exceeds demand at low prices and is therefore completely inelastic—as may be the case for natural gas in particular. In Argentina, China, and Poland, demand for natural gas is considered to be constrained by supply. Within a certain range, therefore, an increase in natural gas prices may not have any significant effect on natural gas consumption. Factoring out emission reductions resulting from natural gas price increases in these three countries would have only a minor effect on the overall estimate for global carbon emissions reductions. Since, in the case of the Russian Federation, the share of natural gas in total energy consumption is as large as that of petroleum products and coal, it is not unrealistic to assume that natural gas price increases will lead to reduced natural gas consumption.

We assume a constant own price elasticity of demand, $-\epsilon$ ($\epsilon > 0$), with an inverse demand function,

$$p(q) = cq^{-1/\epsilon} \quad (9.2)$$

where q is consumption of fossil fuel, p is the domestic unit price of q, and c is a constant determined by the initial equilibrium. If (p_1, q_1) is the initial equilibrium at subsidized prices, p_1, and $(p_w, \boldsymbol{q_w})$ is the equilibrium that would prevail if domestic prices were raised to world prices, p_w, the percentage reduction in fossil fuel consumption that results from raising prices from p_1, to p_w, is,

$$(q_1 - q_w)/q_1 = 1 - (p_1/p_w)^\epsilon \quad (9.3)$$

Estimates of carbon emission reductions resulting from subsidy removal on oil, coal, natural gas, and electricity are presented by country in Table 9.7 and by country and fuel in Tables 9.8, 9.9, 9.10, and 9.11. Subsidy removal with no world price effect could result in a 15% reduction in subsidizing countries.

9.3.2 World Price Effects

Large reductions in fossil fuel demand in subsidizing countries may have significant effects on world or regional prices and therefore on demand in non-subsidizing countries. We consider only those world price effects that arise from changes in fossil fuel demand caused by removing subsidies. Although changes in the relative prices of other goods may affect consumption patterns, this is ignored—even though such changes may to some extent affect fossil fuel consumption since fossil fuels are inputs in their production. We assume world prices, p, are determined by supply and demand in the long run, and define linear world demand and supply functions,

$$q^D = a^D - b^D p \\ q^s = a^s + b^s p \quad (9.4)$$

where $q^D = \mathbf{q^D}_1 + \mathbf{q^D}_2$, with $\mathbf{q^D}_1$ and $\mathbf{q^D}_2$ linear demand functions for subsidizing and non-subsidizing countries, respectively, and similarly for $\mathbf{q}^s$.

Price equilibrium in the market is,

$$p = \left(a^D - a^S\right) / \left(b^D + b^S\right) \quad (9.5)$$

Corresponding to p in Fig. 9.2. Emission reductions resulting from subsidy removal (but estimated without taking account of world price effects) correspond to a movement from a to b in Fig. 9.2. This movement is equivalent to an inward shift in the world demand curve in Fig. 9.3, here noted as a change in a^D.

Thus ∂a^D is tons of reduction of carbon or fossil fuel consumption assuming no world price effects. Differentiating (9.5) with respect to $\mathbf{a^D}$ gives,

$$\partial p / \partial a^D = 1 / \left(b^S + b^D\right)$$

which in elasticity form is,

$$(q/p)\partial p / \partial a^D = (q/p) / \left(b^s + b^D\right) = 1 / \left(e^s + e^D\right)$$

9 WORLDWIDE ENERGY SUBSIDIES AND THE IMPACT ... 341

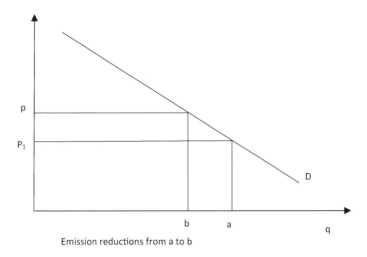

Emission reductions from a to b

Fig. 9.2 Impact of subsidy removal (in subsidizing country)

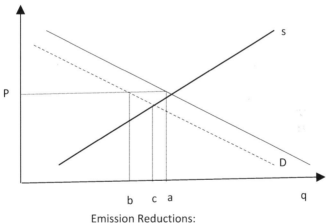

Emission Reductions:
No World Price effect: a to b
World price effect: a to c

Fig. 9.3 Impact of subsidy removal (World Market)

$$= 1/\left(e^s + e_1^D[q_1/q] + e_2^D[q_2/q]\right) \qquad (9.6)$$

$$\partial p/p = \left\{1/\left(e^s + e_1^D[q_1/q] + e_2^D[q_2/q]\right)\right\}\partial a_1^D/q$$

where e_1^D and e_2^D are absolute values of own price elasticities of demand for subsidizing and non-subsidizing countries, respectively, and e^S is weighted average own price supply elasticity for subsidizing and non-subsidizing countries.

The increase in consumption of each fossil fuel resulting from the reduction in world prices is given by,

$$\partial q_i = q_i e_i^D \partial p/p \qquad (9.7)$$

for country i. The net aggregate effect on consumption (i.e., the decrease in consumption resulting from subsidy removal plus the increase resulting from reduced world prices) is,

$$\partial q = \partial a^D + \Sigma_j \partial q_j \qquad (9.8)$$

with $\partial a^D < 0$, represented in Fig. 9.3 as the movement from a to c. The following sections apply this framework to the markets for oil/petroleum products, natural gas, coal, and electricity.

Oil/petroleum products: Two assumptions are made here: first, that there is a perfectly integrated world market for oil/petroleum products in which prices are determined by supply and demand in the long run; second, that a percentage change in the price of crude oil translates into an equivalent percentage change in the prices for refined products.

Assuming a weighted average supply elasticity of 0.5 and a demand elasticity of 0.8 in non-subsidizing countries, world prices of petroleum products (9.6) are estimated to fall 2.7%. Demand elasticity in subsidizing countries is as assumed in the previous section. Increases in petroleum product consumption, and hence carbon emissions, resulting from lower world prices are estimated by Eq. (9.7) and presented in Table 9.10 for oil, coal, and natural gas in aggregate and for oil only in Table 9.7 for each of the subsidizing countries. Net world emission reductions in sample countries are 8.53% when world price effects are incorporated, compared to 9.21% % when world price effects are ignored.

Table 9.7 Greenhouse gas emission reductions from removing subsidies on fossil fuels—oil (2019)

Countries	Domestic consumption—oil 2019, petajoule	Domestic price to border price—oil	Own price elasticity—oil	Emission reduction—oil (million tons, CO_2 eq)	Emission reduction to total emission—oil (%)	Emission increase from regional price's fall—oil (million tons, CO_2 eq)	Net emission reduction—oil (million tons, CO_2 eq)	Net emission reduction to total emission—oil (%)
Armenia	16.78	1.35	0.80			0.02		
Azerbaijan	189.54	0.55	0.25	1.53	14.03	0.07	1.46	13.35
Belarus	243.25	1.08	0.80			0.37		
Estonia	43.62	1.85	0.80			0.07		
Georgia	57.75	1.35	0.80			0.08		
Kazakhstan	455.47	0.40	0.25	7.83	20.57	0.26	7.57	19.89
Kyrgyzstan	60.24	0.74	0.25	0.34	7.37	0.03	0.31	6.70
Latvia	59.76	1.75	0.80			0.09		
Lithuania	101.84	1.70	0.80			0.18		
Moldova	40.34	1.29	0.80			0.06		
Russia	5459.96	0.80	0.20	14.19	4.45	1.73	12.46	3.91
Tajikistan	44.39	0.98	0.25	0.01	0.43	0.02	−0.01	−0.24
Turkmenistan	263.73	0.40	0.25	3.99	20.66	0.13	3.86	19.98
Ukraine	442.04	1.25	0.80			0.66		
Uzbekistan	149.76	1.03	0.80			0.25		
Argentina	931.20	0.84	0.25	2.78	4.19	0.45	2.33	3.51

(continued)

Table 9.7 (continued)

Countries	Domestic consumption—oil 2019, petajoule	Domestic price to border price—oil	Own price elasticity—oil	Emission reduction—oil (million tons, CO_2 eq)	Emission reduction to total emission—oil (%)	Emission increase from regional price's fall—oil (million tons, CO_2 eq)	Net emission reduction—oil (million tons, CO_2 eq)	Net emission reduction to total emission—oil (%)
Brazil	4099.38	1.01	0.80			6.11		
China	22,688.96	0.95	0.25	16.97	1.18	9.76	7.21	0.50
Czech Republic	384.72	1.78	0.80			0.47		
Egypt	1213.34	0.45	0.25	19.20	17.93	0.73	18.48	17.25
India	8706.53	1.12	0.80			13.12		
Indonesia	2921.80	0.47	0.25	38.14	17.10	1.51	36.62	16.43
Mexico	2896.84	0.98	0.25	1.13	0.47	1.64	−0.51	−0.21
Poland	1219.11	1.63	0.80			1.76		
Saudi Arabia	3664.83	0.57	0.30	49.51	15.34	2.62	46.89	14.52
Slovak Republic	143.17	1.86	0.80			0.22		
South Africa	1138.49	1.48	0.80			1.66		
Venezuela	515.80	0.02	0.20	32.03	54.97	0.32	31.71	54.43
Algeria	785.33	0.30	0.25	15.36	26.22	0.40	14.97	25.54
Angola	176.85	0.35	0.25	3.84	23.34	0.11	3.73	22.66
Bahrain	73.83	0.34	0.25	1.07	23.49	0.03	1.04	22.82
Bangladesh	201.35	0.87	0.25	0.66	3.41	0.13	0.53	2.73
Bolivia	165.13	0.50	0.25	1.95	15.91	0.08	1.87	15.23

Brunei	30.59	0.43	0.25	0.36	19.21	0.01	0.35	18.54
Colombia	586.47	0.45	0.25	7.23	17.93	0.27	6.95	17.25
Ecuador	416.82	0.59	0.25	4.35	12.47	0.24	4.12	11.79
Gabon	13.41	0.87	0.25	0.05	3.33	0.01	0.04	2.66
Iran	2911.50	0.33	0.25	45.98	24.02	1.30	44.68	23.34
Kuwait	386.78	0.26	0.25	12.17	28.64	0.29	11.88	27.97
Libya	277.51	0.03	0.25	16.90	58.84	0.19	16.71	58.16
Malaysia	1234.28	0.43	0.25	15.92	19.21	0.56	15.36	18.54
Nigeria	867.47	0.44	0.25	11.49	18.35	0.42	11.07	17.67
Oman	319.21	0.57	0.25	2.46	12.96	0.13	2.33	12.28
Pakistan	779.63	0.89	0.25	1.90	2.90	0.44	1.46	2.22
Qatar	331.41	0.48	0.25	3.08	16.70	0.12	2.95	16.02
Taipei	1551.12	0.95	0.25	0.70	1.15	0.41	0.29	0.48

(continued)

Table 9.7 (continued)

Countries	Domestic consumption—oil 2019, petajoule	Domestic price to border price—oil	Own price elasticity—oil	Emission reduction—oil (million tons, CO_2 eq)	Emission reduction to total emission—oil (%)	Emission increase from regional price's fall—oil (million tons, CO_2 eq)	Net emission reduction—oil (million tons, CO_2 eq)	Net emission reduction to total emission—oil (%)
Trinidad and Tobago	46.62	0.99	0.25	0.01	0.23	0.02	−0.02	−0.45
Sample—subsidizing countries				333.12	9.21	24.44	308.68	8.53
Sample—No subsidizing countries						128.78		
Sample total[a]						153.22	179.90	

[a]The sample includes data for 87 countries
Source Authors' Calculations

Natural gas: The natural gas market is more regional in nature than the oil market. For the purposes of this paper, we distinguish the following natural gas markets:

- The United States, Canada, and Mexico;
- Western and Eastern Europe, the former Soviet Union, and Algeria;
- Rest of the world.

The first two markets account for more than 80% of production, consumption, and trade. Furthermore, almost all trade is intra-market. We therefore assume that general equilibrium price effects will not affect prices in other regional markets. World subsidies on natural gas are primarily in the former Soviet Union and therefore our analysis will be confined to relevant European market only. Large reductions in natural gas demand in subsidizing countries may have significant effects on gas prices and consequently on demand in the corresponding regional gas market. We assume that regional prices are determined by supply and demand in the long run. We further assume that a percentage change in the price of natural gas translates into an equivalent percentage change in gas prices in all sectors of consumption.

Assuming a weighted average supply elasticity of 0.5 and a demand elasticity of 0.8 for non-subsidizing countries, regional prices are estimated to fall by 9.3%. Again, demand elasticities for subsidizing countries are the same as discussed earlier. Similarly, increases in natural gas consumption, and thus carbon emissions, due to the reduction in regional prices are derived using Eq. (9.7) and presented in Table 9.8 for each of the subsidizing countries. Net emission reductions in the subsidizing countries are still substantial—as much as 29.8% of global carbon emissions.

Coal: World coal markets are not as integrated as world oil markets, in part because of significant domestic protection in the form of producer subsidies and trade barriers. World coal trade is only 10% of world production, but is intercontinental. The United States and Australia are the largest exporters, followed by South Africa, Canada, Poland, and the Russian Federation. The largest import markets are Western Europe and Japan. Subsidy removal can be expected to have some general equilibrium price effects, but the corresponding demand effects, although difficult to quantify, are muted by protectionism. Consequently, although

Table 9.8 Greenhouse gas emission reductions from removing subsidies on fossil fuels—gas

Countries	Domestic consumption—natural gas 2019, petajoule	Domestic price to border price—gas	Own price elasticity—gas	Emission reduction—gas (million tons, CO_2 eq)	Emission reduction to total emission—gas (%)	Emission increase from regional price's fall—gas (million tons CO_2 eq)	Net emission reduction—gas (million tons, CO_2 eq)	Net emission reduction to total emission—gas (%)
Armenia	61.14		0.80			0.36		
Azerbaijan	180.29	0.23	0.25	7.11	30.57	0.54	6.56	28.24
Belarus	177.05	0.25	0.25	10.33	29.37	0.82	9.51	27.03
Estonia	10.10	0.63	0.25	0.10	10.83	0.02	0.08	8.49
Georgia	69.91		0.80			0.35		
Kazakhstan	350.39		0.80			2.89		
Kyrgyzstan	6.37		0.80			0.04		
Latvia	13.55	0.58	0.25	0.32	12.56	0.06	0.26	10.23
Lithuania	64.59	0.55	0.25	0.30	14.01	0.05	0.25	11.67
Moldova	31.23	0.89	0.25	0.15	2.93	0.12	0.03	0.59
Russia	7724.51	0.16	0.20	264.24	30.89	15.99	248.26	29.02
Tajikistan	4.21		0.80			0.03		
Turkmenistan	437.40		0.80			3.76		
Ukraine	564.33	0.89	0.25	1.44	2.93	1.15	0.29	0.59
Uzbekistan	763.22		0.80			6.94		
Argentina	847.17	0.58	0.25	11.86	12.68	2.18	9.67	10.35
Brazil	521.42	9.99	0.80			5.26		

China	7492.77		0.80		42.89	
Czech Republic	221.25	1.99	0.80		1.23	
Egypt	597.35		0.80		8.18	
India	1437.25	0.18	0.20	23.95	1.55	22.40	26.96
Indonesia	692.30		0.80		6.63	
Mexico	515.06	3.59	0.80	28.83	10.43		
Poland	468.68	0.95	0.25	0.37	0.74	−0.37	−1.17
Saudi Arabia	1099.27		0.80	1.16	13.20		
Slovak Republic	122.57	0.95	0.20	0.09	0.15	−0.06	−0.76
South Africa	79.82		0.80	1.11	0.34		
Venezuela	138.50		0.80		2.44		
Algeria	814.44	0.05	0.20	38.10	1.58	36.52	43.25
Bahrain	98.22	0.22	0.20	7.44	0.53	6.92	24.55
Bangladesh	428.98	0.18	0.20	15.83	1.03	14.80	26.96
Brunei	20.41	0.47	0.20	0.57	0.08	0.49	12.04
Hungary	250.54	0.53	0.20	2.17	0.34	1.83	10.02
Iran	4508.86	0.01	0.20	241.50	7.32	234.18	59.78

(continued)

Table 9.8 (continued)

Countries	Domestic consumption—natural gas 2019, petajoule	Domestic price to border price—gas	Own price elasticity—gas	Emission reduction—gas (million tons, CO_2 eq)	Emission reduction to total emission—gas (%)	Emission increase from regional price's fall—gas (million tons CO_2 eq)	Net emission reduction—gas (million tons, CO_2 eq)	Net emission reduction to total emission—gas (%)
South Korea	912.43	0.98	0.20	0.46	0.42	2.06	−1.60	−1.45
Libya	105.30	0.05	0.20	8.00	45.12	0.33	7.66	43.25
Luxembourg	26.09	0.56	0.20	0.18	10.95	0.03	0.15	9.08
Malaysia	823.44	0.47	0.20	8.84	13.91	1.19	7.65	12.04
Pakistan	801.44	0.18	0.20	17.78	28.83	1.15	16.63	26.96
Slovenia	24.95	0.80	0.20	0.08	4.43	0.03	0.04	2.56
Taipei	166.98	0.46	0.20	6.61	14.21	0.87	5.74	12.34
Turkeiye	1082.24	0.29	0.20	18.58	21.91	1.59	17.00	20.04
Sample—subsidizing countries				686.39	31.78	41.5	644.89	29.86
Sample—No subsidizing countries						371.45		
Sample total[a]						412.95	273.44	

[a]The sample includes data for 87 countries
Source Authors' Calculations

we estimate general equilibrium effects under the assumption of a fully integrated world coal market with no domestic protection, we keep in mind that the increase in world demand that results from a decline in world coal prices may be reduced by domestic protection.

Assuming a weighted average supply elasticity of 0.5 and a demand elasticity of 0.8 in non-subsidizing countries, Eq. (9.6) implies a 20% fall in coal prices. Again, the demand elasticity in the subsidizing countries is the same as discussed earlier. Increases in coal consumption, or carbon emissions, due to the fall in world coal prices are estimated by Eq. (9.7) and presented in Table 9.9 for each of the subsidizing countries. Thus, net world emission reductions are 10.5% if world price effects are incorporated, compared to 20.8% if world price effects are ignored.

Aggregate effects in oil, natural gas, and coal markets: The aggregate effect of changes in all three markets would be to reduce emissions in subsidizing countries by 14.7%, assuming unchanged world prices. Accounting for reductions in world prices, emission reductions in subsidizing countries would be 9.12%.

Emissions Reductions from Removing Subsidies on Electricity: Emissions reductions from the removal of energy subsidies on electricity consumption are reported in Table 9.11. This takes into the fact that electricity production utilizes oil, natural gas, and coal. Electricity subsidy removal in the sample subsidizing countries results in 20.7% reduction in global carbon emissions with no world price effects (partial equilibrium) and 15.1% reductions in the same taking into consideration world price effects (general equilibrium).

9.4 Welfare Costs of Fossil Fuel Subsidies

In the long-run, removing fossil fuel subsidies will improve welfare, assuming no changes in world prices. If subsidies are removed and world prices do fall, the welfare of fossil fuel exporters may decline. The model used here to estimate welfare effects is limited to changes in the fossil fuel markets and ignores effects from potential changes in the relative prices of other goods. Changes in welfare are first estimated on the assumption of constant world prices. (If only a single "small country" eliminated subsidies, this assumption would be realistic.) Welfare effects are then estimated for subsidizing and non-subsidizing countries taking into account

Table 9.9 Greenhouse gas emission reductions from removing subsidies on fossil fuels—coal (2019)

Countries	Domestic consumption—coal 2019, petajoule	Domestic price to border price—coal	Own price elasticity—coal	Emission reduction—coal (million tons, CO_2 eq)	Emission reduction to total emission—coal (%)	Emission increase from regional price's fall—coal (million tons, CO_2 eq)	Net emission reduction—coal (million tonnes, CO_2 eq)	Net emission reduction to total emission—coal (%)
Armenia	0.11			0.00	5.70	0.00	0.00	0.64
Azerbaijan		0.79	0.25		17.11			
Belarus	31.29	0.79	0.80	0.20	5.70	0.18	0.02	0.64
Estonia	5.34	0.79	0.25	0.33	5.70	0.29	0.04	0.64
Georgia	10.09	0.79	0.25	0.06	5.70	0.05	0.01	0.64
Kazakhstan	377.14	0.71	0.25	10.62	8.11	6.62	4.00	3.06
Kyrgyzstan	23.92	0.71	0.25	0.35	8.11	0.22	0.13	3.06
Latvia	1.54	0.79	0.25	0.01	5.70	0.01	0.00	0.64
Lithuania	7.93	0.79	0.25	0.05	5.70	0.04	0.01	0.64
Moldova	4.26	0.79	0.25	0.02	5.70	0.02	0.00	0.64
Russia	1297.79	0.79	0.20	19.85	4.58	17.53	2.33	0.54
Tajikistan	27.02	0.71	0.25	0.35	8.11	0.22	0.13	3.06
Turkmenistan		0.71	0.80		23.72			
Ukraine	264.44	0.79	0.25	5.24	5.70	4.65	0.59	0.64
Uzbekistan	32.40	0.71	0.25	0.69	8.11	0.43	0.26	3.06
Argentina	15.33	1.34	0.80			0.64		
Brazil	303.74	1.34	0.80			9.88		
China	24,040.58	0.60	0.60	2087.75	26.38	960.85	1126.90	14.24
Czech Republic	87.54	0.79	0.25	3.14	5.70	2.79	0.35	0.64

Egypt	106.57	1.00	0.30	0.00	0.01	0.65	−0.64	−6.06
India	4474.64	0.65	0.60	368.74	22.50	198.96	169.78	10.36
Indonesia	1027.82	0.67	0.30	31.86	11.48	16.85	15.01	5.41
Mexico	105.15	1.34	0.80			6.92		
Poland	399.70	0.79	0.20	7.93	4.58	7.00	0.93	0.54
Saudi Arabia		0.79	0.80		17.11			
Slovak Republic	34.91	0.79	0.25	0.58	5.70	0.52	0.07	0.64
South Africa	750.65	0.99	0.25	1.07	0.30	18.09	−17.02	−4.76
Venezuela		1.34	0.80			0.03		
Algeria	3.58	1.00	0.30	0.00	0.01	0.03	−0.03	−6.06
Australia	125.64	0.90	0.30	4.95	2.98	10.09	−5.14	−3.09
Austria	18.44	0.83	0.30	0.59	5.49	0.65	−0.06	−0.58
Bangladesh	153.22	0.65	0.30	1.89	11.97	0.96	0.93	5.90
Belgium	55.65	0.83	0.30	0.63	5.49	0.69	−0.07	−0.58
Bulgaria	15.11	0.79	0.30	1.39	6.80	1.24	0.15	0.73
Denmark	4.55	0.79	0.30	0.25	6.80	0.22	0.03	0.73
Finland	24.01	0.79	0.30	1.01	6.80	0.90	0.11	0.73
France	51.03	0.83	0.30	1.58	5.49	1.75	−0.17	−0.58
Germany	257.47	0.83	0.30	12.03	5.49	13.31	−1.28	−0.58

(continued)

Table 9.9 (continued)

Countries	Domestic consumption—coal 2019, petajoule	Domestic price to border price—coal	Own price elasticity—coal	Emission reduction—coal (million tons, CO_2 eq)	Emission reduction to total emission—coal (%)	Emission increase from regional price's fall—coal (million tons, CO_2 eq)	Net emission reduction—coal (million tonnes, CO_2 eq)	Net emission reduction to total emission—coal (%)
Greece	8.30	0.79	0.30	0.94	6.80	0.84	0.10	0.73
Hungary	10.20	0.79	0.30	0.48	6.80	0.43	0.05	0.73
Iceland	4.54	0.79	0.30		6.80			
Iran	32.48	0.78	0.30	0.39	7.31	0.32	0.07	1.24
Ireland	17.50	0.79	0.30	0.30	6.80	0.27	0.03	0.73
Israel	0.74	0.67	0.30	2.27	11.48	1.20	1.07	5.41
Italy	33.59	0.79	0.30	1.75	6.80	1.57	0.19	0.73
Japan	870.06	0.60	0.30	60.50	14.20	25.87	34.63	8.13
South Korea	331.43	0.60	0.30	42.72	14.20	18.27	24.46	8.13
Luxembourg	1.78	0.83	0.30	0.01	5.49	0.01	−0.00	−0.58
Malaysia	78.99	0.67	0.30	10.62	11.48	5.62	5.01	5.41
Morocco	0.83	1.00	0.30	0.00	0.01	1.61	−1.61	−6.06
Netherlands	36.60	0.83	0.30	1.38	5.49	1.53	−0.15	−0.58
New Zealand	27.64	0.90	0.30	0.18	2.98	0.37	−0.19	−3.09
Nigeria	1.41	0.99	0.30	0.00	0.36	0.01	−0.01	−5.71
Norway	25.01	0.79	0.30	0.20	6.80	0.18	0.02	0.73
Pakistan	370.68	0.65	0.30	5.89	11.97	2.99	2.90	5.90
Portugal	0.45	0.79	0.30	0.34	6.80	0.30	0.04	0.73
Romania	25.35	0.79	0.30	1.45	6.80	1.29	0.15	0.73

Singapore	7.70	0.67	0.30	0.21	11.48	0.11	0.10	5.41
Slovenia	2.06	0.79	0.30	0.30	6.80	0.26	0.03	0.73
Spain	20.22	0.79	0.30	1.25	6.80	1.11	0.13	0.73
Sweden	29.30	0.79	0.30	0.44	6.80	0.39	0.05	0.73
Switzerland	3.80	0.83	0.30	0.02	5.49	0.02	−0.00	−0.58
Taipei	222.98	0.60	0.30	21.04	14.20	9.00	12.04	8.13
Thailand	327.65	0.67	0.30	7.59	11.48	4.01	3.58	5.41
Turkeiye	452.79	0.79	0.30	11.21	6.80	10.01	1.20	0.73
United Arab Emirates	74.41	0.79	0.30	0.48	6.80	0.43	0.05	0.73
United Kingdom	84.81	0.79	0.30	1.56	6.80	1.40	0.17	0.73
Vietnam	683.07	0.67	0.30	23.47	11.48	12.41	11.06	5.41
Sample—subsidizing countries				2762.20	20.77	1368.8	1393.40	10.48
Sample—No subsidizing countries						208.65		
Sample total[a]						1577.44	1184.76	

[a]The sample includes data for 87 countries

Source Authors' Calculations

Table 9.10 Greenhouse gas emission reductions from removing subsidies on fossil fuels—total (oil, coal, and natural gas)

Countries	Emission reduction (million tons CO_2eq)—no world price effect	Emission reduction (%)—no world price effect to total subsidizing countries	Emission reduction (million tons CO_2 eq)—world price effect	Emission reduction (%)—world price effect, to total subsidizing countries
Armenia	0.00	0.01	0	0
Azerbaijan	8.64	25.17	8.02	23.37
Belarus	10.53	18.76	9.53	16.98
Estonia	0.42	4.16	0.11	1.11
Georgia	0.06	0.60	0.01	0.07
Kazakhstan	18.45	8.88	11.57	5.57
Kyrgyzstan	0.69	7.25	0.44	4.62
Latvia	0.33	4.47	0.26	3.55
Lithuania	0.35	2.98	0.25	2.19
Moldova	0.17	2.03	0.03	0.38
Russia	298.29	18.06	263.04	15.92
Tajikistan	0.36	4.71	0.12	1.62
Turkmenistan	3.99	5.74	3.86	5.55
Ukraine	6.68	3.88	0.88	0.51
Uzbekistan	0.69	0.62	0.26	0.23
Argentina	14.63	8.91	12.00	7.30
Brazil				
China	2104.72	21.08	1134.11	11.36
Czech Republic	3.14	3.27	0.35	0.37
Egypt	19.20	8.44	17.83	7.84
India	392.69	16.56	192.18	8.10
Indonesia	70.00	11.74	51.64	8.66

Mexico	1.13	0.26	−0.12
Poland	8.30	2.83	0.19
Saudi Arabia	49.51	9.91	9.39
Slovak Republic	0.67	2.23	0.01
South Africa	1.07	0.24	−3.87
Venezuela	32.03	35.11	34.77
Algeria	53.46	37.24	35.84
Angola	3.84	18.67	18.12
Australia	4.95	1.29	−1.33
Austria	0.59	0.91	−0.10
Bahrain	8.51	26.01	24.31
Bangladesh	18.37	19.81	17.53
Belgium	0.63	0.69	−0.07
Bolivia	1.95	9.17	8.78
Brunei	1.11	16.77	13.83
Bulgaria	1.39	3.58	0.38
Colombia	7.23	9.36	9.01
Denmark	0.25	0.84	0.09
Ecuador	4.35	12.09	11.44
Finland	1.01	2.42	0.26
France	1.58	0.53	−0.06
Gabon	0.05	1.62	1.29

(continued)

Table 9.10 (continued)

Countries	Emission reduction (million tons CO_2eq)—no world price effect	Emission reduction (%)—no world price effect to total subsidizing countries	Emission reduction (million tons CO_2 eq)—world price effect	Emission reduction (%)—world price effect, to total subsidizing countries
Germany	12.03	1.84	−1.28	−0.20
Greece	0.94	1.63	0.10	0.17
Hungary	2.65	5.72	1.88	4.06
Iran	287.86	48.90	278.92	47.38
Ireland	0.30	0.88	0.03	0.09
Israel	2.27	3.70	1.07	1.74
Italy	1.75	0.56	0.19	0.06
Japan	60.50	5.68	34.63	3.25
South Korea	43.19	7.31	22.86	3.87
Kuwait	12.17	13.52	11.88	13.20
Libya	24.90	53.51	24.37	52.38
Luxembourg	0.19	2.01	0.15	1.57
Malaysia	35.38	14.80	28.02	11.72
Morocco	0.00	0.00	−1.61	−2.40
Netherlands	1.38	0.94	−0.15	−0.10
New Zealand	0.18	0.53	−0.19	−0.55
Nigeria	11.49	8.72	11.06	8.39
Norway	0.20	0.57	0.02	0.06
Oman	2.46	3.54	2.33	3.36
Pakistan	25.57	13.57	20.99	11.14
Portugal	0.34	0.78	0.04	0.08
Qatar	3.08	3.53	2.95	3.38

Romania	1.45	2.01	0.22
Singapore	0.21	0.44	0.21
Slovenia	0.37	2.76	0.56
Spain	1.25	0.53	0.06
Sweden	0.44	1.24	0.13
Switzerland	0.02	0.06	−0.01
Taipei	28.35	10.99	7.01
Thailand	7.59	2.95	1.39
Trinidad and Tobago	0.01	0.05	−0.09
Turkiye	29.79	8.05	4.91
United Arab Emirates	0.48	0.27	0.03
United Kingdom	1.56	0.45	0.05
Vietnam	23.47	8.22	3.87
Sample—subsidizing countries[a]	3781.71	14.69	9.12
		2346.97	
% of sample total[a]		11.92	7.40
% of world total[b]		11.05	6.86

[a]The sample includes data for 87 countries
[b]% of total emissions in Table 9.4. Greenhouse Gas Emissions from Fuel Combustion 2019—Total (Oil, Coal and Natural Gas)
Source Authors' Calculations

Table 9.11 CO_2 emission reductions from removing subsidies on fossil fuels—electricity

Countries	Domestic consumption—electricity 2019, petajoule	Domestic price to border price—electricity	Own price elasticity—electricity
Armenia		0.65	0.80
Azerbaijan	13.21	0.40	0.25
Belarus		0.71	0.55
Estonia	26.34	1.20	0.66
Georgia		0.67	0.76
Kazakhstan	259.04	0.34	0.41
Kyrgyzstan		0.16	0.29
Latvia	23.95	1.51	0.69
Lithuania	37.95	1.62	0.57
Moldova		0.52	0.54
Russia	2720.29	0.71	0.20
Tajikistan		0.41	0.28
Turkmenistan	44.91	0.13	0.59
Ukraine	420.08	0.49	0.44
Uzbekistan	190.79	0.25	0.78
Argentina	452.30	0.28	0.25
Brazil	1849.49	1.27	0.80
China	23,482.14	0.63	0.48
Czech Republic	210.38	1.85	0.73
Egypt	562.82	0.37	0.42
India	4720.04	0.65	0.68
Indonesia	934.44	0.62	0.34
Mexico	1008.27	0.89	0.35
Poland	505.48	1.14	0.56
Saudi Arabia	1058.42	0.42	0.42

Slovak Republic	90.69	1.19	0.49
South Africa	673.69	0.82	0.59
Venezuela	203.98	1.91	0.33
Algeria	225.29	0.25	0.22
Angola	48.00	0.18	0.29
Bahrain	118.02	0.45	0.22
Bangladesh	284.03	0.56	0.23
Brunei	14.06	0.49	0.23
Canada	1906.79	0.77	0.80
Chile	266.75	0.99	0.80
Colombia	243.23	0.98	0.42
Ecuador	92.93	0.65	0.25
Ghana	50.20	0.48	0.80
Hungary	145.14	0.78	0.53
Iceland	65.62	0.72	0.72
Iran	940.92	0.04	0.22
South Korea	1885.74	0.70	0.67
Kuwait	217.12	0.28	0.42
Libya	64.31	0.04	0.24
Malaysia	571.33	0.49	0.23
Morocco	120.42	0.79	0.80
Nigeria	98.19	0.53	0.33
Norway	416.25	0.96	0.77
Oman	121.67	0.63	0.60
Pakistan	408.77	0.65	0.24

(continued)

Table 9.11 (continued)

Countries	Domestic consumption—electricity 2019, petajoule	Domestic price to border price—electricity	Own price elasticity—electricity
Qatar	157.18	0.25	0.54
Sri Lanka	53.06	0.28	0.80
Taipei	886.28	0.77	0.25
Thailand	695.43	0.77	0.74
Trinidad and Tobago	30.34	0.38	0.75
Turkeiye	911.55	0.38	0.53
United Arab Emirates	452.58	0.65	0.78
Vietnam	753.24	0.56	0.59
Sample—subsidizing countries			
Sample—No subsidizing countries			
Sample total[a]			

Countries	Emission reduction—electricity (million tons, CO_2)	Emission reduction to total emission—electricity (%)	Emission increase from regional price's fall—electricity (million tons, CO_2)	Net emission reduction—electricity (million tons, CO_2)	Net emission reduction to total emission—electricity (%)
Armenia	0.39	29.14	0.13	0.27	19.88
Azerbaijan	2.50	20.26	0.36	2.15	17.36
Belarus	4.92	16.99	1.83	3.08	10.66
Estonia			0.45		
Georgia	0.35	26.11	0.12	0.23	17.30
Kazakhstan	37.49	35.72	5.02	32.46	30.93

Kyrgyzstan	0.87	41.21	0.80	37.86
Latvia			0.07	
Lithuania			0.15	
Moldova	1.00	29.88	0.06	23.59
Russia	53.77	6.59	0.21	4.27
Tajikistan	0.41	22.36	18.91	19.11
Turkmenistan	14.51	69.94	0.06	63.07
Ukraine	21.45	26.71	1.43	21.59
Uzbekistan	26.46	66.62	4.11	57.57
Argentina	11.12	27.82	3.60	24.87
Brazil			1.18	
China	1038.52	19.83	6.05	14.25
Czech Republic			292.08	
Egypt	33.49	34.27	4.06	29.35
India	297.94	25.40	4.80	17.52
Indonesia	33.90	15.08	92.40	11.11
Mexico	5.33	3.90	8.94	−0.12
Poland			5.50	
Saudi Arabia	72.01	30.34	8.94	25.53
Slovak Republic			11.42	
South Africa	25.58	11.09	0.33	4.25
Venezuela			15.78	
Algeria	10.58	26.83	1.01	24.22
Angola	1.82	39.39	1.03	36.00
Bahrain	3.72	16.16	0.16	13.59

(continued)

Table 9.11 (continued)

Countries	Emission reduction—electricity (million tons, CO_2)	Emission reduction to total emission—electricity (%)	Emission increase from regional price's fall—electricity (million tons, CO_2)	Net emission reduction—electricity (million tons, CO_2)	Net emission reduction to total emission—electricity (%)
Bangladesh	5.02	12.75	1.06	3.96	10.06
Brunei	0.51	14.95	0.09	0.42	12.28
Canada	16.31	19.26	7.85	8.46	9.99
Chile	0.32	0.87	3.46	−3.14	−8.40
Colombia	0.14	0.92	0.75	−0.61	−3.95
Ecuador	0.49	10.16	0.14	0.35	7.25
Ghana	2.59	44.40	0.54	2.05	35.13
Hungary	1.35	12.57	0.66	0.69	6.43
Iceland		20.87			12.54
Iran	82.35	51.84	4.05	78.30	49.29
South Korea	66.10	20.85	24.47	41.63	13.13
Kuwait	18.68	40.93	2.20	16.48	36.12
Libya	11.76	54.37	0.59	11.17	51.64
Malaysia	17.58	15.10	3.14	14.44	12.41
Morocco	5.00	17.42	2.66	2.34	8.16
Nigeria	2.43	18.91	0.50	1.93	15.03
Norway	0.06	2.78	0.18	−0.12	−6.10
Oman	3.74	24.33	1.07	2.67	17.38
Pakistan	4.56	9.79	1.29	3.27	7.02
Qatar	12.69	53.15	1.49	11.20	46.89
Sri Lanka	6.30	63.72	0.92	5.38	54.45
Taipei	10.20	6.39	4.65	5.55	3.48
Thailand	15.84	18.03	7.57	8.27	9.42
Trinidad and Tobago	2.67	51.56	0.45	2.22	42.90

Turkeiye	53.84	40.25	8.15	45.69
United Arab Emirates	19.75	28.27	6.32	13.43
Vietnam	44.07	28.54	10.47	33.59
Sample—subsidizing countries	2102.49	20.67	564.45	1538.04
Sample—No subsidizing countries			308.69	
Sample total[a]			873.14	1229.35

34.16	
19.22	
21.76	
15.12	

[a]The sample includes data for 87 countries
Source Authors' Calculations

the impacts on world prices estimated in the previous section. Welfare calculations are based on consumer and producer surpluses of fossil fuel consumption and production, an approach which assumes full employment of resources, and should therefore be considered a long-run approximation. Demand and supply elasticities are as presented earlier.

Case 1. No change in world prices:

Welfare measured as the sum of consumer and producer surplus at subsidized fossil fuel prices

$$W_p = \int_0^{q_1} D\delta q - \int_0^{x_1} S\delta x + p_w(x_1 - q_1) \tag{9.9}$$

where D is the inverse demand function, S is the inverse supply function, q_1 is domestic consumption at subsidized price p, x_1, is domestic production at price p, and p_w world price. Welfare at non-subsidized world price (p_w) is,

$$W_{p_w} = \int_0^{q_w} D\delta q - \int_0^{x_w} S\delta x + p_w(x_w - q_w) \tag{9.10}$$

where q_w is domestic consumption and x_w is domestic production at world price p_w. Thus, change in welfare from subsidy removal is,

$$\Delta W = W_{p_v} - W_p = -\int_{q_w}^{q_1} D\delta q + p_w(q_1 - q_w)$$

$$- \int_{x_1}^{x_w} S\delta x + p_w(x_w - x_1) > 0 \tag{9.11}$$

This welfare effect is illustrated in Fig. 9.4 for fossil fuel exporters and importers, with the shaded area (+) representing welfare gain. Approximating ΔW by assuming linear demand and supply functions in the relevant range gives,

$$\Delta W = 0.5(q_1 - q_w)(p_w - p) \\ + 0.5(x_w - x_1)(p_w - p)$$

$$\Delta W = 0.5\{(q_1 - q_w)/q_1\}\{(p_w - p)q_1\}$$
$$+ 0.5\{(x_w - x_1)/x_1\}\{(p_w - p)x_1\} \quad (9.12)$$

where the first factor in the first term is percentage change in consumption from subsidy removal, the second factor in the first term is total subsidies, and the first factor in the second term is percentage change in production from subsidy removal. The latter is non-zero if domestic prices are below world prices due to either price ceilings and/or producer subsidies. We assume that the last term is zero—i.e., that subsidy removal engenders no supply response in a subsidizing country.

Total welfare gains in subsidizing countries from removing fossil fuel subsidies are more than US$172 billion. Welfare gains are largest for the Russian Federation at approximately US$26 billion (Table 9.12), a figure which amounts to 15% of world welfare gains with no world price effects. China and India register the second and third largest welfare gains. Welfare gains by fuel are largest in aggregate for oil, natural gas, and coal followed by electricity.

Case II. Change in world prices:

We assume that all subsidizing countries remove subsidies in the same time period. Thus, welfare at subsidized prices (p) is as given by Eq. (9.9) since there is no change in world prices before subsidy removal. When subsidies are removed, world prices fall from p_w to $p_{w'}$ and domestic prices are adjusted to $p_{w'}$. Welfare at non-subsidized prices, $p_{w'}$, is,

$$W_{p_{w'}} = \int_0^{q_{w'}} D\delta q - \int_0^{x_{w'}} S\delta x + p_{w'}(x_{w'} - q_{w'}) \quad (9.13)$$

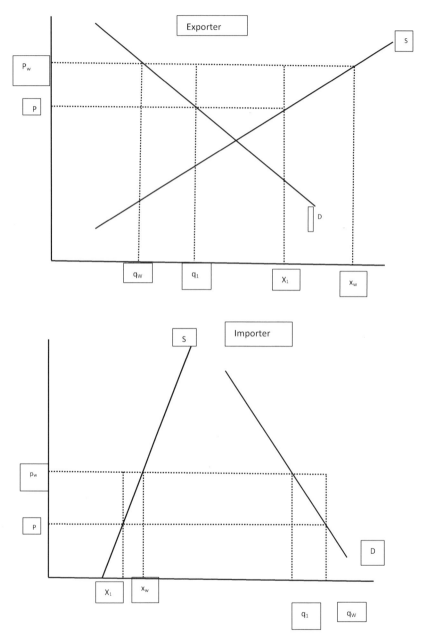

Fig. 9.4 Welfare gain from subsidy removal

Table 9.12 Welfare impacts of subsidy removal

Countries	Welfare gains—no world price effect, million US$—oil, coal and natural gas	Welfare gains—world price effect, million US$—oil, coal and natural gas	Welfare Gains—no world price effect, million US$—electricity
Armenia	0.00	0.11	
Azerbaijan	517.50	881.06	28.95
Belarus	309.19	517.75	
Estonia	3.32	−117.39	
Georgia	0.28	9.35	
Kazakhstan	877.66	−249.60	1104.07
Kyrgyzstan	19.13	0.11	
Latvia	5.63	24.90	
Lithuania	32.60	123.94	
Moldova	0.93	46.10	
Russia	16,652.14	27,254.53	940.22
Tajikistan	1.53	−21.56	
Turkmenistan	499.99	679.19	495.38
Ukraine	22.05	−498.01	1029.85
Uzbekistan	1.78	−25.94	1741.74
Argentina	179.58	732.11	1650.36
China	7264.45	−38,378.46	31,136.71
Czech Republic	2.45	−451.53	
Egypt	1806.57	2756.30	2200.46
India	3647.55	−3450.55	7625.12
Indonesia	4119.49	−7354.39	970.33
Mexico	3.80	2387.83	77.34
Poland	10.98	−913.04	
Saudi Arabia	3639.57	6247.19	3390.00

(continued)

Table 9.12 (continued)

Countries	Welfare gains—no world price effect, million US$—oil, coal and natural gas	Welfare gains—world price effect, million US$—oil, coal and natural gas	Welfare Gains—no world price effect, million US$—electricity
Slovak Republic	1.56	193.86	
South Africa	0.06	−4949.58	244.87
Venezuela	4230.37	4498.02	
Algeria	4959.93	10,944.15	824.25
Angola	410.77	528.07	281.41
Australia	1.00	−13,970.23	
Austria	0.41	17.55	
Bahrain	334.01	421.56	190.03
Bangladesh	829.60	1686.81	292.15
Belgium	1.23	50.88	
Bolivia	199.70	316.90	
Brunei	63.11	−297.49	19.30
Bulgaria	0.50	−174.37	
Canada	0	0	1565.25
Chile	0	0	0.46
Colombia	873.21	1281.78	0.88
Denmark	0.15	4.33	
Ecuador	326.26	631.75	59.46
Finland	0.80	−18.93	
France	1.13	0.02	
Gabon	0.86	11.59	
Germany	5.69	−886.33	
Ghana	0	0	210.45
Greece	0.28	−115.05	

Hungary	110.64	348.76	74.17
Iceland	0.15	0	69.14
Iran	28,853.11	57,545.54	8539.06
Ireland	0.58	1.09	
Israel	0.08	−0.92	
Italy	1.12	31.96	
Japan	137.84	1002.26	
South Korea	53.14	1707.25	2113.85
Kuwait	1247.55	1492.19	1159.68
Libya	2767.45	3712.61	612.17
Luxembourg	9.96	46.68	
Malaysia	2554.73	1778.56	792.15
Morocco	0.00	0.81	81.16
Netherlands	0.81	34.83	
New Zealand	0.22	−53.67	
Nigeria	1344.40	1946.48	157.02
Norway	0.84	20.55	7.58
Oman	267.77	500.66	199.91
Pakistan	1572.58	3469.88	254.47
Portugal	0.02	0.43	
Qatar	436.26	669.77	1144.78
Romania	0.85	−132.30	
Singapore	0.82	8.85	
Slovenia	1.84	4.01	

(continued)

Table 9.12 (continued)

Countries	Welfare gains—no world price effect, million US$—oil, coal and natural gas	Welfare gains—world price effect, million US$—oil, coal and natural gas	Welfare Gains—no world price effect, million US$—electricity
Spain	0.68	19.24	
Sri Lanka	0	0	441.12
Sweden	0.98	23.65	
Switzerland	0.08	3.62	
Taipei	147.98	1854.54	237.75
Thailand	35.07	212.53	534.42
Trinidad and Tobago	0.02	38.47	176.46
Turkiye	1343.54	2380.68	4162.25
United Arab Emirates	2.49	70.79	805.23
United Kingdom	2.83	25.43	
Vietnam	73.11	−406.81	1705.33
Sample—subsidizing countries	92,828.32	68,733.72	79,346.70
Sample—No subsidizing countries		30,008.66	
Sample total[a]		98,742.38	

[a]The sample includes data for 87 countries
Source Authors' Calculations

where $q_{w'}$ is domestic consumption and $x_{w'}$ is production at new world prices $p_{w'}$. Change in welfare from subsidy removal is by linear approximation of D and S,

$$\begin{aligned}\Delta W = W_{pw'} - W_p =\ &0.5\{(q_1 - q_{w'})/q_1\}\{(p_{w'} - p)q_1\} \\ &+ 0.5\{(x_{w'} - x_1)/x_1\}\{(p_{w'} - p)x_1\} \\ &+ (p_{w'} - p_w)(x_1 - q_1)\end{aligned} \quad (9.14)$$

The two first terms are like those in case 1, but with new world prices and corresponding quantities of consumption and production. The last term represents an exporting country's welfare loss from lower world prices, or an importer's welfare gain from lower prices. Welfare gains (+) and losses (−) are illustrated by shaded area in Fig. 9.5, for both exporters and importers. Welfare gains or losses resulting from the import or export effects of changed world fossil fuel prices are based on border prices for crude oil, natural gas, and coal, and not on end-user prices. If a country produces no fossil fuels, then the second term is zero and $x_1 = 0$.

Reduced world prices for fossil fuels do not leave welfare in non-subsidizing countries unaffected. Welfare change by linear approximation of D and S is,

$$\begin{aligned}\Delta W = W_{pw'} - W_{pw} =\ &0.5(q_{w'} - q_w)(p_w - p_{w'}) \\ &+ 0.5(x_w - x_{w'})(p_w - p_{w'}) \\ &+ (p_{w'} - p_w)(x_w - q_w)\end{aligned} \quad (9.15)$$

Or

$$\begin{aligned}\Delta W = W_{pw'} - W_{pw} =\ &0.5 e_D (\partial p_w/p_w)^2 q_w p_w \\ &+ 0.5 e_s (\partial p_w/p_w)^2 x_w p_w \\ &+ (p_{w'} - p_w)(x_w - q_w)\end{aligned} \quad (9.16)$$

where e_D and e_S are the absolute values of demand and supply elasticities.

Table 9.12 reports these calculations for the sample subsidizing countries. These calculations assume full employment of resources. In fact, subsidy removal may have significant short-run adjustment costs. It may not therefore be politically acceptable over a short time horizon unless some external inducement for subsidy removal is provided to subsidizing countries. This issue is considered in the next section.

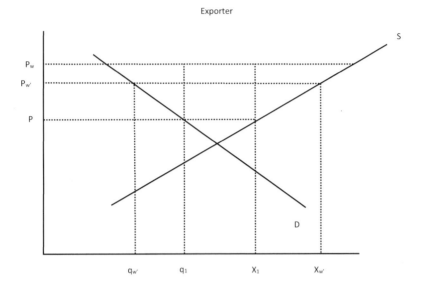

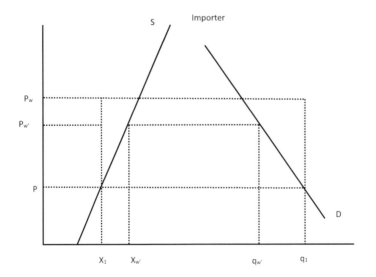

Fig. 9.5 Welfare effect of subsidy removal

9.5 POTENTIAL FOREIGN INDUCEMENT FOR REMOVAL OF SUBSIDIES

Suppose OECD countries decide to reduce carbon emissions from fossil fuel consumption by some percentage below current levels. This may be achieved in several ways, one of which is to impose a carbon tax in the OECD countries. An alternative is to achieve equivalent reductions by paying countries that subsidize fossil fuels to remove such subsidies. While removing such subsidies would improve welfare in the long run, there might be short-run adjustment costs and distributional consequences; without compensation subsidy removal is unlikely to politically acceptable. It is, therefore, of some interest to determine: first, the level of OECD carbon taxes needed to achieve emission reductions equivalent to those achieved by subsidy removal; second, OECD willingness to "buy" equivalent reductions from subsidizing countries.

Estimated carbon emission reductions from subsidy removal are lowered if world price constancy is not assumed. Similarly, if OECD countries unilaterally reduce carbon emissions, we may expect emission increases in non-OECD countries in response to reduced world prices. Furthermore, the effect on carbon emissions per dollar of carbon tax in any one OECD country will be decreased if such a tax is also imposed in all OECD countries, since world prices may fall. The net global effect of an OECD carbon tax on carbon emissions from fuel use is,

$$\partial q = \left\{ e^S / \left(e^S + e_1^D [q_1/q] + e_2^D [q_2/q] \right) \right\} \partial a^D \qquad (9.17)$$

derived from (9.4) and (9.5), with

$$\partial a^D = \partial q_2 = \left(1 - (p/p_t)^e \right) q_2 \qquad (9.18)$$

where p and p_t are weighted average fossil fuel prices in OECD countries before and after carbon tax, respectively, and e is price elasticity of demand (e_2^D) as in Eq. (9.3). Table 9.13 presents a range of three cases in which price elasticity of fossil fuel demand in OECD countries varies from 0.6, through 0.8, to 1.0; supply elasticities for all countries and demand elasticities for all non-OECD countries are the same as in the earlier section. Using Eq. (9.8), and thus accounting for world price effects, world emission reductions from subsidy removal are estimated for each of the demand elasticities of non-subsidizing (primarily OECD) countries. Next, a carbon tax is estimated that provides a value for p/p_t in (9.18) such that the value of ∂q in (9.17) is equal to world emission reductions resulting from subsidy removal. A range for ∂q is presented in Table 9.13. The lower bound assumes a demand effect in non-OECD countries from lower world prices, which would be the case if lower world prices were to translate into lower end-user prices. The upper bound assumes no demand effect in non-OECD countries, which would be the case if prices were fixed in subsidizing countries: lower world prices would not translate via the market to lower end-user prices, and there would be no demand effect in those countries. To achieve as substantial a reduction in emissions worldwide as subsidy removal, a carbon tax would need to be in the range of US$155–341 (see Table 9.13). Total emission reductions in OECD countries are about 19% assuming no world price effects. Nordhaus (1991), using a survey of cost estimates of carbon reductions in several countries and regions, derives a marginal cost curve according to which a US$60 carbon tax would reduce emissions by 20%. A demand elasticity of 0.8 is therefore quite consistent with Nordhaus' marginal cost curve. Estimations presented in Table 9.13 suggest that a substantial carbon tax is necessary to reduce emissions by the same amount as subsidy removal. But, subsidy removal may not be politically acceptable in the short run without some form of external compensation for adjustment costs. OECD countries might therefore consider such compensation a lower cost strategy for reducing emissions than the relatively high carbon tax estimated in this section.

9.6 Summary and Conclusions

Substantial fossil fuel subsidies prevail in a handful of large carbon emitting countries. Total subsidies in sample countries are estimated to be more than US$1.38 trillion. Removing such subsidies would substantially reduce national carbon emissions in some countries and reduce global carbon emissions by 15%, assuming no change in world prices, and by 9%, accounting for changes in world prices. Welfare gains from subsidy removal worldwide would be more than US$172 billion assuming no change in world prices, even ignoring the benefits from curtailment of greenhouse gases emissions and abatement of local pollution. Total welfare gains from removing fossil fuel subsidies when accounting for world price changes would still be some US$148 billion in subsidizing countries. Net fossil fuel importers in Western Europe, United States, and Japan would experience a welfare gain of approximately US$30 billion in

Table 9.13 Carbon tax equivalent to subsidy removal: oil, gas, coal and electricity sectors

(a) Carbon tax in OECD countries[a]—oil, coal, and natural gas

	Case 1	Case 2	Case 3
Elasticity of demand in OECD	0.6	0.8	1.0
World emission reduction from subsidy removal (million tons)	1332.12	1228.81	1119.51
OECD carbon tax (US $ / PJ) for equivalent world emission reduction accounting for world price effect	9,662,825	6,934,491	5,285,650
World emission reductions from OECD carbon tax accounting for world price effect (million tons)	1315.79–1348.86	1213.11–1244.92	1104.78–1134.64
OECD reductions assuming no world price effects (%)	19.13	19.17	18.85

(b) Tax converted to equivalent units under alternate demand elasticity scenarios

	US$/PJ	US$/ton of oil	US$/ton of coal	US$/million BTU
Case 1 e = 0.6	9,662,825.00	341.08	283.20	10.20
Case 2 e = 0.8	6,934,491.00	244.78	203.24	7.32
Case 3 e = 1	5,285,650.00	186.57	154.91	5.58

(continued)

Table 9.13 (continued)

	US$/PJ US$/ton of oil	US$/ton of coal		US$/million BTU
		Case 1	Case 2	Case 2
(c) Carbon tax in OECD countries[a]—electricity				
Elasticity of demand in OECD		0.6	0.8	1.0
World emission reduction from subsidy removal (million tons)		886.41	801.43	711.55
OECD carbon tax (US$/kwh) for equivalent world emission reduction accounting for world price effect		0.10	0.07	0.05
World emission reductions from OECD carbon tax accounting for world priceeffect (million tons)		759.80–915.73	678.80–830.17	596.95–738.68
OECD reductions assuming no world price effects (%)		26.22	25.42	24.11

[a]The sample includes data for 87 countries (39 OECD countries)
Source Authors' calculations

the event of subsidy removal dampening world energy prices. Equivalent reductions in carbon emissions could be achieved by an OECD carbon tax on the order of US$155–341 per ton. It should be noted that neither the subsidy removal nor an equivalent carbon tax would be sufficient to stabilize global carbon emissions at 2025 levels. To achieve that objective, stronger economic policy responses would be required.

Annex

See Tables 9.14 and 9.15.

Table 9.14 CO_2 emissions from fuel combustion by sector[a] 2019—total

Countries	Million tonnes of CO2	Percentage of world emissions	Cumulative percentage
China	9876.50	29.38	29.38
United States	4744.45	14.11	43.49
India	2309.98	6.87	50.36
Russia	1640.33	4.88	55.24
Japan	1056.19	3.14	58.38
Germany	644.11	1.92	60.29
South Korea	585.71	1.74	62.04
Iran	583.52	1.74	63.77
Indonesia	583.41	1.74	65.51
Canada	571.01	1.70	67.20
Saudi Arabia	495.15	1.47	68.68
South Africa	433.57	1.29	69.97
Mexico	419.39	1.25	71.21
Brazil	410.99	1.22	72.44
Australia	380.74	1.13	73.57
Turkeiye	366.42	1.09	74.66
United Kingdom	342.24	1.02	75.68
Italy	309.32	0.92	76.60
France	293.87	0.87	77.47
Poland	287.38	0.85	78.33

(continued)

Table 9.14 (continued)

Countries	Million tonnes of CO2	Percentage of world emissions	Cumulative percentage
Viet Nam	282.28	0.84	79.17
Taipei	255.96	0.76	79.93
World[b]	33,621.53	100	
Non-OECD total	21,001.24	62.46	
OECD total	11,317.68	33.66	

[a]Sectors include electric and heat production, other energy industry own use, manufacture industries and construction, transport, residential, commercial and public services
[b]Data for over 190 countries and regions
Source International Energy Agency. (2021). Greenhouse Gas Emissions from Energy Highlight https://www.iea.org/data-and-statistics/data-product/greenhouse-gas-emissions-from-energy-highlights#overview

Table 9.15 IEA—fossil fuel subsidies 2019, real 2020 million USD

Countries	Oil	Coal	Gas	Electricity	Total
Argentina	2709.08		878.13	59.12	3646.32
Armenia					
Azerbaijan	834.49		517.25	455.13	1806.86
Belarus					
Brazil					
China	18,840.34			15,168.74	34,009.08
Czech Republic					
Egypt	9830.66		53.86	6683.15	16,567.67
Estonia					
Georgia					
India	21,126.68		765.14	11,135.52	33,027.34
Indonesia	15,303.35				15,303.35
Kazakhstan	2632.99	1831.88	316.63	911.57	5693.07
Kyrgyzstan					
Latvia					
Lithuania					
Mexico	18.53			2886.19	2904.72
Poland					

(continued)

Table 9.15 (continued)

Countries	Oil	Coal	Gas	Electricity	Total
Moldova					
Russia			10,203.28	8475.63	18,678.91
Saudi Arabia	15,254.34		4633.33	5321.33	25,209.00
Slovak Republic					
South Africa					
Tajikistan					
Turkmenistan	963.71		1754.63	269.70	2988.04
Ukraine				2145.72	2145.72
Uzbekistan	748.39		3024.18	1471.25	5243.82
Venezuela	3776.35		667.62	3921.79	8365.76
Algeria	7342.68		1934.57	1869.45	11,146.69
Angola	632.44			185.71	818.15
Bahrain	207.07			527.27	734.34
Bangladesh	10.03		692.58	1279.85	1982.47
Bolivia	630.15		41.88		672.03
Brunei	155.87				155.87
Colombia	526.72				526.72
Ecuador	2498.78				2498.78
El Salvador	23.18			316.90	340.08
Gabon	25.83		0.54		26.37
Ghana	115.21		11.70		126.92
Iran	14,785.00		17,039.66	56,119.47	87,944.13
Iraq	5047.65		218.72	1382.90	6649.28
South Korea		53.82			53.82
Kuwait	1016.95		439.14	2335.71	3791.81
Libya	4532.82		179.29	984.67	5696.78
Malaysia	1107.76				1107.76
Nigeria	1654.00			50.97	1704.97
Oman	101.11				101.11
Pakistan	164.88		1414.42		1579.30
Qatar	271.96		0.33	180.24	452.53
Sri Lanka	281.49			81.14	362.63
Taipei	2.47				2.47
Thailand	424.57				424.57
Trinidad and Tobago	438.18			83.53	521.71
United Arab Emirates	139.14		6306.60	463.53	6909.27
Vietnam	0.09	308.78			308.87
Sample total	134,174.94	2194.48	51,093.48	124,766.18	312,229.09

Source International Energy Agency, Fossil Fuel Consumption Subsidies, 2010–2020 https://www.iea.org/data-and-statistics/data-product/fossil-fuel-subsidies-database

References

Bohi, Douglas R. (1981). "Analyzing Demand Behavior—A Study of Energy Elasticities." Resources for the Future. The Johns Hopkins University Press, Baltimore.

Boudekhdekh, Karim (2022). "A Comparative Analysis of Energy Subsidy in the MENA Region." *Economic Insights—Trend and Challenges*, XI(2):37–56.

Churchill, A.A. and Saunders, R.J. (1991). "Global Warming and The Developing World." *Finance and Development* (June).

Hoeller, P. and Wallin, M. (1991). "Energy Prices, Taxes and Carbon Dioxide Emissions." Working Paper No. 106. Economics and Statistics Department. Public Economics Division. OECD, Paris.

International Energy Agency. (2021). Greenhouse Gas Emissions from Energy Highlight. https://www.iea.org/data-and-statistics/data-product/greenhouse-gas-emissions-from-energy-highlights#overview.

International Energy Agency (2022). Fossil Fuel Consumption Subsidies, 2010–2020. https://www.iea.org/data-and-statistics/data-product/fossil-fuel-subsidies-database

Larsen, Bjorn and Shah, Anwar (1992a). "Tradeable Carbon Emissions Permits and International Transfers." Presented at the 15th Annual International Conference of the International Association for Energy Economics, Tours, France, May 18–20.

Larsen, Bjorn and Shah, Anwar (1992b). "Combating the 'Greenhouse Effect.'" *Finance and Development*, 29(4):20–23.

Larsen, Bjorn and Shah, Anwar (1992c). "World Fossil Fuel Subsidies and Global Carbon Emissions." Policy Research Working Paper No. WPS 1002, October, World Bank, Washington, DC.

Larsen, Bjorn and Shah, Anwar (1994a). "Energy Pricing and Taxation Options for Combating the "Greenhouse Effect"." In Akihiro Amano, Brian Fisher, et al. (eds.), *Climate Change: Policy Instruments and Their Implications*. Proceedings of the Tsukuba Workshop of the Intergovernmental Panel on Climate Change Working Group III. Tsukuba, Japan, January 17–20. Cambridge University Press, UK.

Larsen, Bjorn and Shah, Anwar (1994b). "Global Tradable Carbon Permits, Participation Incentives, and Transfers" (with Bjorn Larsen). *Oxford Economic Papers*, 46(0):841–856. http://imagebank.worldbank.org/servlet/WDSContentServer/IW3P/IB/1994b/06/01/000009265_3970716141051/Rendered/PDF/multi_page.pdf.

Larsen, Bjorn and Shah, Anwar (1995). "Global Climate Change, Energy Subsidies and National Carbon Taxes" (with Bjorn Larsen). In Lans Bovenberg and Sijbren Cnossen (eds.), *Public Economics and the Environment in An Imperfect World*. Boston, London, Dordrecht: Kluwer Academic Publishers.

Nordhaus, William D. (1991). "To Slow or Not to Slow: The Economics of The Greenhouse Effect." *The Economic Journal*, 101(July):920–937.

Shah, Anwar and Larsen, Bjorn (1992a). "Carbon Taxes, the Greenhouse Effect and Developing Countries." Working Paper Series No. 957. The World Bank. Washington, DC.

Shah, Anwar and Larsen, Bjorn (1992b). "Global Warming, Carbon Taxes and Developing Countries." Proceedings of the 1992 Annual Meetings of the American Economic Association January 3. New Orleans, USA.

Summers, Lawrence H. (1991). "The Case For Corrective Taxation." *The National Tax Journal*, XLIV(3, September):289–292.

PART V

Combating Corruption

CHAPTER 10

Combating Corruption in the Oil and Gas Sector

Anwar Shah

10.1 Introduction

Quite a few of resource rich countries fare poorly on good governance and economic performance. They experience the highest incidence of corruption in all sectors. They lag in political and economic freedom. They are beset with autocratic regimes with little respect for human rights. On Polity IV ratings on democracy scale of 0–10, the average hydrocarbon country scores 0. In fact, hydrocarbons are statistically significant negative predictor of the level of democracy controlling for income (Siegle, 2007, p. 35). High oil and gas revenues are often associated with poor governance, lack of economic and social development, lack of respect for basic human rights and poverty amid plenty. In Nigeria with all its oil wealth, 90% of its citizens live on less than $2 per day (see also, Lovei and McKechnie, 2000). In some of these countries, benefits of oil

A. Shah (✉)
Brookings Institution, Washington, DC, USA
e-mail: shah.anwar@gmail.com

largesse accrue to a small group of ruling elites and the public is deprived of opportunities for economic and social development (Ehiemua, 2015). They often suffer from "Dutch disease" or "resource curse" with resource wealth being associated with poor economic performance. A growing body of recent literature deals with the evidence and tools to address the resource curse (see DAI, 2007; Chebab et al., 2022; Donwa, 2015; Eriksen and Soreide, 2017; Al-Kasim et al., 2013; Auty, 2007; Gallagher and Rozner, 2007).

This chapter provides a brief overview of the literature as to how corruption contributes to the above malaise. The chapter provides a conceptual framework on corruption. It reviews issues relating to corruption in oil and gas industries. It discusses various global initiatives undertaken to combat corruption in this sector and discusses next steps for the effectiveness of such efforts. A final section draws conclusions from the analysis in this chapter.

10.2 Corruption and Its Drivers: Fundamental Concepts

10.2.1 Fundamental Concepts

Corruption is defined as exercise of official powers against public interest or the abuse of public office for private gains. Public sector corruption is a symptom of failed governance. Here, we define "governance" as the norms, traditions, and institutions by which power and authority in a country is exercised—including the institutions of participation and accountability in governance and mechanisms of citizens' voice and exit and norms and networks of civic engagement; the constitutional-legal framework and the nature of accountability relationships among citizens and governments; the process by which governments are selected, monitored, held accountable, and renewed or replaced; and the legitimacy, credibility, and efficacy of the institutions that govern political, economic, cultural, and social interactions among citizens themselves and their governments.

Concern about corruption—the abuse of public office for private gain—is as old as the history of government. In the 4th Century BCE, Kautaliya, who served as an advisor responded to the Indian King's concerns about corruption by arguing that corruption is inevitable. He wrote that "Just as it is not possible to not to taste honey placed on the surface of the tongue, even so it is not possible for one dealing with the money of the king not to taste the money in however a small quantity." He further advised the king that corruption would be difficult to detect. He stated "Just as fish moving under water cannot possibly be found out either as drinking or not drinking water, so government servants employed in the government work cannot be found out (while) taking money (for themselves)." In 350 B.C.E., Aristotle suggested in *The Politics* that "... to protect the treasury from being defrauded, let all money be issued openly in front of the whole city, and let copies of the accounts be deposited in various wards."

In recent years, concerns about corruption have mounted in tandem with growing evidence of its detrimental impact on growth and development (Amundsen, 2012). As a result of this growing concern, there has been universal condemnation of corrupt practices, leading to the removal of some country leaders. Moreover, many governments and development agencies have devoted substantial resources and energies to fighting corruption in recent years. Even so, it is not yet clear that the incidence of corruption has declined perceptibly, especially in highly corrupt resource rich countries. The lack of significant progress can be attributed to the fact that many programs are simply folk remedies or "one size fits all" approaches and offer little chance of success. For programs to work, they must identify the type of corruption they are targeting and tackle the underlying, country-specific causes, or "drivers," of dysfunctional governance. This calls for examination of structure of government among other factors. For corruption in oil and gas industry, it also requires a rigorous examination of industry relationships with the executive and legislative branches of the state.

10.2.2 *The Many Forms of Corruption in Oil and Gas Industry*

Corruption is not manifested in one single form; indeed, it typically takes at least four broad forms.

Petty, administrative, or bureaucratic corruption. Many corrupt acts are isolated transactions by individual public officials who abuse their office, for example, by demanding bribes and kickbacks, diverting public funds, or awarding favors in return for personal considerations. Such acts are often referred to as petty corruption even though, in the aggregate, a substantial number/amount of public resources may be involved. For oil and gas industry, administrative corruption is of lesser importance but may still be observed for "greasing the wheels" in public-private interface in proposals, registration, procurement, and access to local public services.

Grand corruption. The theft or misuse of vast amounts of public resources by state officials-usually members of, or associated with, the political or administrative elite-constitutes grand corruption. For oil and gas sector, grand corruption permeates through the entire value chain from legislative framework to production and distribution. In developing countries, foreign private sector participation is considered is the predominant source of such corruption.

Legislative and executive capture. Collusion by private actors with public officials or legislative members or politicians for their mutual, private benefit is referred to as legislative and/or executive branch capture. That is, the private sector "captures" the state legislative, executive, and judicial apparatus for its own purposes. Legislative and executive capture coexists with the conventional (and opposite) view of corruption, in which public officials extort or otherwise exploit the private sector for private ends. This form of corruption is prevalent in both industrial and developing countries.

Patronage/paternalism and being a "team player." Using official position to aid clients having the same geographic, ethnic, and cultural origin so that they receive preferential treatment in their dealings with the public sector including public sector contracts and employment. Also providing the same assistance on a quid pro quo basis to colleagues belonging to an informal network of friends and allies. Geographic, ethnic, and culture-based patronage is commonplace in developing countries and "team player" form of corruption is fact of everyday life in industrial countries. This type of corruption is of lesser significance in the oil and gas sector.

It is also known that corruption is country-specific; thus, approaches that apply common policies and tools (that is, one-size-fits-all approaches) to countries in which acts of corruption and the quality of governance vary widely are likely to fail. One needs to understand the local circumstances that encourage or permit public and private actors to be corrupt. Finally, we know that if corruption is about governance and governance is about the exercise of state power, then efforts to combat corruption demand strong local leadership and ownership if they are to be successful and sustainable.

10.2.3 Many Facets of Corruption in the Oil and Gas Sector

An important distinguishing aspect of oil and gas sector corruption is that mostly it comes in the form of grand corruption resulting from collusive practices among stakeholders in industrial and developing countries. Table 10.1 highlights the incidence of corruption in the oil and gas sector at various decision points in the legislative-executive-private sector interface in provision of public infrastructure.

10.3 WHAT DRIVES CORRUPTION? CONCEPTUAL PERSPECTIVES

Public sector corruption, as a symptom of failed governance, depends on multitude of factors such as the quality of public sector management, the nature of accountability relations between the government and citizens, the legal framework, and the degree to which public sector processes are accompanied by transparency and dissemination of information. Efforts to address corruption that fail to adequately account for these underlying "drivers" are unlikely to generate profound and sustainable results (Shah and Schacter, 2004). To understand these drivers, a conceptual perspective is needed to understand why corruption persists and what can be a useful antidote. At the conceptual level, several interesting ideas have been put forward. These ideas can be broadly grouped together in three categories (a) Principal-agent or agency models; (b) new public management perspectives; and (c) neo-institutional economics frameworks.

10.3.1 Principal-Agent Models

This is the most widely used modeling strategy. A common thread in these models is that the government is led by a benevolent dictator, the principal, who aims to motivate government officials (agents) to act with integrity in the use of public resources (see Becker, 1968, 1983). One such view, the so-called crime and punishment model by Gary Becker (1968), states that self-interested public officials seek out or accept bribes so long as the expected gains from corruption exceed the expected costs (detection and punishment) associated with corrupt acts. Thus, according

Table 10.1 Many facets of corruption in the oil and gas sector

Decision point	Incidence of corruption	Type of corruption
Legislative framework	Lobbying legislators for pet projects, log rolling, political campaign financing, perks and privileges in exchange for legislative favors, preparation of draft legislation by lobbyists	Grand corruption, executive and legislative capture (universal phenomenon)
Policy making	Policy framework to suit oil and gas interest groups and bribe payers	Executive capture, grand corruption, most countries
Tax policy regime	Hidden tax expenditures to favor of oil and gas industry. Kickbacks in return for favorable terms. Lax enforcements	Executive capture, grand corruption, most countries
Exploration rights	Licensing, permits, contract awards in return for bribes	Executive capture, grand corruption, most countries
Production and development	Bribe payments to receive approvals, procurement, contracts and theft of oil and gas or underpayment of revenues	Executive capture, grand corruption, most developing countries
Transport and trade	Bribes to receive access or to falsify trade volumes and value, smuggling	Grand corruption
Refining and marketing	Bribes in return for lax registration and favorable procurement	Grand corruption
Corporate finance	Bribe payments in return for tax evasion, money laundering	Mostly grand corruption, most developing countries
Public-private partnerships	Collusion, re-negotiation of contracts during and after execution	Grand corruption most countries

| Public procurement | Kickbacks rigged bidding and tendering. Mechanisms: intermediary collusion, consulting fees, false services, overpayment, payment to a front company, Swiss and Cayman Island bank accounts, collusion among competitors,. Procurement cycle: registration, manipulating specifications, insider info, quality reduction, clearance from inspectors, expediting payments | Mostly grand corruption, most countries |

Source Author's conceptualization

to this view, corruption could be mitigated by (a) reducing the number of transactions over which public officials have discretion; (b) reducing the scope of gains from each transaction; (c) increasing the probability for detection; and (d) increasing the penalty for corrupt activities. Moreover, since it is costly to increase detection, but not to increase penalties (at least assuming detection is accurate), the most efficient way to eliminate corruption is to impose very high penalties with a relatively low probability of detection. Klitgaard (1988) restates this model to emphasize the unrestrained monopoly power and discretionary authority of government officials. According to him, corruption equals monopoly plus discretion minus accountability. To curtail corruption under this framework, one must have a rules-driven government with strong internal controls and with little discretion to public officials. This model gained wide acceptance in public policy circles and served as a foundation for empirical research and policy design to combat administrative, bureaucratic, or petty corruption. Experience in highly corrupt countries, however, contradicts the effectiveness of such an approach as the rules enforcers themselves add extra burden of corruption and lack of discretion is also thwarted by collusive behavior of corruptors. In fact, lack of discretion is often cited as a defense by corrupt officials who partake in corruption as part of a vertically well-knit network enjoying immunity from prosecution.

Another variant of principal-agent models integrates the role of legislators and elected officials in the analysis. In this variant, high level government officials—represented by legislators or elected public officials—institute or manipulate existing policy and legislation in favor of particular interest groups—representing private sector interests and entities or individual units of public bureaucracy competing for higher budgets—in exchange of rents or side payments. In this framework, legislators weigh the personal monetary gains from corrupt practices and improved chances of re-election against the chance of being caught, punished, and losing an election with a tarnished reputation. Factors affecting this decision include campaign financing mechanisms, information access by voters, the ability of citizens to vote out corrupt legislators, the degree of political contestability, electoral systems, democratic institutions and traditions, and institutions of accountability in governance. Examples of such analyses include Rose-Ackerman (1975, 1978), Van Rijckeghem and Weder (2001), Acconcia et al. (2003). This conceptual framework is useful in analyzing political corruption or state capture in the oil and gas sector.

10.3.2 New Public Management Frameworks

The new public management (NPM) literature, on the other hand, points to a more fundamental discordance among the public sector mandate, its authorizing environment and the operational culture and capacity. According to NPM, this discordance contributes to government acting like a runaway train and government officials indulging in rent-seeking behaviors with little opportunity for citizens to constrain government behavior. This viewpoint calls for fundamental civil service and political reforms to create a government under contract and accountable for results. Public officials will no longer have permanent rotating appointments but instead they could keep their jobs if they fulfilled their contractual obligations (see Shah, 1999, 2005, 2007, 2015). This framework has some relevance for an analysis of corruption in the oil and gas sector as with grand corruption there are serious failures in achieving service delivery and economic development goals.

10.3.3 Neo-Institutional Economics (NIE) Frameworks

Finally, Shah (2006) has utilized the transactions costs approach of the neo-institutional economics (NIE) to present a newer perspective on the causes and cures of corruption. Shah argues that corruption results from opportunistic behavior of public officials (agents) given that citizens (as governors and principals) are either not empowered or face high transaction costs to hold public officials accountable for their corrupt acts. The principals have bounded rationality—they act rationally based upon the incomplete information they have. To have a more informed perspective on public sector operations, they face high transaction costs in acquiring and processing the information. On the other hand, agents (public officials) are better informed. This asymmetry of information allows agents to indulge in opportunistic behavior which goes unchecked due to high *transactions costs* faced by the principals and a lack of or inadequacy of countervailing institutions to enforce accountable governance. Thus, corrupt countries have inadequate mechanisms for contract enforcement, weak judicial systems, and inadequate provision for public safety. This raises the transactions costs in the economy further raising the cost of private capital as well as the cost of public service provision. The problem is further compounded by path dependency (i.e., a major break with the past is difficult to achieve as any major reforms are likely to be blocked by influential interest groups), cultural and historical factors and mental models where those who are victimized by corruption feel that attempts to deal with corruption will lead to further victimization, with little hope of corrupt actors being brought to justice. These considerations lead principals to the conclusion that any attempt on their part to constrain corrupt behaviors will invite strong retaliation from powerful interests. Therefore, citizen empowerment (e.g., through decentralization, open government with citizens' right-to-know, citizens' charter, bill of rights, electoral finance, and other forms of civic engagement) assumes critical importance in combating corruption because it may have a significant impact on the incentives faced by public officials to be responsive to public interest.

10.3.4   Measuring the Incidence of Corruption

Measuring the country-specific and global incidence of corruption is a complex task and oil and gas sector corruption adds multiple layers of greater complexity. Existing cross-country indexes of corruption and governance quality such the Corruption Perception Index (CPI) of Transparency International and World Bank's Worldwide Governance Indicators (WGIs) are perception-based indexes often capturing the perceptions of a few influential Western institutions or "experts" and their empirical validity for cross-country, and country-specific time series incidence of corruptions could not be established. In some cases, political bias of these indicators can be established by a careful analysis of Western leanings of political regimes in a country (Thomson and Shah, 2005). Across countries and in time series data, they use different factors and weights. Therefore, even for the measurement of perceptions, their methodologies are not valid for cross-country and country-specific time series comparisons (Thompson and Shah, 2005; Thomas, 2006; Iqbal and Shah, 2008; Arndt, 2008; Ivanyna and Shah, 2011; Pelizzo et al., 2017). Figures 10.1, 10.2, and 10.3 illustrate their time and country assessment inconsistencies making their ranking highly suspect.

Given the invalidity of the macro aggregate perceptions-based indicators, measurement of the incidence of corruption and more importantly incidence of corruption in the oil and gas sector is better assessed by

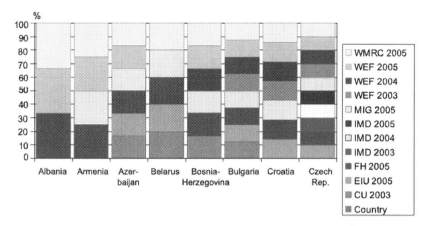

Fig. 10.1 Inconsistency of weights across countries: Transparency International's Corruption Perception Index—2005 weights (*Notes* WMRC—The Worlds Markets Research Center, WEF—The World Economic Forum, MIG—Grey Area Dynamics Ratings by the Merchant International Group, IMD—The International Institute for Management Development, FH—The Freedom House, EIU—The Economist Intelligence Unit, CU—The State Capacity survey by the Center of International Earth Science Information Network. *Source* Arndt [2008])

investigative journalism as demonstrated by the cases of Petrobras in Brazil in 2014, JP Morgan Chase in Nigeria in 2011, Prime Minister Najib/Goldman Sachs 1MDB scandal in Malaysia; or Kazakhgate in Kazakhstan in 2003 (Dossim, 2014; Moise, 2020). Investigative journalism could be assisted by the strong "Right to Know" legislation facilitating transparency in governance.

10.4 What Is Special About Corruption in the Oil and Gas Sector

Oil and gas sector has a complex value chain with high probability of corruption being undetected. Even if great deal of information may be made available, it is very difficult to assess the fair value of various contracts due to high risk and uncertainty associated with exploration and development and technical complexity of various tasks. Also, comparative data is of little use. In view of this, corruption can take place in plain

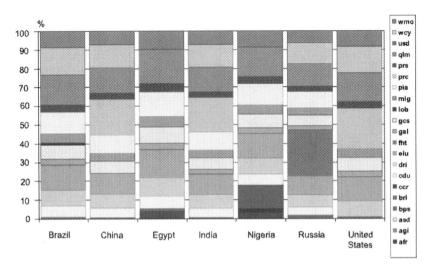

Fig. 10.2 Inconsistency of weights across countries used by the Worldwide Governance Indicators (WGIs) (*Notes* WMO—Global Insight Business Conditions and Risk Indicators, WCY—Institute for Management and Development World Competitiveness Worldbook, USD—USAID, QLM—Business Environment Risk Intelligence, PRS—Political Risk Services International Country Risk Guide, PRC—Political Economic Risk Consultancy Corruption in Asia Survey, PIA—World Bank Country Policy and Institutional Assessments, MIG—Merchant International Group, LOB—Latinobarometro, GCS—World Economic Forum Global Competitiveness Report, GAL—Gallup International, FNT—Freedom House, EIU—Economist Intelligence Unit, DRI—Global Insight Global Risk Service, CDU—Columbia University State's Capacity Survey, CCR—Freedom House Countries at Crossroads, BRI—Bertelsman Foundation, BPS—Business Enterprise Environment Survey, ASD—Asian Development Bank Country Policy and Institutional Assessments, AGI—United Nations Economic Commission for Africa, AFR—Afrobarometer, ADB—African Development Bank Country Policy and Institutional Assessments. *Source* Arndt [2008])

sight without any recognition by even informed citizenry. Most of the corruption in this sector is grand corruption undertaken by politically powerful and dense international networks. The so-called Big Men facilitate interactions of this network in across border transactions. Strong and well-financed lobby groups build long-term relationships with members of the legislature though campaign finance and with influential politicians

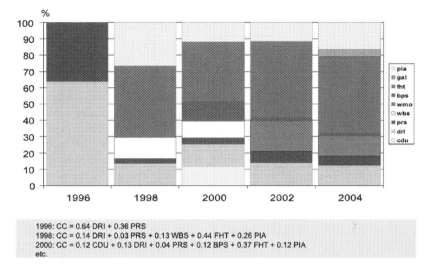

Fig. 10.3 Inconsistency of weights over time—Albania as an example (*Notes* PIA—World Bank Country Policy and Institutional Assessments, GAL—Gallup International, FHT—Freedom House Nations in Transition, BPS—Business Enterprise Environmental Survey, WMO—Global Insight Business Conditions and Risk Indicators, WBS—World Business Environment Survey, PRS—Political Risk Services International Country Risk Guide, DRI—Global Insight Global Risk Service, CDU—Columbia University State Capacity Project. *Source* Arndt [2008])

to secure their interests (McPherson and MacSearraigh, 2007; Riyadi, 2020; Sayne and Gillies, 2016).

Grand corruption in this sector typically has international dimensions where Western stakeholders act as bribe payers and political leaders and officials in resource rich developing countries act as bribe recipients. Elf, a French conglomerate, frequently made regular bribe payments to the African heads of the state with full knowledge of the French Government. Nigerian Economic and Financial Commission has alleged that US firm Halliburton is involved in bribe payments to Nigerian officials.

In this sector, the West's strategic, security, economic, and commercial interests collide with its moral compass preventing any corrective action. France, until recently, explicitly recognized bribe payments to foreign leaders and officials as a legitimate cost of doing business in oil rich countries and allowed it to be a tax-deductible expense. USA is seen to be lax in prosecuting Saudi officials for violations of its laws.

Even when grand corruption is detected, it is difficult to prosecute. US prosecuted a New York businessman Mr. Giffen in 2007 under the Foreign Corrupt Practices Act for US$80 million in bribe payments to the Kazakhstan President for securing interests of Shell Oil Company. Mr. Giffen pleaded not guilty stating that the bribe was paid on encouragement from the US Government. Judge William Pauley III of the US District Court New York gave the verdict of not guilty and wrote in his judgment that Mr. Giffen is a great patriot who advanced US economic interests abroad.

10.5 Global Initiatives in Combating Corruption in Oil and Gas Industry

In view of the global interest, several initiatives have been undertaken to address corruption in oil and gas sector by seeking enhanced transparency.

10.5.1 Fiscal Transparency Code by the IMF

IMF's Fiscal Transparency Code (Revised 2019) seeks transparency of public finances and has four pillars: Pillar I requires comprehensive, timely, and reliable reporting on government's financial position and performance; Pillar 2 seeks clarity in budgetary policy and accuracy in fiscal forecasting; Pillar III requires disclosure of disclosure and management of risks to public finances; Pillar IV is specifically concerned with resource revenue management and advises the government to provide a transparent framework for the ownership, contracting, taxation, and utilization of natural resource endowments. A Handbook covering Pillar 4 is currently under preparation. The IMF conducts Fiscal Transparency Evaluations (FTEs) using the Fiscal Transparency Code standards upon request from member governments.

IMF's fiscal transparency is a welcome initiative in informing legislatures, oversight bodies, markets, and citizens with important information to hold government to account. It would be also helpful in shedding light on the use of resource revenues. The IMF code simply aims at fiscal transparency, and it does not make any claims regarding combating corruption, and it is unlikely to have significant impact on grand corruption in the oil and gas sector as true contracting details showing bribes, if any, are neither required and nor likely to be revealed by corrupt governments.

10.5.2 *Extractive Industries Transparency Initiative (EITI)*

The EITI was launched in 2003 with an ambitious goal of remedying the resource curse through revenue transparency and public dialogue. However, its criteria were modest and required member countries to publish "extractive payments by companies to governments as well as revenues received by governments from companies." However, over the years EITI mission objectives and associated standards have undergone significant upward revisions. Most recent, EITI Standard 2019, requires making financial disclosure and open data a routine part of government and corporate reporting to provide information to stakeholders in a more timely, relevant, and accessible form, that supports its use in analysis and decision-making. It also requires member countries to publish new contracts effective January 2021 and introduced requirements for environmental and gender impact assessment and commodity trading.

EITI membership is voluntary. As of August 2022, EITI has 58 member countries representing a very small percentage of world oil production as none of the top ten oil producing countries (USA, Russia, Saudi Arabia, Canada, Iraq, China, United Arab Emirates, Brazil, Kuwait, Iran) are represented. Also, most of the bribe-paying countries are not represented. The EITI transparency in practice provides a small coverage of the value chain and its revenue reporting does not go much beyond what is already available from the IMF and other sources. The current impact of this initiative is unknown as we lack any reliable and credible impact assessment. However, given its small coverage of relevant oil producing countries and the value chain and absence of bribe payers, one can safely conclude that even if effective, this initiative is unlikely to make a significant dent on combatting corruption in the oil and gas sector. Its mission creep into environmental and gender impact assessments further will divert its focus from its original goal of curtailing corruption (For a comprehensive review of EITI initiative, see Moise, 2020; Ejiogu et al., 2018; Olujobi, 2020; Moses, 2018; Kolstad and Wiig, 2009).

10.5.3 Publish What You Pay (PWYP) Initiative

This is a complementary initiative to EITI. It aims "...to build a global movement of civil society organizations making oil, gas and mineral governance open accountable, sustainable, equitable and responsive to all people." More specifically in the oil and gas sector, PWYP aims "..to ensure that EITI reports provide detailed information on fiscal regimes and tax payments by extractive companies so the data can be reconciled with other sources – such as alternative reports, journalists investigations and supreme audit institutions – to provide real fiscal transparency" (PWYP website). PWFP has expounded lofty goals but so far little to show in terms of its achievements.

10.5.4 USA Foreign Corrupt Practices Act (FCPA)

The Foreign Corrupt Practices Act (FCPA), enacted in 1977, prohibits the payment of bribes to foreign officials to assist in obtaining or retaining business. The FCPA can apply to prohibited conduct anywhere in the world and extends to publicly traded companies and their officers, directors, employees, stockholders, and agents. Agents can include third party agents, consultants, distributors, joint-venture partners, and others.

The FCPA also requires issuers to maintain accurate books and records and have a system of internal controls sufficient to, among other things, provide reasonable assurances that transactions are executed, and assets are accessed and accounted for in accordance with management's authorization.

The sanctions for FCPA violations can be significant. The SEC may bring civil enforcement actions against issuers and their officers, directors, employees, stockholders, and agents for violations of the anti-bribery or accounting provisions of the FCPA. Companies and individuals that have committed violations of the FCPA may have to disgorge their ill-gotten gains plus pay prejudgment interest and substantial civil penalties. Companies may also be subject to oversight by an independent consultant.

The SEC and the Department of Justice are jointly responsible for enforcing the FCPA. The SEC's Enforcement Division has created a specialized unit to further enhance its enforcement of the FCPA (see https://www.sec.gov/spotlight/foreign-corrupt-practices-act.shtml).

FCIA is by far the most welcome and effective instrument in curtailing US source corruption in the oil and gas sector worldwide. The statistics on enforcement actions show that during the period 1977 to June 2022, 312 total actions were taken, out of which 47 related to bribe activities in the oil and gas sector. US$13.4 billion dollars monetary sanctions were paid to foreign governments in FCPA-related enforcement actions in all sectors including oil and gas (https://fcpa.stanford.edu/total-sanction.html).

In conclusion, while several global initiatives aim to address corruption in the oil and gas, with the sole exception of FCPA, none has shown any perceptible impact on the incidence of corruption. The design of most of these initiatives are not consistent with their objectives and therefore one does not expect future results to be different from the past impacts. As noted earlier, people empowerment through the right-to-know legislation and independent media with skills in investigative journalisms is critical to uncovering corruption and none of the international initiatives are aimed at developing and strengthening these institutions in oil-rich countries. In view of this, high incidence of corruption is likely to persist in the oil and gas sector in the foreseeable future.

10.6 The Way Forward: Pathways to Reform

Combating corruption in the oil and gas sector is one of the most challenging tasks facing reformers worldwide. Given the enormous complexity of the underlying transactions, uncovering corruption is an enormously difficult even in the most ideal public policy environment. To assist in this effort, creating an enabling environment for people's right-to-know and freedom for investigative journalism to pursue all leads without fear of harassment, persecution, and risks to life and liberty should be the first order of priority for policymakers. A second order of priority is to establish an authorizing environment that holds powerful individuals and entities to account for corrupt practices through timely and fair dispensation of justice. This would require difficult legislative and judicial reforms. Since corruption in the oil and gas sector, in a large part, originates from the stakeholders in industrial countries, it would also require strong judicial interventions by these countries along the lines of the US Foreign Corrupt Practices Act. International advocacy groups concerned with corruption in the oil and gas sector would be well advised to focus their activities on both the open government issues in resource rich countries as well as creating disincentives for stakeholders in bribe-paying countries to disengage them from corrupt practices to advance their economic and political agendas.

REFERENCES

Acconcia, Antonio, Marcello D'Amato and Riccardo Martina (2003). Corruption and Tax Evasion with Competitive Bribes. CSEF Working Papers 112, University of Salerno, Centre for Studies in Economics and Finance, Italy.

Al-Kasim, Farouk, Tina Soreide and Aled Williams (2013). Corruption and Reduced Oil Production: An Additional Resource Curse Factor? *Energy Policy*, **54**: 137–147.

Amundsen, Inge (2012). Corruption, Corruption Prevention and Good Governance, Christian Michelsen Institute, Norway, presentation, Elmina, Ghana, November 8.

Arndt, C. (2008). What Is Happening to the Level of Corruption in ECA and World. PowerPoint Presentation at the World Bank, Washington, DC (mimeo).

Auty, R.M. (2007). Rent Cycling Theory, the Resource Curse and Development Policy. In DAI, Washington, *From Curse to Cures. Developing Alternatives*, V.11, issue 1, Spring 2007, pp. 7–13. Washington, DC.

Becker, Gary Stanley (1968). Crime and Punishment: An Economic Approach. *Journal of Political Economy*, **76**(2): 169–217.

Becker, Gary Stanley (1983). A Theory of Competition Among Pressure Groups for Political Influence. *Quarterly Journal of Economics*, **XCVII**(3), 371–400.

Chebab, Daouia, N. Mazlan, M. Rabbani, L. Chin and A. Ogiri (2022). The Role of Corruption in Natural Resources-Financial Development Nexus: Evidence from MENA Region. *Institutions and Economies*, **14**(2): 1–29.

DAI, Washington (2007). From Curse to Cures. *Developing Alternatives*, V.11, issue 1, Spring 2007. Washington, DC.

Donwa, P., C. Mgbame and O. Collins (2015). Corruption in the Oil and Gas Industry: Implications for Economic Growth. *European Scientific Journal*, **11**(22): 212–230.

Dossim, Satpayev (2014). Corruption in Kazakhstan and the Quality of Governance. Institute of Developing Economies Discussion Paper No. 475, Tokyo, Japan.

Ehiemua, Solomon (2015). Nigeria Crude Oil: Sources of Corruption and Economic Disparity in the Nation. *European Journal of Research in Social Sciences*, **3**(4): 76–83.

Ejiogu, A., C. Ejiogu and A. Ambituuni (2018). The Dark Side of Transparency: Does the Nigeria Extractive Industries Transparency Initiative Help or Hinder

Accountability and Corruption Control? *British Accounting Review* (in-press). https://doi.org/10.1016/j.bar.2018.10.004

Eriksen, Birthe and Tina Soreide (2017). Chapter in volume edited by Philippe Le Billon and Aled, *Williams: Corruption, Natural Resources and Development: From Resource Curse to Political Ecology*. New York: Edward Elgar

Gallagher, Mark and Steve Rozner (2007). In DAI, Washington, *From Curse to Cures. Developing Alternatives*, V.11, issue 1, Spring 2007, pp. 28–34. Washington, DC.

Iqbal, Kazi and Anwar Shah (2008). Truth in Advertisement: How Do Governance Indicators Stack Up? World Bank Institute, Washington, DC, https://wwww.worldbank.org/wbi/public/finance.

Ivanyna, Maksym and Anwar Shah (2011). Citizen-Centric Governance Indicators: Measuring and Monitoring Governance by Listening to the People. *CESifo Forum*, 1: 59–70.

Klitgaard, Robert E. (1988). *Controlling Corruption*. Berkeley, California: University of California Press.

Kolstad, I. and A. Wiig (2009). Is Transparency the Key to Reducing Corruption in Resource-Rich Countries? *World Development*, 37(3): 521–532. https://doi.org/10.1016/J.WORLDDEV.2008.07.002

Lovei, Lazlo and Alastair McKechnie (2000). The Cost of Corruption for the Poor—The Energy Sector, Public Policy for the Private Sector, Note No. 207, World Bank, Washington, DC.

McPherson, Charles and Stephen MacSearraigh (2007). Corruption in the Petroleum Sector. In J. Edgardo Campos and Sanjay Pradhan (eds.), *The Many Facets of Corruption*, chapter 6, pp. 191–220. Washington, DC: World Bank

Moise, Gian Marco (2020). Corruption in the Oil and Gas Sector: A Systematic Review and Critique of the Literature, Science Direct. *The Extractive Industries and Society*, 7: 217–236.

Moses, Olayinka (2018). Effectiveness of the Extractive Industries Transparency Initiative, Ph.D Dissertation, University of Wellington, Victoria.

Olujobi, Olusola (2020). Nigeria's Upstream Petroleum Industry Anti-corruption Legal Framework: The Necessity for Overhauling and Enrichment. *Journal of Money Laundering Control*, Emerald Insight, pp. 1–27. https://www.emerald.com/insight/1368-5201.htm

Pelizzo, Boris and S. Janenova (2017). Objective or Perception-Based? A Debate on the Ideal Measure of Corruption. *Cornell International Law Journal*.

Riyadi, Bambang (2020). Culture of Abuse of Power Due to Conflict of Interest to Corruption for Too Long on the Management Form Resources in Oil and Gas in Indonesia. *International Journal of Criminology and Sociology*, 9: 246–254.

Rose-Ackerman, S. (1975). The Economics of Corruption. *Journal of Public Economics*, **4**(February): 187–203.
Rose-Ackerman, S. (1978). *Corruption: A Study in Political Economy*. New York, San Francisco, London: Academic Press.
Sayne, Aaron and Alexandra Gillies (2016). Initial Evidence of Corruption Risks in Government Oil and Gas sales. Natural Resources Governance Institute Briefing Paper, June.
Shah, Anwar (1999). Governing for Results in a Globalized and Localized World. *The Pakistan Development Review*, **38**(4, Part I): 385–431.
Shah, Anwar (2005). On Getting the Giant to Kneel: Approaches to a Change in the Bureaucratic Culture. In A. Shah (ed.), *Fiscal Management, Public Sector Governance and Accountability Series*, pp. 211–29. Washington, DC: World Bank.
Shah, Anwar (2006). Decentralized Governance and Corruption. In E. Ahmad and G. Brosio (eds.), *Handbook of Fiscal Federalism*, pp. 478–498. Cheltenham, UK: Edward Elgar.
Shah, Anwar (2007). Tailoring the Fight Against Corruption to Country Circumstances. In Anwar Shah (ed.), *Performance Accountability and Combating Corruption*, chapter 7, pp. 233–254. Washington, DC: World Bank
Shah, Anwar (2015). Decentralized Provision of Infrastructure and Corruption. In Jonas Frank and Jorge Martinez-Vasquez (eds.), *Decentralization and Infrastructure in the Global Economy: From Gaps to Solutions*, chapter 14, pp. 418–454. New York: Routledge Press.
Shah, Anwar and Mark Schacter (2004). Combating Corruption. Look Before You Leap. *Finance and Development*, **41**(4, December): 40–43.
Siegle, Joseph (2007). The Governance Root of the Natural Resource Curse. In DAI, Washington, *From Curse to Cures. Developing Alternatives*, V.11, issue 1, Spring 2007, pp. 35–43. Washington, DC.
Thomas, Melissa (2006). What Do Worldwide Governance Indicators Measure? Johns Hopkins University, Washington, DC (mimeo).
Thompson, Theresa and Anwar Shah (2005). Transparency International's Corruption Perception Index: Whose Perceptions Are They Anyway? World Bank Institute, Washington, DC. https://wwww.worldbank.org/wbi/public/finance.
Van Rijckeghem, Caroline and Beatrice Weder (2001). Bureaucratic Corruption and the Rate of Temptation: Do Low Wages in Civil Service Cause Corruption? *Journal of Development Economics*, **65**: 307–331.

Author Index

A
Acconcia, Antonio, 394
Acquatella, J., 231
Afzal, A., 85
Ahmad, Junaid, 181
Ahmadov, Ingilab, 162
Alberton, Mariachiara, 193, 206
Al-Kasim, Farouk, 388
Ambituuni, A., 401
Amundsen, Inge, 389
Aragon, Nazli, 206
Aristi, Juan, 163
Arndt, C., 396
Aslanli, Kenan, 162
Astrov, Vasily, 162
Aubert, Cecile, 187
Auty, R.M., 388

B
Baffes, John, 300
Bagattini, Gustavo, 163
Bahl, Roy, 122, 125, 126, 132
Ballard, C.L., 113
Barrett, S., 256
Bates, R., 293
Becker, Gary Stanley, 108, 391
Benoit Bosquet, 123
Bernow, S., 292, 296, 302
Berkelaar, Arjan and Jennifer Johnson-Calari, 164
Bernstein, Jeffrey, 299
Bird, Richard, 181
Bishop, Grant, 130
Boadway, Robin, 4, 5, 122, 123, 125, 175
Boessenkool, Kenneth J., 145
Boex, Jamieson, 147
Bohi, Douglas R., 338
Bohm, P., 256
Born, Rubens H., 239
Boudekhdekh, Karim, 329
Briere, M., 164
Browning, Edgar K., 284
Bruce, Neil, 5
Budina, Nina, 162

C

Capcelea, Arcadie, 198, 199
Cartaxo, Tiago de Melo, 206
Carvalho, José Carlos, 219
Chebab, Daouia, 388
Chin, L., 388
Churchill, A.A., 258, 259, 325
Coase, R.H., 85, 86
Collins, O., 388
Cullenward, Danny, 259

D

Dalmazzone, Silvana, 174, 177–180, 182
D'Amato, Marcello, 394
Das, Udaibir, 164
Davis, Jeffrey, 155, 162, 163
Davis, J.M., 147
Devlin, J., 163
Donwa, P., 388
Dossim, Satpayev, 397

E

Eckardt, Sebastian, 136
Eckaus, Richard S., 259
Ehiemua, Solomon, 388
Ejiogu, A., 401
Ejiogu, C., 401
Ellis, Gregory, 5
Engel, Kirsten H., 199
Epstein, J.M., 304
Eric S. Maskin, 85
Eriksen, Birthe, 388
Eskeland, G., 92, 109
Esty, Daniel C., 174, 177, 180–182

F

Fasano, Ugo, 162
Fedelino, A., 147
Feehan, James, 145

Feldstein, Martin, 289, 309, 313, 317
Fisher, B., 256
Flatter, Frank, 4, 5, 123
Flavin, C., 259
Frankel, J., 154
Franklin, Aimee, 163
Fredriksson, Per G., 179, 180, 199
Freire, Nelson, 237

G

Gabeira, Fernando, 238
Gallagher, Mark, 388
Galligan, Brian, 190
Gillies, Alexandra, 399
Glomsrod, S., 292, 296, 302
Gómez, J.J., 231
Gonzales, Adrian, 154
Goulder, Lawrence H., 256, 284

H

Haghiri, Fateme, 154
Harrington, W., 109
Henderson, J.V., 179
Heston, Alan, 296
Heuty, Antoine, 163
Hoeller, P., 338
Hoel, Michael, 262, 303, 305
Hofman, Bert, 123
Holland, Kenneth, 190
Horn, Murry, 182

I

Ilahi, N., 163
International Energy Agency, 287
Iqbal, Kazi, 396
Ivanyna, Maksym, 396

J

Jaffe, A.B., 179

Jatobá, Jorge, 227, 228
Jenkins, G., 88, 260
Johnsen, T., 292, 296, 302
Jorgenson, D.W., 256, 299, 308

K
Kaiser, Kai, 123
Karl-Göran, Mäler, 85
Karsalari, Abbas Rezazadeh, 154
Keleman, Daniel, 198, 206
Klotsvog, F.N., 140
Knill, Christoph, 206
Kolstad, I., 401
Kosmo, Mark, 111, 114
Krupnick, A., 91, 109, 112, 115, 116
Kunce, Mitch, 179
Kuroda, M., 256
Kushnikova, I.A., 140

L
Laffont, Jean-Jacques, 187
Lamech, R., 88
Larsen, Bjorn, 10, 11, 94, 96, 97, 256, 259, 263, 269, 274, 277, 287, 288, 292, 293, 301, 302, 304, 307, 325, 326
Le, Angie, 11
Lebdioui, Amir, 163
Lenschow, Andrea, 206
Lerda, J.C., 231
Levinson, Arik, 180
Lewin, M., 163
Lichtenberg, E., 114
List, John, 178–180
Litvack, Jennie, 181
Lovei, Lazlo, 387
Lucke, Matthias, 162

M
MacSearraigh, Stephen, 399

Magat, W., 109
Maler, K.G., 262
Marçal, Cláudia, 225
Marron, D., 292, 296, 302
Martimort, David, 183
Martina, Ricardo, 394
Mason, Charles, 178
Mathieu, Paul, 155, 161, 162
Mazlan, N., 388
McConnell, Virginia, 179
McKechnie, Alastair, 387
McLure, Charles E Jr., 125, 126, 140, 145
McPherson, Charles, 399
Mehrara, Mohsen, 154
Metcalf, Gilbert E., 275
Mgbame, C., 388
Millimet, Daniel, 179, 180, 199, 204
Moise, Gian Marco, 397, 401
Moore, E., 293
Moore, Samuel, 163
Morton, F., 190
Moses, Olayinka, 401
Mubazi, J., 256
Mulder, Yinqium Christian, 164
Munasinghe, Mohan, 91

N
Nalini, Renato, 235, 237, 240
Natural Resources Governance Institute, 162
Nordhaus, William D., 258, 376

O
Oates, Wallace, 174–181, 204–206, 211–213, 226, 303, 306
Ogiri, A., 388
Olson, M., 175
Olujobi, Olusola, 401

P

Pearce, David, 85, 256
Pelizzo, Boris, 396
Percival, Robert, 199
Petersen, Christian, 160, 161
Peterson, S.R., 179
Portney, Paul R., 180, 181, 213, 303, 306
Post, Diahanna, 211
Postel, S., 259
Poterba, James M., 91, 256, 278, 284, 293

Q

Qiao, Baoyun, 6, 7

R

Rabbani, M., 388
Rabe, Barry C., 275
Radian Corporation, 295
Revesz, Richard, 179, 182
Riyadi, Bambang, 399
Rose-Ackerman, Susan, 394
Rozner, Steve, 388

S

Sachs, Jeffrey D., 121, 154
Saeed, Khalid Adnan, 154
Sarnoff, Joshua, 182
Saunders, R.J., 258, 259, 325
Sayne, Aaron, 399
Scancke, Martin, 146
Schwab, Robert, 178, 179
Scott Barret, 85
Shabsigh, G., 163
Shah, Anwar, 6–8, 10–13, 94, 96, 97, 122–128, 130, 136, 138, 175, 182, 192, 256, 259, 263, 269, 274, 277, 279, 281, 283, 287, 288, 292, 293, 299–302, 304, 307, 325, 326, 391, 395, 396
Shah, Sana, 182
Shoven, J.B., 113
Siegle, Joseph, 387
Smith, S., 116, 161, 164
Soreide, Tina, 388
Stavins, Robert N., 179, 182
Summers, Lawrence H., 256, 259, 325
Summers, Robert, 296
Sy, Amadou, 164

T

Terkla, D., 113
Thomas, Melissa, 396
Thompson, Theresa, 396
Tietenberg, T., 262
Toffel, Michael, 206, 211
Tsani, Stella, 162
Tumennasan, Bayar, 122, 125, 126, 132

U

Udeh, John, 143

V

Valdes, Rodrigo, 163
Van der Ploeg, Frederick, 154
Van Rijckeghem, Caroline, 394
van Selm, Bert, 155, 161, 162
Vennemo, H., 292, 296, 302
Victor, David, 259
Viscusi, W.K., 108
Vogel, David, 206, 211
Von Ritter, Konrad, 198, 199

W

Wakerman-Lin, John, Paul, 155, 161, 162

Wallin, M., 338
Warner, Andrew, M., 121, 154
Weder, Beatric, 394
Weitzman, M.L., 102, 103
Whalley, John, 113, 256, 279, 281
Wilcoxen, P.L., 256, 308
Williams, Aled, 388
Williamson, Oliver, 182

World Bank, 255, 260, 275, 396
World Resources Institute, 275, 287, 306

Z

Zeckhauser, R., 108
Zilberman, D., 114

Subject Index

A

Abatement of local pollution, 12, 376
Abatement technology standards, 98, 100, 112
Accountable fund governance, 8, 168
Agenda 21, 241, 243, 244
Air pollution, 6, 87, 91, 95, 115, 116, 176, 192, 194, 198, 199, 201, 204, 222, 228, 238, 248
Alberta Heritage Fund, 159
Anti-corruption, 12
A race to the top, 9, 179
Asian Development Bank, 398
Asset management, 8, 135, 166
Asset allocation, 8, 164–166
Assignment of environmental functions, 8, 173–175, 190, 192, 194, 199, 200, 214, 249
Auctions, 48, 49, 51, 68
Auction systems, 49, 50, 52, 67, 68, 73, 75
Australia, 8, 127, 174, 190–192, 206, 265, 269, 271, 273, 277, 328, 330–333, 335, 347, 353, 357, 370, 379
Authorizing environment, 13, 395, 403
Azerbaijan, 130, 131, 133, 134, 146, 162, 334, 343, 348, 352, 356, 360, 362, 369, 380

B

Brazil, 4, 9, 122, 128, 129, 131, 211, 215, 216, 219, 220, 226–233, 235–241, 243–249, 264, 267, 270, 272, 276, 328, 330, 332, 334, 344, 348, 352, 356, 360, 363, 379, 380, 397, 401
Brazilian System of Environmental Management, 215
Bribe-paying countries, 13, 401, 404
Bribe-recipient countries, 399

C

Canada, 8, 71, 76, 127–129, 132, 143, 144, 147, 154–156, 161, 167, 174, 190, 191, 193, 194, 197, 199, 206, 260, 263, 265, 269, 271, 273, 277, 328, 330, 332, 333, 335, 347, 361, 364, 370, 379, 401
Capital income taxation, 28, 69
Capital inputs, 20, 27, 30, 36, 37, 44, 45, 80, 299, 300
Capitalization, 67
Capital losses, 24, 32, 38, 45, 47, 53
Capturing resource rents, 66
Carbon dioxide emissions, 11, 97
Carbon Tax Center, 260
Cash flow, 20, 27, 29, 33, 41–44, 46, 52, 53, 59, 66–68, 70, 79
Cash flow and cash flow equivalent taxation, 52, 53, 66
Cash flow tax, 5, 20–22, 33, 52, 53, 67, 68, 71, 75
Centralized sharing arrangements, 136
Chile's Copper Stabilization Fund, 162
China, 122, 131, 140, 147, 264, 266, 270, 272, 275, 276, 328, 330–334, 339, 344, 349, 352, 356, 360, 363, 367, 369, 379, 380, 401
Citizen empowerment, 396
Civil society oversight, 8, 164, 166, 168
Closely-held firms, 33
Coal, 6, 11, 71, 97, 115, 116, 201, 255–257, 259, 286, 287, 291, 293–295, 300, 303, 319, 326, 327, 329, 331, 332, 338, 339, 342, 347, 351, 352, 367, 373, 377, 378, 380, 381
Common citizenship, 7, 8, 168

Comparative advantage of policy instruments under uncertainty, 103
Concessions, 62, 72
Conference of the Parties on Climate Change (COP26), 246
Conservation of resources, 30
Constraints on Environmental policies, 110
Content taxes, 96, 97, 99, 100, 112, 115
Coordination failures, 10
Corporate income taxes, 11, 28, 113, 145, 258, 281, 288–290, 294, 295, 307, 314
Corrective taxation, 6, 259
Corruption, 4, 6, 7, 12, 13, 33, 154, 155, 184, 188, 190, 249, 388–392, 394–398, 401, 403
Corruption, incidence, 6, 154, 387, 389, 391, 392, 396, 403
Corruption in the oil and gas sector, 13, 391, 392, 395–397, 400, 401, 403
Cost effectiveness, 88, 111
Cost of abatement, 89, 99–101, 104–107, 111
Costs of imperfect rent taxes, 57
Cross-border externalities, 9, 207
Current inputs, 24, 25, 30, 36, 37, 40, 44, 45, 60, 78, 80

D

DAI, Washington, 388
Decentralized governance, 9, 174, 207
Deforestation, 229, 231, 232, 235, 248, 269, 274, 277
Democratic participation, 9, 207
Depreciation, 20, 24, 32, 36, 38, 40–45, 53, 54, 67, 77, 78

SUBJECT INDEX   415

Desertification, 228, 233, 241, 243, 244, 248
Designing natural resources revenue funds, 164
Developing countries, 5, 11, 12, 23, 49, 51, 65, 71–73, 76, 91, 92, 102, 109–117, 121, 122, 146, 154, 163, 174, 188, 189, 239, 256, 258, 259, 261, 263, 278, 279, 281, 287, 299, 304, 307, 390–392, 399
Differential welfare costs, 308
Discount factor, 36, 39, 40
Discounting, 39–41
Discount rate, 40, 41, 49, 178
Discretionary policies, 76
Distorting effects of taxes, 63
Distributional impact, 86–88, 90, 91, 94
Distributional implications of carbon taxes, 278
Drinking water, 91, 176, 194, 199, 200, 204, 205, 212, 222, 389
Dutch disease, 6, 7, 154, 388

E
Economic growth, 6, 133, 154, 162, 256, 294, 299, 304
Economic instruments to combat global climate change, 3, 260–262, 275
Economic instruments to combat local pollution, 258
Economic interests, 249, 400
Economic profits, 24, 26, 27, 30, 35, 40, 42
Economic rents, 4, 23, 25, 27, 36, 41, 51, 54, 55, 57–61, 65, 71, 76, 113, 114, 123, 135
Economics of a national carbon tax, 256, 275

Economic stabilization, 7, 148, 164, 168
Economic welfare, 11, 57, 326
Efficiency-based argument, 4, 65
Efficiency costs of carbon taxes, 281, 293
Efficiency gains, 5, 91, 147
Efficiency value of environmental taxation, 111
Effluent and emissions charges, 96, 100
Electricity, 11, 18, 229, 326, 329, 333, 334, 336, 338, 339, 342, 351, 360, 367, 369, 370, 372, 377, 378, 380, 381
Emission (abatement standards), 5, 98, 101, 104, 106, 108
Energy Information Administration, 287
Energy prices, 12, 111, 259, 286, 287, 295, 300, 303, 326, 379
Energy pricing, 259, 293, 307, 325
Energy subsidies, 3, 4, 11, 12, 111, 259, 297, 326, 327, 329, 351
Enforcement considerations, 100, 107, 108
Environmental functions, 8, 9, 173, 175, 191, 192, 199, 207, 216, 221, 227
Environmental funds, 228
Environmental licensing, 214, 216, 218, 220, 221, 224–226, 234, 239
Environmental policy, 10, 84, 92–94, 97, 100, 102, 103, 110, 111, 114, 117, 177–179, 191, 199, 211, 212, 214, 217, 219, 226, 230, 231, 234, 238, 242, 248, 249
Environmental protection, 5, 7, 9, 83, 139, 149, 178, 191, 193, 198,

199, 205–207, 223, 249, 256, 259
Environmental quality, 6, 9, 84–86, 88, 91–93, 107, 109, 110, 113, 174, 176, 179, 207, 212, 213, 217, 219, 230, 248
Environmental regulation and pre-existing market distortions, 114
Environmental security, 193
Environmental standards, 109, 178, 179, 212–214, 234, 238, 239
Environmental taxes, 112–114, 140
Environment, global, 3, 11, 255, 325
Environment, local, 3, 9, 11, 127, 176, 207, 242, 259, 296, 298, 308, 325
Equal yield reductions, 286, 288, 290
Equitable development, 249
Equity-based argument, 4
Equity, intergenerational, 6, 7, 154, 155, 157, 159, 161, 164
Equity, interjurisdictional, 6
Equity participation, 55, 56, 68, 74, 76
Equivalence of policy instruments under perfect information, 94
Ethics rules, 8
European Union, 8, 174, 175, 190, 205, 211–213
Evaluation of resource revenue sharing practices, 146
Ex-ante rent, 31, 48–52, 62
Expenditure responsibility, 124, 127, 128
Export taxes, 21, 30, 54, 55, 61, 62, 70, 72, 73
Ex-post rent, 31, 48, 50, 52, 54
Externalities, 5, 58, 70, 83, 84, 99, 123, 147, 174–176, 178, 189, 212, 214, 231, 258, 275, 281, 293, 303, 308

Extractive industries, 4
Extractive Industries Transparency Initiative (EITI), 12, 401, 402

F
Federal Arsenic Rule, 9, 207
Federalism, 4, 8, 9, 76, 175, 180, 181, 191, 193, 198, 206, 207, 211–213, 240, 244, 246, 248
Financing costs, 38
Fiscal equalization, 7, 125, 128, 144, 147–149
Fiscal policy, 84, 98, 102, 135, 136, 145, 146, 160, 162, 164
Fiscal stabilization, 146, 162, 163
Fiscal sustainability, 6, 153, 155
Fiscal system reform, 3
Fiscal Transparency Code, 400
Flexible instrument, 11, 308
Foreign income, 35
Foreign inducement, 375
Foreign subsidiaries, 23
Forward shifting, 278–280
Fossil fuel pricing regimes, 326, 327
Fossil fuel subsidies, 11, 12, 293, 294, 325–327, 331, 334, 338, 351, 367, 376, 380
Fund governance, 7, 164
Future rents, 28, 39, 40

G
Gas, 3, 6, 11, 12, 18–21, 71, 97, 122, 124, 130, 132–136, 139–142, 146, 154, 156, 157, 160, 255, 257, 259, 286, 291, 293–295, 319, 326, 327, 329, 338, 339, 342, 347, 348, 351, 367, 373, 377, 380, 381, 387–392, 395–397, 400, 402, 403
Germany, 8, 117, 174, 190, 193, 206, 265, 268, 269, 271, 273,

SUBJECT INDEX 417

274, 277, 328, 330–333, 335, 353, 358, 370, 379
Global carbon emissions, 12, 259, 260, 326, 339, 347, 351, 376, 379
Global carbon tax, 11, 256, 260, 262, 263, 270, 307
Global climate change, 4, 11, 192, 197, 203, 206, 212, 255, 256, 258, 325, 326
Global warming, 96, 97, 222, 256, 258, 260–263, 275, 306, 307
Good governance, 7, 164, 387
Governance, 7, 8, 122, 146, 154, 160, 163, 174, 175, 183, 193, 206, 247, 387–391, 394–397, 402
Grand corruption, 4, 12, 154, 390–393, 395, 398–401
Green federalism, 206
Greenhouse gas emissions, 135, 176, 197, 205, 206, 245, 247, 259, 262, 307, 327, 328, 330–332
Greenhouse gas(es), 12, 96, 176, 243, 246, 256, 258, 260, 294, 376
Green taxes, 5, 83, 244

H
Heritage fund, 7, 159

I
Impact of carbon taxes, 256, 281, 294, 299, 301
Incentives for innovation in abatement technology, 100
Incidence of corruption, 6, 154, 387, 389, 391, 392, 396, 403
Incidence of personal and corporate income taxes, 279, 282
Inclusive development, 249
Income-based taxes, 71

Indirect taxes, 64, 94, 99, 100, 133
Indonesia, 11, 70, 71, 74, 122, 123, 131, 136, 137, 256, 264, 267, 270, 272, 276, 286–288, 291–298, 307, 328, 330–334, 344, 349, 353, 356, 360, 363, 369, 379, 380
Industrial countries, 13, 154, 157, 206, 263, 390, 403
Industrial performance, 256, 294, 299
Industrial policy, 29, 55
Inefficiency, 50, 75, 76, 84, 95, 102, 125, 128, 185, 212, 223, 230, 233, 236, 315
Inflow rules, 8, 166
Intergovernmental conflicts, 9, 193, 246
International agreements, 9, 192, 207, 220, 240, 261–263
International aspects, 76
Inter-state equalization, 7, 147, 149
Inventories, 24, 25, 32, 38, 39, 45, 47, 78

J
Joint base sharing arrangements, 142
Judicial reforms, 13, 403
Jurisdictional issues, 76

K
Kyoto Protocol, 206, 243

L
Lease, 17, 21, 22, 24, 39, 48–52, 55, 62, 63, 65, 68, 158, 167
Leasing arrangements, 58, 62
Leasing fees, 72
Legislative and executive capture, 390
Legislative compliance, 7, 165
Legislative reforms, 13

L

Local pollutants, 294, 297, 298, 302, 308
Loss offsetting, 21, 22, 29, 34, 63

M

Marginal effective tax rate, 63, 77, 78, 80
Marketable pollution quotas, 5, 93, 94
Market failure, 5, 174, 230
Martins de Castro, Deborah I., 214
Measurement of rents, 26, 27, 37
Measuring rents, 32, 36
Mechanisms for taxation of resource rents, 47
Missing markets, 84
Monitoring and implementation problems, 33
Monopoly power, 30, 61, 73, 394
Municipal systems, 217

N

National Action Plan to Combat Desertification (PAN), 243
National carbon tax, 11, 256, 275, 307, 308
National Fund of the Republic of Kazakhstan, 160
Natural resources, management, 3, 122, 139, 246
Natural resources, revenue funds, 3, 140, 157, 164, 165, 168
Natural resources, revenue sharing, 3, 6, 121–123, 125, 127, 140–142, 148
Natural resources, taxation, 3, 4, 17, 23, 30, 31, 137, 139
Neo-institutional economics, 8, 174, 182, 183, 391, 395
New public management frameworks, 395

Nigeria, 122, 132, 142, 143, 147, 154, 161, 162, 264, 266, 270, 272, 276, 328, 336, 345, 354, 358, 361, 364, 371, 381, 387, 397, 399
Non-renewable resource rents, 24
Non-renewable resource revenues, 6, 7, 159
Non-renewable resources revenues funds (NRRFs), 3, 7, 8, 156, 157, 160, 163, 164
Non-tax measures, 21
Norway, 7, 131, 134, 135, 148, 154–157, 162, 165–167, 260, 263, 296, 336, 354, 358, 361, 364, 371
Norway Government Pension Fund Global, 156, 157, 167

O

OECD countries, 4, 9, 174, 259, 261, 263, 275, 327, 338, 375–378
Oil, 3, 6, 11, 12, 18–21, 70, 71, 74, 122, 124–126, 130–136, 139–143, 145–147, 155–158, 160–163, 166, 167, 177, 229, 255, 286, 327, 339, 342, 347, 351, 367, 373, 377, 378, 387–392, 395–397, 400–403
Oman' Oil Funds, 162
One-size-fits-all approach, 390
Organizational theory, 182, 187
Outflow rules, 8, 160, 161, 164, 165, 167
Output losses, 11, 300, 308

P

Pakistan, 131, 139, 256, 264, 267, 270, 272, 276, 278–282, 286–288, 291–293, 295–297,

299–302, 308, 328, 330, 336, 345, 350, 354, 358, 361, 364, 371, 381
Parliamentary oversight, 7, 165
Paternalism, 390
Pathways to reform, 403
Patronage, 390
Petroleum products, 291, 293–295, 319, 327, 338, 339, 342
Petty, administrative and bureaucratic corruption, 389
Pigouvian taxes, 87–92, 94, 96, 115
Policies, command and control, 5
Policies, market-based incentives, 5, 94
Policy implications, 64
Political cohesion, 7, 149
Political economy, 155, 181
Political interests, 8, 181, 186, 224
Political party competition, 9, 207, 238
Political science, 8, 174, 175, 177
Polluter-pay principle, 235, 237
Polluters, 85, 86, 94, 96, 98, 100, 101, 104, 107, 109, 117, 174, 222
Present value, 4, 25–28, 31, 36, 39–44, 46, 48, 49, 52, 53, 59, 60, 63, 75, 78, 79, 101, 167, 178, 299
Principles of taxing resource rents, 30
Production sharing, 55, 58, 73, 74, 76, 133
Profit tax, 20, 133, 142
Progressive, 90, 92, 279–281
Property rights, 4, 17, 20, 21, 24, 26, 35, 39, 41, 65–67, 76, 85, 86, 93, 122, 126, 305
Property taxes, 72, 127, 128, 137, 145
Public choice, 8, 174, 175, 181

Public sector, 4, 5, 17, 21–23, 28–31, 50, 54, 65–70, 74, 86, 110, 122, 388, 390, 391, 395
Public sector equity participation, 55, 68
Publish What You Pay (PWYP) Initiative, 402

R
Rationale for public policies, 84
Regressive, 11, 72, 90, 91, 278, 280, 281, 307
Regulators, 100, 101, 103, 107, 108, 183–189, 204
Renewable-resource rents, 24, 25
Renewable resources, 18, 19, 31, 32, 36, 38, 39, 45, 46, 51, 53, 58, 60, 62, 63, 67, 217
Rent collection, 28, 48, 73
Rent-maximizing decision rules, 44
Rent-seeking behavior, 6, 154, 395
Repatriation of profits, 23
Research and development, 33, 98, 177, 218
Resource depletion, 20, 36, 38, 53, 153
Resource exploitation, 23, 26, 55, 58–60, 63, 67, 122, 123
Resource exploitation and development, 50
Resource rents, 4, 7, 23, 30, 35, 48, 50–52, 55–58, 62, 65, 71, 72, 76, 124–128, 136, 149, 154
Resource revenue dependence, 6
Resources revenue sharing principles, 122
Resource wealth, 7, 126, 388
Revenue-neutral, 11, 256, 290, 294, 307, 309–312, 314, 315, 318
Revenue potential of carbon taxes, 275

Revenue sharing, 3, 6, 121–123, 125, 127–131, 136, 140–143, 146–148, 158, 167
Right to know, 13, 403
Rio Earth Summit, 243, 244, 246
Risk, allocation, 5
Risk pooling and financing, 29
Risk tolerance, 8, 165
Royalties, 54, 55, 58–60, 70, 71, 76, 77, 126–128, 135, 139, 142, 145, 146, 149, 158, 167
Royalty structures, 59
Russia, 122, 123, 131, 132, 139, 140, 147, 166, 263, 328, 330–334, 343, 348, 352, 356, 360, 363, 369, 379, 381, 401
Russian Oil Stabilization Fund, 162

S

Safe Drinking Water Act, 205
Severance taxes, 20, 54, 69, 78, 127, 145
Shared revenues, 128, 136
Shut-down costs, 75
Societal consensus, 9, 207, 239
Soil pollution, 231, 233
State Environmental Systems, 217
State of Alaska Permanent Fund, 158
State of Alaska USA, 7
State Oil Fund of Azerbaijan Republic, 160
States as laboratories, 180, 181
Stumpage fees, 20, 69
Subnational agreements, 9, 207
Sustainable development, 122, 148, 163, 205, 219, 231, 235, 239, 244, 246, 247, 249
Sustainable Development Goals (SDGs), 244, 245

T

Taxation of Foreigners, 29
Tax avoidance, 35, 275
Tax harmonization, 125, 147
Tax liabilities, 21–23, 31, 33, 34, 44, 67, 79
Tax treatment of losses, 21
The Convention on Biological Diversity (CBD), 241
The Convention on Climate Change, 243
The Foreign Corrupt Practices Act (FCPA), 1977, 402, 403
The Government Council, 216
The greenhouse effect, 4, 259, 261, 306
The Mechanism for Clean Development (MDL), 243
The National Environmental Council (CONAMA), 216, 217, 225, 230
The National Environment System, 214, 218, 219
The National Water Agency (ANA), 218, 219
The National Water Resources Council (CNRH), 218, 219
The National Water Resources Management System (SNGRH), 218, 219
Time Horizon, 26, 39, 62, 75, 153–155, 177, 373
Tradable permits, 3, 11
Transition economies, 160
Transparency, 7, 8, 12, 135, 155, 161, 162, 164–166, 168, 391, 397, 400–402
Treatment of inflation, 40

U

Uncertainty, 6, 7, 25, 31, 34, 84, 102, 103, 105–107, 123, 125,

SUBJECT INDEX

126, 155, 180, 182, 258, 263, 306, 397

United Arab Emirates, 132, 145, 146, 154, 156, 328, 330, 333, 337, 355, 359, 362, 365, 372, 381, 401

United Nations Framework Convention on Climate Change, 242

United States, 12, 23, 96, 107, 109, 113, 114, 127, 132, 145, 190, 204, 211, 243, 291, 292, 295–297, 328, 330–333, 347, 376, 379

Urban Institute and Brookings Institution, Tax Policy Center, 275

U.S. Energy Information Administration, 287

US Foreign Corrupt Practices Act, 13, 400, 402, 403

V

Value chain, 6, 7, 12, 154, 390, 397, 401

Venezuela's Macroeconomic Stabilization Fund, 162

Victims' compensation, 12

W

Water pollution, 114, 117, 192, 194, 200, 221, 222

Welfare gain, 12, 213, 286, 290, 293, 294, 311, 315, 320, 326, 366–370, 372, 373, 376

Welfare loss, 105, 113, 284, 286, 288–290, 293, 294, 297, 319, 373

World Markets, 30, 61, 73, 123, 341, 342

World prices, 12, 74, 112, 294, 326, 327, 329, 331, 332, 338–340, 342, 351, 366, 367, 373, 375–378

Worldwide energy subsidies, 12